ENOUGH FOR ONE LIFETIME

Wallace Carothers, Inventor of Nylon

Reproduced with permission from Helen Carothers.

ENOUGH FOR ONE LIFETIME

Wallace Carothers, Inventor of Nylon

Matthew E. Hermes
University of Wyoming

History of Modern Chemical Sciences

American Chemical Society
and the Chemical Heritage Foundation

ISBN: 0–8412–3331–4
ISSN: 1069–2452

The paper used in this publication meets the minimum requirements of American National Standard for Information Sciences—Permanence of Paper for Printed Library Materials, ANSI Z39.48–1984.

PRINTED IN THE UNITED STATES OF AMERICA

Dedication

To Anne Greene and Cas Speare

About the Cover

"Enough for One Lifetime" is a quote from one of Wallace Carothers' letters in which he states that if he can nail down not only a synthetic rubber but also a synthetic silk it will be enough for one lifetime.

The cover photograph on the left is reprinted with permission from the Hagley Museum and Library Collection. The one on the upper right is reprinted with permission from Tort Gelvin Machetanz. The one on the lower right is reprinted with permission from Barbara Osborn.

About the Author

The fate of Wallace Carothers led Matthew E. Hermes to a second career in writing. Hermes wanted a life in chemistry from the moment he first encountered the periodic table as a high school student in Brooklyn, New York. He earned his B.S. degree from St. Johns University in New York, then studied under Professor William Bailey at the University of Maryland where he received his Ph.D. in 1959.

Around the laboratories at DuPont's Central Research Department, Hermes first heard the sketchy tales of the inventor of nylon. After serving DuPont's Central Research, Plastics, Elastomers, and Fabrics and Finishes Department with modest impact, he left DuPont in 1979 to serve as Director of Research of Virginia Chemicals, a subsidiary of the then Celanese Corporation.

In 1983, Hermes became Corporate Research Scientist at the U.S. Surgical Corporation, where he led a fast-paced and successful effort to develop and introduce more than 1000 surgical suture products. In the course of this effort, he began to be fascinated by the mysterious Carothers and his imprint on the field of synthetic fibers. Determined to prepare Carothers' biography, he earned a Master of Arts in Liberal Studies in 1992 from Wesleyan University. There he began the first draft of *Enough for One Lifetime: Wallace Carothers, Inventor of Nylon*, while studying with Professor Anne Greene.

Hermes has lived in Steamboat Springs, Colorado, since 1993. He served as Adjunct Professor of Chemistry at the University of Wyoming. As new methods of information distribution and exchange began to affect how the population of one million square miles of the rural west compete with its urban counterparts, he became a consultant in telecommunications. He is currently President of the F.V. Hayden Institute, which specializes in rural telecommunications research.

Hermes holds 22 U.S. patents and has published more than a dozen scientific papers.

Contents

Preface   xiii

Tarkio College   1

At Urbana, the Best of the Best   21

Paris and Harvard   45

The Road to Pure Science   59

DuPont Hires Carothers   75

Arrow of Discovery   89

We Will Have Quite a Lot of Things
   in My Division To Show You   103

Elmer Bolton's Chemical Department   121

A Synthetic Rubber and a Synthetic Silk:
   "Enough for One Lifetime"   135

"1932 Looks Pretty Black to Me Just Now"   145

A Struggle Over Research Objectives   157

The Invention of Nylon   181

Carothers Disappears   191

A Year at Wawaset Park   211

A Gathering of Seven Friends   227

A Second, Superb Idea   235

Carothers Marries   249

At the Philadelphia Institute   259

In the Alps with Carothers   271

Isobel Carothers Dies   279

Wallace Carothers Swallows Cyanide and Dies   285

DuPont and Ira Carothers   295

What Is the Family Illness?   309

Appendix A: Chemistry Cited   321

Acknowledgments   327

Bibliography   331

Index   337

Preface

This is a story of invention and chemistry and the ineluctable fate of the inventor of nylon. A young Harvard chemist, Wallace Carothers was lured to the DuPont Company in 1928 to lead a program they chose to call basic research. In a very few years, he invented for them neoprene and nylon, the world's first synthetic rubber and fiber. Carothers took an infant science called polymer chemistry, defined it, and gave it maturity. But in the process, the still-young Carothers aged to the childlike dependence of the very old. Though sheltered in the care of psychiatrists, wife, and friends, he took poison and died painfully two days after his 41st birthday on April 29, 1937.

Do not be misled by a text that jumps ahead to 1938, to New York City on the eve of a new world war, to a crowded meeting hall awaiting the arrival of Katherine Hepburn. For what occurred on October 27, 1938, 18 months after Wallace Carothers died, was the beginning of a revolution for which the inventor Carothers was responsible.

Before the fact, before even seemingly modest events of our history occur, in the precise instant before our worlds forever change, we sit complacent, ignorant, unaware that revolution is racing upon us, consuming the last moments before our awakening. DuPont's Dr. Charles M. A. Stine spoke to a quiet audience that day, a crowd awaiting Miss Hepburn's appearance later in the program.

The DuPont chemist disclosed his company's new textile yarn, a fiber he called nylon. Stine stood before *The New York Herald Tribune*'s eighth annual Forum on Current Problems at the New York World's Fair. Outside the building, across New York, and over all the civilized lands, the world shook with the implications of Neville Chamberlain's acquiescence to Hitler's partition of Czechoslovakia. That was the world's story in October 1938. And in three days, Orson Welles would narrate his Halloween tale, *The War of the Worlds*. It would panic an edgy populace, still reeling from the aftershock of the decade of Depression and fearful that an instant of peace won from Hitler by the British prime minister would not prevent an all-encompassing world conflict.

Dr. Stine told a scientific story. DuPont's new filaments were the world's first absolutely manmade fiber. They came to being from "coal, air and water."

The audience still looked on, dully. But when they heard Stine hint that a major stocking mill had replaced silk and was now knitting nylon hosiery and that these new stockings felt better, didn't bag or snag, and when they snagged they didn't run, that passive audience stood and applauded.[1]

Eighteen months later, on the morning of May 15, 1940, newspapers heralded Hitler's climactic *blitzkrieg* across the Low Countries. They told confused stories of German armor and air power sweeping toward France. They drew maps showing how closely the Germans at Rotterdam now stood to fortress England. As Hitler completed his dramatic conquest, women and men across this country snaked in line to reach the hosiery counters in upscale department stores. The opening hour on May 15 brought bedlam. Swarms of customers paid premium prices—$1.15–$1.35 per pair—to sweep nearly 5 million pairs of nylon stockings from the suppliers' racks that first day of sales. In city after city the new nylons were sold out by sundown.[2]

Fortune magazine quickly made a startling and precise prediction. DuPont had "staked out a vast new area of revolutionary plastics and fibers," the business magazine reported.

> ...nylon breaks the basic elements like nitrogen and carbon out of coal, air and water to create a completely new molecular structure of its own. It flouts Solomon. It's an entirely new arrangement of matter under the sun, and the first completely new synthetic fiber made by man. In over four thousand years, textiles have seen only three basic developments aside from mechanical mass production: mercerized cotton, synthetic dyes and rayon. Nylon is a fourth.

Within a year, nylon went to serve the needs of war. Its introduction to American consumers was delayed until enough woven para-

[1]The *New York Times*, October 28, 1938; *Time*, November 7, 1938.

[2]*Fortune* 1940, 22, 56. *Fortune*'s use of lower case "nylon" as opposed to capitalizing the word and indicating it as a trademark resulted from a conscious decision by DuPont to use the term nylon generically to refer to a family of proteinlike, fiber-forming synthetic materials.

chute fabric and braided parachute cord and outerwear and tenting and tire yarn and glider towrope had been made. The vanguard of the U.S. Army floated to earth in Normandy carried by and covered with nylon. And by 1945 the long conflict came to an end. For two years after the war, restless women battled at the hosiery counters across the United States until DuPont could supply the stocking mills with enough of the new fiber.[3] An American GI with a pair of "nylons" found he possessed an almost universal medium of exchange on the devastated European continent.

DuPont sells $4.5 billion of nylon each year now, and the profits from this single material have sustained the DuPont Company, the extended duPont family and all the Company's stockholders for more than half a century.[4,5]

But the real importance of nylon lies not with its staggering commercial success, nor with the riches accrued to DuPont and the duPont's. The invention of nylon marked the beginning of the modern era of scientific design of materials matched to an ever more sophisticated and challenging description of the "needs" of mankind. Chemists at DuPont and in laboratories across this country and Europe learned from DuPont's design of the nylon molecule that chemistry could improve on nature's building blocks. Chemists could make better elastomeric materials than natural rubber and more versatile structural elements than maple and oak and spin finer and stronger filaments than any from the cocoon of the silkworm.

The careers and livelihoods of many of today's chemists draw directly from the fundamental understanding and laboratory achievement behind DuPont's introduction of nylon. The DuPont laboratories proved in the 1930s, by choosing the polyhexamethylene adipamide molecule over hundreds of other candidates, that careful manipulation of molecular structure could be driven to yield specialized materials, with properties on demand. Modern chemists model

---

[3]Hounshell, David; Smith, John Kenly. *Science & Corporate Strategy, DuPont R&D 1902–1980;* Cambridge University Press: Cambridge, England, 1988; p 272.

[4]*Chemical & Engineering News* **1993**, *August 2, 12.*

[5]Hounshell, David; Smith, John Kenly. *Science & Corporate Strategy, DuPont R&D 1902–1980;* Cambridge University Press: Cambridge, England, 1988; p 272.

chemical structures by computer now. They design on the screen molecules they expect will bind to DNA, interact with enzymes, catalyze petroleum decomposition, attack cancers, and tangle with the complex virus called HIV. In the laboratory, chemists craft molecules, fashioning them to the exact shape, length, and chemical sequence they anticipate will perform predictable roles.[6] The sophisticated tools of our science bear scant relation to the crude devices used in the first efforts to make high polymers. But the goals remain in place. From carefully devised laboratory syntheses we can develop the critical properties of advanced synthetic materials.

■ ■

Nylon, and the materials revolution behind the miracle fiber, came from the mind of Wallace H. Carothers. Dr. Stine, who introduced nylon to the world in 1938, had introduced Carothers to DuPont a decade earlier. But by the time Stine disclosed nylon in New York, his young protege had been dead for more than a year.

This book extracts Wallace Carothers' life from the clinging, silent mists that began to curl and flow and hide his story on the chilly morning after he died in the spring of 1937. The silence after suicide is the singular offense of no one. The dead man's role is complete now. The survivors move quietly on, somehow trusting that their amorphous sense of guilt, shame, and remorse will be served best in solitude.

Carothers' rise to scientific creation and to significant prominence was a symbol of hope underlying the stillness of the Great Depression and the inevitable portent of approaching war. But his own despair and the death he chose trace the inevitable course of deeply embedded disease: of clinical depression, of alcoholism. Depression and alcohol played out in Wallace Carothers in the 1930s, a decade in which their victims and medical science strained to fathom and overcome their grip. The help and understanding, the treatment and support we might offer Wallace Carothers today, came more than a decade too late.

■ ■

I have lived in Wallace Carothers' long scientific shadow for nearly 40 years. I learned organic chemistry and earned my Ph.D. in

---

[6]Hermes, Matthew E. U.S. Patent 5 248 761, September 28, 1993.

polymer chemistry from Carothers' direct successors. And I continue to practice research and invention in polymer chemistry with sufficient skill.

In 1959, I joined DuPont's Central Research Department, where Wallace Carothers had once invented nylon. But I found at Central Research no trace of Carothers, no tradition surrounding his brilliant inventions from the 1930s. The success of nylon still fueled DuPont's obsessive quest for another dramatic research breakthrough—for a new nylon—more than 20 years after he was gone. But of Carothers there remained virtually nothing—a few whispers, some talk among the younger scientists about Carothers' mysterious death, but that was all.

Once, long ago in 1946, DuPont honored Wallace Carothers. At an elaborate ceremony they named a laboratory building at its Experimental Station near Wilmington after the inventor. Carothers' wife, Helen, and his only child, Jane, who was born seven months after Carothers died, attended the ceremony. DuPont has modified this research building many times since 1946.

Today, only a dusty photograph of the inventor hanging in the small lobby serves to remind the chemists entering the Carothers Laboratory, how DuPont came to earn its fortune in nylon and the world came to gain its materials revolution. Only the black and white photograph honors the inventor who came to the end, unable to manage his own life.

Chapter 1

Tarkio College

A young man and a young woman from the city of Des Moines, Iowa, stepped down from a Chicago, Burlington and Quincy railroad train at Tarkio, Missouri, in September 1915, to join the freshman class at Tarkio College. Each was 19, a year older than the other entering students. Each had graduated from Capitol Cities Commercial College in Des Moines and, in traveling to Tarkio, they shared a special mission. Back in Des Moines the commercial college provided one year of business training for the young people of Iowa's biggest city. The school grandly claimed placement of more than 1000 students a year in good positions.[1]

These two students from Des Moines formed part of a plan to revitalize Tarkio's own lagging Commercial Department. Few of Tarkio's young people chose to study the less rigorous business curriculum. Rev. Dr. J. A. Thompson, a Presbyterian minister and president of Presbyterian Tarkio, wished to make the business program more attractive by adding to the staff. Rev. Thompson learned, perhaps through his Des Moines trustee Frederick McMillan, that the vice-president of the commercial college there, Ira Carothers, was a devout and active member of the United Presbyterian Church. Rev. Thompson asked for help. Ira Carothers promised that two members of his newly graduated class would matriculate at Tarkio. They would work part-time to help staff the small Commercial Department there. Alice Gifford was one of the students he sent to Tarkio. Ira's oldest child, his son Wallace Hume Carothers, was the other.

---

[1] *Des Moines Evening Tribune*, September 7, 1914.

Wallace Carothers' first semester was the first time in years the Commercial Department at Tarkio did not lose money.[2] And from the beginning the other Tarkio students called Carothers "Prof."

■ ■

Ira Carothers carried more than a century of Scotch–Presbyterian heritage into his discussions with the Tarkio College trustees. Ira's grandfather, Andrew Carothers, farmed a 208-acre tract of land in Pennsborough Township, Cumberland County, Pennsylvania, obtained by Andrew's father, Rodger Carothers, in 1765. Andrew served in the War of 1812 and was married in the United Secession Presbyterian Church at Newville, just west of Carlisle, Pennsylvania, in 1814.

Andrew Carothers had nine children. The third, John Carothers, was born in 1820 and later married Nancy Andrew. These Carothers migrated west to the black dirt Illinois town of Stronghurst near Burlington, Iowa, on the Mississippi. They named their first son, born in Stronghurst on October 3, 1869, Ira Hume.[3]

Ira married Mary McMullin, a Scotch–Irish descendant of Ulstermen, in Burlington in 1895. Wallace Hume Carothers was born there the next year. In 1901, Ira took Mary, Wallace, second son John, and infant daughter Isobel further west to Des Moines where Ira began his long tenure at the Capital Cities Commercial College.

No denominational body carries the dedication to political independence, ardor for freedom, and firmness of moral values more than the Scotch–Irish Presbyterians. But by the very establishment of the local Presbyters as the organizational powers and moral guides of the church community, the Presbyterian Church has guaranteed its history of passionate division, and of passionate and loyal devotion.[4]

Ira Carothers was the picture of a Presbyterian elder, precise and correct in appearance. He seems both trim and grim in his photo-

---

[2] *The Tarkiana, 1917;* published by the junior class of Tarkio College, 1916; Vol. 8, reveals the characteristics of the school and its students. Barbara Osborn of Los Altos Hills, California, daughter of Ora (Tort) Gelvin Machetanz of the Tarkio class of 1917, allowed me to use this yearbook. The Commercial Department is described on page 62.

[3] Suzie Wallace Kyle Terrill, daughter of Wallace Hume Carothers' sister, Elizabeth, provided an undated genealogical outline of the Carothers family prepared by Russell Rankin, a classmate of Wallace Carothers at Tarkio.

[4] Presbyterianism and characteristics of its fragmentaion in the United States are described by Lewis S. Mudge, Secretary of the Presbyterian Church in the United States, in the *Encyclopedia Britannica,* 14th ed.; Chicago, IL, 1939; Vol. 18, p 446.

Carothers Family (left-to-right back row, Wallace and John; middle row, Mary and Ira; and front row, Isobel). (Suzie Terrill.)

graphs, balding, even as a young man, always with pince-nez, starched collar, and precisely knotted tie. His family appeared regularly at his side in the United Presbyterian church.

■ ■

Wallace's mother reported his American boyhood. Her son and his friends established a club in an old barn behind their house on Arlington Avenue. They wired battery-operated doorbells and wound coils for crystal sets on cylindrical Quaker Oats boxes. Wallace attended North Des Moines High School. He read water meters and clerked in the reference department of the Public Library of Des Moines to earn a little money.[5]

Wallace found science early. The boy sat on the steps of the library overlooking the Des Moines River and read Robert Kennedy Duncan's 1905 report, *The New Knowledge.* Duncan wrote with wide-eyed wonder of the mysteries of the universe: of matter and ether and energy. He assembled the phenomena reported by the turn-of-the-

[5]Louis Cook interviewed Mary and Ira Carothers for the *Des Moines Register,* 1947.

century scientific masters: the apparent periodic behavior of the masses of the then-known 70 elements. He wrote, "This periodic law of the atoms is God's alphabet of the universe." The lack of symmetry in the periodic law puzzled Duncan. The relationship of the periodic table to atomic number—the number of positively charged protons in the nucleus of the atoms—was not yet known. He told of the vivid but unexplained colors of these elements detected in the gaseous state in the spectra cast on Earth by distant stars. He related the strange behavior of electrons. He called them corpuscles, "the fulcrum for the lever of thought, the philosophers desire," the particles whose electric charge and mass he carefully derived for his lay readers.[6]

Questions, questions, questions. Early twentieth century chemists and physicists had observed and reported the basic phenomena and pondered what it all implied. But the understanding of the structure of the atom, of its relationship to periodicity, of the nature and universal presence of Duncan's corpuscles would come from the laboratory of the physicist, Duncan predicted. "Here, we are in another world," he wrote. "Instruments of infinite precision surround us, optical, electrical, and magnetic, and the whole atmosphere of the place tingles with accuracy."[7] And as young Wallace Carothers read Duncan's wondering yet already outdated text and constructed his first chemistry laboratory in his bedroom in Des Moines, Rutherford and Millikan completed Duncan's vision and defined for us the atom and the electron.[8]

Ira Carothers wrote of his son as a young man. But only later, after Wallace died. Possibly the saddest duty of a parent is the preparation of a memorial biography of a son or daughter. Perhaps no such remembrance can withstand elevation, for at one time, Ira Carothers had characterized his son as a "slow learner."[9]

> As a growing boy he had a zest for work as well as play. He enjoyed tools and mechanical things, and spent much time in experimenting. He was a great admirer of Thomas A. Edison and seriously considered electrical engineering for a profession.

---

[6]Duncan, Robert Kennedy. *The New Knowledge;* A. S. Barnes & Co.: New York, 1905; p 29.

[7]Ibid. p 46.

[8]*See* Millikan, Robert A. *The Electron;* University of Chicago: Chicago, IL, 1917, for the scientific picture of the electron and its relationship to atomic structure, which evolved from the contemporary experiments described by the awe-struck Duncan.

[9]*Des Moines Register,* November 5, 1931.

His school work was characterized by thoroughness, and high school classmates testify that when called on to recite his answers revealed careful preparation. It was his habit to leave no task unfinished or done in a shoddy manner. To begin a task was to complete it. No doubt this quality of painstaking thoroughness can account for his later success.

Very early in life he displayed a love of books. From the time when *Gulliver's Travels* interest [*sic*] a boy, on through Mark Twain's books, Life of Edison, and on up to the masters of English Literature, he was a great reader. He possessed a melodious singing voice that might have developed under training into something very worth while. Though he had no technical training in music he was a great lover of the great masters. His collection of phonograph records included the works of Bach, Beethoven, Brahms, Schubert and others.[10]

Yet with all this, after Wallace graduated from North in 1914, the quiet, shy son enrolled at his father's college to learn shorthand and bookkeeping. Wallace Carothers' attendance at the commercial college clashes indeed with the intellectual and technical promise Ira reported after Wallace's death. It would seem Ira Carothers restricted his talented son to the limited educational standards of his own position during Wallace's year at commercial college and then sent this finished product off to Tarkio as an offering to please the leaders of his church.

Tarkio's Arthur Pardee described Carothers' parents as rigid, but "nice, responsible people.... His family had really very little educational background. I know his father throughout Wallace's college career didn't really sense the possibilities of the field to which his son wanted to devote himself."[11]

Wallace Carothers went off submissively to Tarkio, acceding to an arrangement whereby he would be able to work for his expenses and perhaps a scholarship. He was a pawn, dominated by his father's efforts to curry favor with the Presbyterian churchmen. Wallace Carothers accepted his exile to Tarkio in September 1915 as he accepted Sunday morning worship and evening prayer and Bible

---

[10]Ira Carothers to Roger Adams, December 2, 1937. (Adams Papers, University of Illinois Archives, Urbana, IL.)

[11]Arthur Pardee to Roger Adams, February 19, 1938. (Adams Papers, University of Illinois Archives, Urbana, IL.)

study in his home. All of these were part of a ritual to be tolerated, a sacrifice of his family by his father to the Church.

■ ■

In the mid-1880s, the elders of the United Presbyterian Church established a college in the new, northwest Missouri town of Tarkio, in Atchison County. The county lay on the fringe of the tall prairie grasslands still extending across Iowa and eastern Kansas, Nebraska, and the Dakotas. The United Presbyterian ministers serving God in the small towns on the plains stretching west to the foothills of the Rockies transformed a small rise of land at Tarkio into their own institution of higher learning, a liberal arts college to educate the minds while protecting the souls of the young plainsmen and women. By 1915, the prairie was in corn and corn-fed cattle—cattle bred in Missouri and cattle still coming in by the thousands of head from the western range—through St. Joseph and Kansas City.[12]

And by the fall of 1915, only 58 men and 56 women enrolled at the four-year college. They came to Tarkio from small hamlets: North Bend, Nebraska; Riverton, Iowa; Sparland, Illinois; Fort Morgan, Colorado; and Webber, Kansas. These were small towns whose very formation was the offspring of the spiderweb of rail lines that moved goods and people across the plains. Of the 114 students, 49 came from Tarkio itself. Until Wallace Carothers arrived in 1915, Russell Rankin from Des Moines was the only student at Tarkio who could say, "I live in a city."[13]

Frances Gelvin of nearby Maitland, Missouri, had enrolled originally at Wellesley College in Massachusetts, but "some eager member of the board of the college came to Maitland to say that all the other Gelvin girls had gone to Tarkio—so," Frances Gelvin wrote. We can picture a Presbyterian Rankin, the most wealthy man in northwest Missouri, with 20,000 acres of rich, rolling land, visiting cattleman Gelvin, to deal with his young daughter's fanciful notion that she should go East.[14]

At Tarkio, Frances Gelvin remembered, "My half sister, Tort, and I were the rich girls. We wore designer clothes, came from a wealthy family, had a Cadillac coupe with a glass vase with flowers. Two of my other half-sisters married into the 'reigning families' the Stevensons

---

[12] *Missouri Historical Review* **1953**, *57*, 212.

[13] *The Tarkiana, 1917;* published by the junior class of Tarkio College, 1916; Vol. 8.

[14] *Missouri Historical Review* **1953**, *57*, 212.

and the Rankins.''[15] Tort looks round and plain in the yearbook pictures from 1916. In contrast, Frances Gelvin had long, dark hair and languid eyes. In studio portraits taken in the early 1920s she affects great glamour, but as she turns her head, we see in profile her nose is large, and overall, her looks are more ordinary.[16]

At Tarkio in September 1915, Wallace Carothers met Wilko Machetanz—everyone called him Mach—and roomed with him for the next two years. Mach, who was from the west-central Ohio town of Kenton, came to Tarkio because his aunt Bertha taught German there. Mach was two years ahead of Wallace Carothers; he graduated in the summer of 1917, served in the army in the Great War, studied law at Yale, and became a landowner and judge in Visalia, California. Mach married Tort Gelvin, his prosperous classmate from Maitland, in 1924. Mach and Tort shouldered the burden of Wallace Carothers' remembrance for the rest of their lives. Carothers became famous for a fleeting moment; then one night he was dead. Mach and Tort passed on their memories, intact, to their children.

Mach's son, Fred Machetanz, said his father and Wallace Carothers shared a pessimistic outlook on life. His father was gloomy; Carothers' outlook was even more bleak. Tarkio College in 1915 had just added considerably to its library collection and now boasted of 8000 volumes. The two young men studied literature by memorizing it. ''I remember my father quoting Carothers' favorite poem,'' Fred

ORA GELVIN *K. L. S.*
Maitland, Missouri
Secretary K. L. S. 1915; Winner Intersociety
Oratorical Contest 1916; Phoenix Staff.

Tort Gelvin from Tarkiana. (Barbara Osborn.)

---

[15]Frances Gelvin Spencer to Matthew E. Hermes, July 19, 1990.

[16]Barbara Osborn showed photographs of Frances Gelvin Spencer taken in Columbus, OH, in 1928.

Frances Gelvin—"Kid"   *Kappa*
    Maitland, Missouri
"Tee-hee-hee! Were woman but constant, she were
perfect."

Frances Gelvin from Tarkiana. (Barbara Osborn.)

Wallace Carothers and Wilko Machetanz. (Barbara Osborn.)

Machetanz said. "Carothers loved Swinburne and would quote the
second stanza of *The Garden of Proserpine*."[17]

> I am tired of tears and laughter,
> And men that laugh and weep;
> Of what may come hereafter
> For men that sow to reap
> I am weary of days and hours,
> Blown buds of barren flowers,

[17]Matthew E. Hermes interview with Fred Machetanz, September 25, 1989.

Desires and dreams and powers,
And everything but sleep.

In Carothers' freshman year of 1915, Mach and Tort Gelvin introduced Carothers to Tort's half-sister, Frances. "He would quote Swinburne from time to time," Frances Gelvin wrote, and she continued, remembering the next verses of *Proserpine* 75 years later:

That dead men rise up never;
That even the weariest river
Winds somewhere safe to sea.

"But it appeared a bit self conscious and derisive," she continued, writing of Wallace Carothers' infatuation with the deadly, pessimistic poetry, "as though we should not concern ourselves with a poet of such negligible stature as Swinburne.... He loved German poems and songs and admired Nietzsche. That should tell a lot. He sang snatches of German songs and quoted poems. Teasingly, he said to me in German, 'You are my flower.' I remember his singing along in the chapel with the rest of us,

The valley stands so thick with corn,
That they laugh and laugh...

on and on in a repetitive crescendo. I do not believe he was singing praises to high heaven, but just enjoyed the sound of music. Incidentally, I did not share W's agnosticism."[18]

Frances Gelvin and Wallace Carothers paused and posed for an outdoor photograph that freshman year. Carothers nearly smiles and looks down at Frances as if he were speaking to her. He wears an overcoat over his jacket. His tie is knotted neatly in an Edwardian collar. Frances looks down, demurely. She is smiling, too. The editors of Tarkio's yearbook framed the photograph in a heart, placed it on a page filled with photographs of the school's boy–girl steadies, and wrote beneath the picture, " 'Prof' Proposes."[19] Frances Gelvin Spencer saw a copy of the photograph three-quarters of a century after its first appearance.

"[I] fear people will get the wrong idea of girl-friend–boy-friend." she wrote.

---

[18]Frances Gelvin Spencer to Matthew E. Hermes, September 23, 1990.

[19]*The Tarkiana, 1917;* published by the junior class of Tarkio College, 1916; Vol. 8, p 149.

Wallace Carothers and Frances Gelvin from Tarkiana. (Barbara Osborn.)

Very different in the teens from 1990. Tarkio was rigid in its code. There was only one girl on campus one could point the finger at, if one were a finger-pointer or a stone-caster. Over all the College there seemed to be a pall of Fundamentalism. Everyone took Greek, Bible, Ethics. Many of the students were going to be missionaries and ministers. We had to attend chapel every day and [we] sang Psalms. Most all were United Presbyterians. I was reared in that faith, but I felt like a square peg in a round hole. Resentfully, at home we had to learn the Presbyterian catechism and one of the Psalms on Sunday.

"The town of Tarkio seemed to have the flavor of the college. The U. P. Church was the place to be," Frances Gelvin insisted.[20] Wallace met his parents as they stepped off the train from Des Moines. They

---

[20]Frances Gelvin Spencer to Matthew E. Hermes, July 19, 1990.

were in Tarkio to see Wallace and meet the Rev. Thompson. Ira Carothers found his son was now smoking cigarettes. Ira threatened to reduce his contribution to Wallace's tuition.

"Mr. C.," remembered Frances Gelvin Spencer in 1990, "appeared to me to be a character right out of Dickens' novels—he could have understudied Scrooge. Of course this is only my impression. He might have been a worthy man in some respects."

"W. wrote me in a letter," Mrs. Spencer continued, "he just couldn't stand to be in a room with his father. I can readily understand this as Mr. C. appeared to me mean-spirited looking, thin nose, close-together cold gray eyes."[21]

Perhaps Tarkio life deviated little from the uniformity of the United Presbyterian ethic as Frances Gelvin lately proposed. But the student yearbook for 1916 hints at a wider exploration among the young midwesterners. Of Des Moines' Russell Rankin the book suggests:

> He makes hay while the sunshines
> And makes love while the moon shines.

And of Merrill Smith:

> His books do not attract him
> Neither does the gym;
> But up he springs with open eyes
> When you say "joy-ride" to him.

Marjorie McGinnis teases with:

> They say we women like to be loved;
> Just try it and see.

Marie Grimm is:

> A mighty hunter and her prey is man.

Under Wallace's photograph the yearbook editor quoted:

> I wish that I could utter the thoughts that arise in me.

■ ■

---

[21]Frances Gelvin Spencer to Matthew E. Hermes, September 23, 1990.

WALLACE CAROTHERS—"Prof"   *Forum*
    Des Moines, Iowa
  Scrub Faculty.
"I wish that I could utter the thoughts that
arise in me."

Wallace Carothers from Tarkiana.(Barbara Osborn.)

"We didn't have dances. We had what they called 'Grand March-es,' " Frances Gelvin wrote. But the Gelvin Cadillac provided a unique freedom for Tort and Mach and Frances and also for Wallace Carothers.

After two years, Mach graduated. Wallace and Mach slipped out of Tarkio together and toasted his graduation. Later, Mach regretted he had given Wallace Carothers his first drink. Frances Gelvin secured her wish to study at one of the "Seven Sisters" and took a year at Radcliffe. Wilko Machetanz joined the U. S. Army. Tort, with Mach gone, stayed on at Tarkio and took business training from Wallace Carothers. They climbed in her Cadillac and drove together to her home in nearby Maitland. And Wallace Carothers and the young woman he was minding for his friend Machetanz pounded along more than 100 miles of rutted roads to Wallace's home at Des Moines, sitting side-by-side in the swaying, dusty Cadillac, giggling and bouncing on the bench seat. They pulled up at Wallace's home in the elegant touring car and announced they were off to the the-ater; they were going to see Sally Rand perform.

Tort Machetanz, at 96 years of age and blind, talked of the old college years and of young Wallace Carothers. "One night we were out on the football field in front of the men's dormitory. I was playing the ukelele. And the aurora borealis started. It was just gorgeous. Oh, he could sing. I'd play the ukelele and he would just sing all the old songs."

She offers a treasured photograph of her with Wallace taken on a visit to her large Victorian home in Maitland. Tort sits at the grand piano. From the backlit glow we can imagine late summer, late evening. Tort sits with a long-stemmed rose between her lips with the blossom weighting the stem in a long, downward arc. Both hands rest at the keyboard, with fingers raised, not to play but perhaps to listen. Her shoulders are lifted slightly and her head is forward, eyes turned toward Wallace who stands at her left. Her hair is swept back to a bun; loose ends appear as a halo in this 1918 photograph.

Wallace stands at the bass end of the keyboard, left elbow akimbo with his hand resting on the top of the piano. He is impeccably dressed in heavy wool: suit buttoned, vest buttoned with an artfully knotted, striped tie. He wears a formal shirt with French cuffs that protrude marginally from his sleeves. Wallace wears a silk flower in his lapel, his hair is trimmed short and is severely parted on the left. It could be said to be slicked down. Carothers looks at Tort and leans slightly left and forward. He is well proportioned, of medium height, and has broad shoulders. His lips are closed and thin, but it appears he may be about to smile or to sing. He has a vertical cleft in his chin; he accents his face with round eyeglasses with dark but thin rims. They are a fine couple; Wallace Carothers is an entirely presentable guest at the home of the well-to-do Gelvins of Maitland, one of whom took great pains in composing this fine photograph.[22]

■ ■

Carothers wrote in 1932, "My interest in chemistry was started by reading Robert Kennedy Duncan's popular books while a high school student in Des Moines, Iowa, so that after some delay when it was possible for me to go to college I had definitely decided to specialize in chemistry."[23]

Professor Duncan had followed *The New Knowledge* with *The Chemistry of Commerce* in which he detailed the marvelous synthetic silks now being drawn from cellulose solutions in the new viscose process.[24] He wrote a collection of essays he called *Some Chemical Problems of Today.* "The Prizes of Chemistry," he called his lead treatise. "Large fortunes are being accumulated by men of creative genius through the cooperation of corporations anxious...to conduct their operations through the principles of progressive scientific practice." Duncan made a series of marvelously accurate predictions in his 1911 essay. He saw a chemical science that would inevitably lead to developing grain alcohol to replace gasoline for the automobile. He foresaw the invention of the electron microscope, decaffeinated coffee, and the refinement of the fibers of certain southwestern plants into common textiles. He wrote of the desire for a shaving cream to replace soap, for a soap whose residues would not leave sticky curds, and he was certain that a positive study of arteriosclerosis would eliminate old age.

Duncan focused on practical science in his text, science protected by patents and done by adequately trained and remunerated professionals. He scoffed at uncentered work:

---

[22]Matthew E. Hermes interview with Tort Machetanz, December 10, 1989.

[23]Wallace Carothers to Pauline G. Beery, April 4, 1932. (Hagley Museum and Library Collection, Wilmington, DE, 1784.)

[24]Duncan, Robert Kennedy. *The Chemistry of Commerce;* Harper & Brothers: New York, 1907; p 219.

Wallace Carothers and Tort Gelvin. (Tort Gelvin Machetanz.)

> The many important and actual opportunities that lie everywhere at hand for applying scientific knowledge to the manufacturing needs of men make one frankly consider why trained and earnest men should devote laborious days to make diketotetrahydroquinazoline, or some equally academic substance, while on every side these men are needed for the accomplishment of real achievement in a world of manufacturing waste and ignorance.[25]

Duncan's outspoken, practical design of the science fortified Carothers as his first year at Tarkio brought a baptism into the formal study of chemistry.

Tarkio entrusted its chemistry to Arthur Pardee. Pardee graduated from Washington and Jefferson College, Washington, Pennsylvania, with a B.A. degree in 1907. Coincidentally, he had studied under Duncan and had taken away Duncan's clear model of the importance of the chemical sciences in the remodeling of national culture.[26] But turn-of-the-century Tarkio moved forward as the collegiate model of the one-room schoolhouse. The 22-year-old Pardee's initial teaching assignments included not only chemistry, but physics, economics, and government. Arthur Pardee aspired to master chemistry beyond his bachelor's degree. He took two leaves from Tarkio to study for his Ph.D.; during the school year of 1910–1911 he went to Johns Hopkins University, Baltimore, Maryland, which was the fading center of graduate education in chemistry in the country. He returned to Tarkio the following year. Pardee's Ph.D. required completion of certain advanced course work. But the doctorate also demands a long period of independent chemical investigation under the loose direction of a mentor—a senior faculty researcher who outlines the general problem and leaves the detailed work to the emerging scientist. Pardee carried out much of the experimental work for his doctoral thesis back at Tarkio after he returned in 1911. He set up a small laboratory in the basement of the main building to complete his research. That work took four years.

Pardee was small in stature, with a high, squeaky voice and an enormous enthusiasm for science. While he completed his own research, he converted two young men, Leslie Grimm and Leonard Hasche, to focus on the chemical sciences. Both Grimm and Hasche went on to receive their own Ph.D. degrees in chemistry.

---

[25]Duncan, Robert Kennedy. *Some Chemical Problems of Today;* Harpers: New York, 1911; pp 1–18.

[26]Duncan taught at Washington and Jefferson from 1901 to 1906. *Encyclopedia Britannica*, 14th ed.; Chicago, IL, 1939; Vol. 7, p 737

In 1915, Pardee returned to Hopkins for a final year to submit his dissertation in physical chemistry to Professor Harry Clary Jones.[27] And as Pardee left Tarkio for Baltimore, Wallace Carothers arrived.

Tarkio assigned Harry Claire Weamer who, like Pardee, had a bachelor's degree from Washington and Jefferson, to teach chemistry and physics in 1915. Carothers took his first year of college chemistry from Weamer, under the lingering influence of Pardee.

Picture the basement laboratory at Tarkio in the fall of 1915. The small room—with its wooden center bench and oak desks, its chemicals and burners, flasks and test tubes—lacks its master. Pardee is present only by the stories Grimm, Hasche, and the caretaker, Weamer, tell to each other about this marvelous and inspiring teacher. The laboratory is an outpost, a basement hideout at the fringe of the fundamentalist core of Tarkio life, as far away from the mainstream of Tarkio's religious culture on the three main floors of the building as is the Commercial Department stuck under the hot eaves of its roof. The laboratory serves as a home for the agnosticism of youthful science. And a place to study and to tinker and to work out the passionate embrace of an emerging scientist for his science.

The laboratory is a gift for young Carothers. An unexpected dream come to reality, a secluded place, away from parents and authority, a clubhouse, library, shop, and bench where no idea, however oblique, can be damned for its lack of orthodoxy. A place of his own. And, in Wallace's sophomore year, Pardee would be back.

Arthur Pardee hung his new diploma when he returned. It showed he was finally, at 31, Dr. Pardee. And he unpacked his books—the latest texts on the developing field of chemistry brought with him from Hopkins. And his new student, Wallace Carothers, began to read them.

Wallace began his studies with Arthur Pardee with an introduction to organic chemistry. His chemical training now focused on only one of the elements: carbon. Centered at the top of Mendelejeff's periodic table, carbon was recognized in the middle of the nineteenth century as the common ingredient of all the chemical compounds crucially important to living things, and the chemistry of the carbon-containing materials began to be called Organic Chemistry. The name stuck, although the limitation to natural materials quickly vanished as chemists learned that a limitless number of chemical

---

[27]Schwartz, A. T. "The Importance of Good Teaching: The Influence of Arthur Pardee on Wallace Carothers," *Journal of College Science Teaching* **1981**, *February*, 218.

compounds could be made in the laboratory. Chemists want to know the properties of the individual compound—physical properties such as melting or boiling point, solubility, odor, density, refraction of light. The German chemists Carothers read about tasted their compounds. And chemists wanted to know the chemical properties—atomic composition, molecular weight, and reaction with acids or bases, with bromine or iodine, or with water or other classes of organic compounds. And they wanted to know, finally, the structure—the atom-to-atom attachments—that complete a unique description of the compound.

Pardee found that his new student already knew his basement laboratory well. Wallace Carothers had demystified the odd assemblage of rusted iron plates and steel rods called ring-stands, pear-shaped, glass-stoppered vessels called separatory funnels, round-bottom flasks, Erlenmeyer flasks, beakers, side-arm flasks, manometers, asbestos-coated wire screens, Bunsen burners, clamps, and elaborate glass condensers. He could obtain accurate weights on the analytical balance in the glass-windowed, mahogany case, finishing by using tweezers to carefully add tiny weights to the right side of the swing-arm to balance the small sample resting on the left-hand pan.

Carothers used these tools to separate compounds by selectively extracting them from water with solvents such as ether, chloroform, or benzene. He obtained purified compounds by distillation, by crystallization. He set up a bath of mineral oil in a small beaker and inserted a thermometer in the bath. He attached a tiny capillary of glass tubing, sealed at one end, to the thermometer after forcing the open end of the tube into a small pile of pure yellow crystals and tapping the closed end of the tube on the table to drop the powder to the bottom of the tube. Then he heated the oil slowly with the low flame of a Bunsen burner. He watched the solid as the bath warmed. The solid melted quickly and completely to an orange liquid, and Wallace Carothers recorded the temperature. Scientists of the early century used the characteristic melting temperature of each substance as identification of the material.

And on it goes in the study of organic chemistry. Wallace will determine the molecular weight of the compound by a method called boiling-point elevation. Then he will settle on a structure featuring some of the "functional groups" that characterize chemical reactivity.[28] The balance of Carothers' organic chemistry revolved around

---

[28]This discussion of organic chemistry in about 1915 comes from Perkin, W. J.; Kipping, F. S. *Organic Chemistry*, new ed.; W. & R. Chambers: Edinburgh, Scotland, 1911.

detailed descriptions of compounds bearing these functional groups—aliphatic and aromatic hydrocarbons, ethers, esters, acids, alcohols, halides, amines, and nitro compounds—the whole of a large and growing catalog.

"Early in his college career he showed mature judgement and was always regarded by his classmates as a more or less remarkable person," Pardee wrote in 1938.[29]

> He never held himself aloof from his fellow students but had little time for the boisterous enthusiasms of the average underclassman.

> His interest in Chemistry and the physical sciences was immediate and lasting and he immediately outdistanced his classmates in accomplishment. Throughout this entire time he helped (me) in sustaining interest in the work of the Department.

> ...he was not taking enough physical exercise for a student leading such an active intellectual life, but he was ever averse to do anything but hard and sustained mental work.

In timely fashion, Carothers completed Tarkio's chemistry curriculum in his sophomore and junior years. In his junior year he had only one student in the Commercial Department—Tort Machetanz—so he now assisted in the English Department.

The summer of 1918, with the nation immersed in the Great War, brought change to Tarkio. Pardee left Tarkio suddenly and accepted a position at his alma mater, Washington and Jefferson. In a nationwide ceremony, the military made college students sweepingly liable for active duty. Army doctors, however, exempted Wallace Carothers from military service because the Iowan, growing up in a geographical area where the water contained no dissolved iodide ion, had developed a goiter.[30]

Tarkio, with Pardee gone for good, named Carothers, now entering his senior year, its instructor in chemistry. Wallace Carothers

---

[29]Pardee, Arthur. *Contribution to the Biographical Memoir of Wallace Carothers*, February 19, 1938. (Adams Papers, University of Illinois Archives, Urbana, IL.)

[30]Julian Hill interview with Matthew E. Hermes, May 29, 1990. Arthur Pardee to Roger Adams, *Contribution to the Biographical Memoir of Wallace Carothers*, February 19, 1938. (Adams Papers, University of Illinois Archives, Urbana, IL.)

delayed his graduation. The basement laboratory in Tarkio's main building became his laboratory. He taught the school's science for two years. It seems unlikely such an extraordinary student, so challenged by the task of learning, was actually 'delayed' in meeting the requirements for graduation by his teaching responsibility. But there it is. Wallace Carothers, who has come to this outpost of education to trade commercial teaching for the simplest of college training, proceeds ever so slowly and cautiously. He does not choose to risk the unknowns of an independent life. He finally earns his bachelor's degree from the tiny Missouri college in 1920, at the age of 24. He knows chemistry but hesitates yet to call himself a chemist. He fails to recognize his transition from a youth, exuberant at the works of Edward Kennedy Duncan, to a man with embryonic status as a knowledgeable scientist, with the opportunity to take Duncan's vision and make it a reality.

After Wallace Carothers died, Dr. Thompson, the president of Tarkio, recalled Carothers had achieved nothing but "the A grade, the highest grade obtainable" while he attended Tarkio.[31]

---

[31] J. A. Thompson to Arthur Pardee, about late 1937. (Adams Papers, University of Illinois Archives, Urbana, IL.)

Chapter 2

At Urbana, the Best of the Best

Arthur Pardee remembered Wallace Carothers after Pardee left Tarkio. He wrote Carothers, he nudged him, he pushed him to continue his education in chemistry—to go ahead toward the doctorate.

Pardee recognized that German chemists dominated pre-war science. They published in the *Chemishe Berichte* and in *Liebig's Annalen*, and their *Beilstein* was the primary reference source for the physical and chemical properties of the organic compounds of the known chemical universe. The language of chemistry was German. American chemistry was second rate. American chemists were ranked, not by their own abilities, but by the reputation of the German professors who had directed them in postdoctoral years abroad. Studied with Willstätter. Studied with Liebig. Studied with Victor Meyer. And Americans depended on German chemicals—from aspirin to the spectrum of shaded dyestuffs coloring fashionable clothing.

The war brought a sudden embargo to things German, including German chemistry and chemicals. Thrown to self-dependence, American chemistry suddenly flourished. Pardee knew the tilt of the science toward Europe was ended forever. Americans trained in chemistry would fill new jobs as the country developed its own dyestuffs and lacquers and coated fabrics and artificial silk.

Carothers was 24 when he graduated from Tarkio. He had no money and few prospects. His degree earned at the minor Presbyterian college tucked in the northwest corner of Missouri hardly prepared him to work at the frontiers of science or, in fact, in science at all. Industry offered no jobs, and the teaching profession was closed to those not prepared by European study.[1]

---

[1] C. S. Marvel interview with John Mulvaney. (Chemical Heritage Foundation, Philadelphia, PA.)

Pardee avoided sending Carothers to his alma mater, Johns Hopkins. Hopkins was leaderless. Preeminence in his science was now passing from Hopkins into the hands of the Illinois department now led by Roger Adams. Wallace Carothers enrolled at the University at Urbana in the fall of 1920 to work for his doctorate at its Department of Chemistry.[2]

■■

Some of us are obsessed with chemistry. We become addicted to the laboratory, to the multiple brown bottles of chemicals lining the shelves, to the pervasive residual odors of the acrid phenols which, once spilled, soak to the core of the wooden bench top.

We habituate to our offices, heaped with papers and books stacked open, threatening their spines. We leave no place to spread out work; we cover every horizontal surface. We clean up on occasion; the task uncovers old ideas and chemical problems, revealed anew. Treasured thoughts, valuable concepts, saved papers.

Intellectual demons demand time. The undergraduate years of college simply fail to relieve the suckling need for intimacy with the science. Chemistry, in graduate school, allows us to challenge all that has gone before in order to make some small research advance, pushing the wave of chemical knowledge, in one place, higher on the unknown shore.

The university provides for new graduate students: Via an assistantship, it provides advanced course work free of tuition, a rectangle of laboratory space, and a smaller stipend. The new student, in consideration, teaches laboratory chemistry to the university's premed, agricultural, and home economics majors. There is little else to do but chemistry; the university carefully adjusts the assistantship to the bare subsistence level.

For most students attempting advanced chemistry, the chemistry department, the building, quickly reveals itself. Suite after suite of slate-covered benches set in perfect alignment border the halls where the undergraduates are trained. By late afternoon these halls are quiet, orderly, nearly inanimate. The laboratory assistants, all graduate students, have shelved the day's flasks and beakers, acids and bases. One student edges slowly, backward down the hall, wielding his wet swab in steady rhythm, back and forth across the floor.

---

[2]Pardee's role in Carothers' choice of Illinois is mentioned in Schwartz, A. Truman. "The Importance of Good Teaching: The Influence of Arthur Pardee on Wallace Carothers," *Journal of Chemical Science Teaching* **1981**, *February*, 218.

But off in the older sections, in graduate laboratories where an aged veneer of slimy film coats every surface, where jumbles of cork-stoppered, unlabeled flasks sit dusty and undisturbed, the real work of the day is just beginning. The graduate students, now freed from teaching, pull out their advanced texts and work toward their own semester examinations and departmental tests and language proficiencies. They manipulate their chemicals. They push aside the clutter; they light burners and swirl flasks and filter solids and take melting points and proceed with their research.

Lights burn on through the night in the research section. From afar, the chemistry building sits aglow like a liner on a quiet nighttime sea. Its crew, this year's cluster of graduate students, has committed to three or more years of full-time study, research, and teaching. For each, the building is now home.

■ ■

The older, four-story section of the University of Illinois Chemistry Department was completed along the edge of the university in 1901. The building was constructed of wood, except for its brick facing. Heavy lateral fire walls and a 12-inch-thick layer of sand between the ceiling of one level and the wooden floor of the next were the only protection against the chemists' occasional fires.[3]

The attractive 1916 addition stood just as high and was built with the long side facing the campus and two short extensions reaching to the old section and enclosing a large inner courtyard. The new section had a less cluttered look and a two-story limestone arch beneath the awning-shaded windows of the departmental offices at the front of the building. The Illinois chemistry laboratory was huge—almost as large as a football field on each of its four stories—and it had been built with the unclouded goal of providing chemists for the developing chemical industry.

By night, Wallace Carothers completed his organic and physical chemistry course work. Illinois required a rigorous year of classroom work on the way to the Ph.D. The university would award a master's degree for completion of courses.

Carothers' Tarkio laboratory—where as faculty for two years he ruled over a rudimentary facility but lacked even the least of colleagues—was in the past. Now in the fall of 1920, Carothers found he

---

[3]Tarbell, D. Stanley; Tarbell, Ann Tracy. *Roger Adams, Scientist and Statesman*; American Chemical Society: Washington, DC, 1981; pp 39, 41. The Tarbell's biography of Adams stands as a vital source for a picture of the Illinois department where Carothers worked for nearly six years.

was no longer alone; 20 professors and associates and instructors and dozens of assistants worked in the laboratory and, as it happens, all of them were immersed in a new experiment in American scientific education—a demonstration of collegiality and openness, a cooperative blending of all the skills of the department to train each individual student for a successful career in chemistry.

The life and vigor pulsing through the halls of the chemistry building came from the leadership of Roger Adams, head of the organic chemistry division.[4] Professor Adams, a Bostonian descendent of the sixth U.S. president, migrated from the private and uncommunicative tradition of Harvard, where he got his Ph.D. in 1912, to study in Emil Fisher's laboratory in Berlin (where he failed in an attempt to make a small molecule composed of two carbon atoms and four nitrogen atoms, dicyanocarbodiimide) (*see* **1** in Appendix A).[5] He returned to Harvard to teach but was lured by the new building at Illinois in 1916. Adams was a tough negotiator. He agitated Illinois' President James during his hiring; he insisted on a guarantee of tenure in a prearranged time frame. President James rejected Adams' demand and suggested Illinois was not about to alter its policies just for him. But Adams grasped the Illinois opportunity anyway.

By 1920, Adams' report to the trustees of the university recounted one program he held in high regard—"Summer Preps." Systematic preparation of organic chemicals, especially the particularly difficult-to-obtain aromatic derivatives that had traditionally been available only from Germany before the war, was carried out in the department by graduate student labor. To substitute for the wartime shortages, students practiced and perfected making the preparations, kept meticulous records of their time and productivity, and were paid a small hourly wage. They competed openly for the best yield of the purest material made in the shortest time. The chemicals were sold to industrial customers and other universities; the balance sheet in mid-1920 placed a value of $3000 on the stock inventory, with sales of close to $30,000 over the previous three years.

This fine chemicals business, run from the department of chemistry, formed the genesis of the annual publication, *Organic Syntheses.* More importantly, however, Preps was the training ground for hundreds of American organic chemists who developed an entire enterprise: the American synthetic chemicals industry.

As a first-year graduate student pursuing his master's degree, Carothers was not expected to carry out experimental work. But he

---

[4]Ibid., pp 38–58.

[5]Ibid., p 27. I made dicyanocarbodiimide first, some 50 years after Adams' attempts (*see* Marsh, F. D.; Hermes, M. E. *Journal of the American Chemical Society* **1965**, *87*, 1819).

slipped along the halls, drawn toward the research laboratories. Night and day, a persistent medicinal odor perfused throughout the building. Carothers found the source of the aroma; there too, he found his instructor in organic chemistry, Dr. Carl S. Marvel, still dressed in formal shirt and tie but covered by a long, black rubber apron. Marvel's laboratory differed from the rest. The flasks and beakers and filter funnels were much larger. A stone crock lay on the floor. On the bench, a five-liter flask containing a pale, yellow-green oil was stirring while Marvel adjusted the stopcock of a larger addition funnel to feed a steady stream of a deep red liquid to the flask. The red liquid discolored as it reached the oil; a cloud of white vapor roiled from the surface, up through a glass condenser and a rubber tube. The cloud disappeared into a beaker of water and cracked ice.

Marvel worked late. "From eight AM until ten PM," he said. "Then I studied. But I had to get to breakfast by seven the next morning. That's the only time I hurried. And that's how I got my nickname, 'Speed'."[6] Speed could expect to finish this preparation in about five hours. He was treating phenol with the red, corrosive, and poisonous liquid element, bromine.[7] If he spilled bromine on his hands, the liquid would bore through his skin, causing painful burns. Marvel used carbon disulfide—the most flammable and evil-smelling of the organic solvents—as a diluent. The vapors of carbon disulfide will ignite if they contact a hot steam pipe. Marvel absorbed corrosive hydrogen bromide into the ice water, and late into the night, he would carefully distill the corrosive product, *p*-bromophenol (*see* reaction 1 in Appendix A), through a flask he had specially made in the glass shop. Marvel would occasionally spill some *p*-bromophenol. It would solidify and soak into the floors and cabinets and books and papers. Its odor would not leave the building for years.

Speed Marvel was 26, two years older than Carothers. He came to Urbana in 1915 from Illinois Wesleyan at Bloomington. He learned in his year of organic chemistry there under Dr. Alfred Homberger that he "had so much fun making new compounds that there was never a doubt in my mind what I would like to do." Homberger sent Marvel to work for his own advisor, Professor Noyes at Illinois. When the war started, Professor Adams gave a "manufacturing scholarship"

---

[6]Videotaped interview of C. S. Marvel with Professor Gortler of Brooklyn College (*see* "Eminent Chemist Series"; American Chemical Society: Washington, DC, 1984). Available from American Chemical Society, Education Distribution Center, P.O. Box 2537, Kearneysville, WV, 25430, 1–800–209–0423. The tape includes a still photograph of the formally dressed "Speed" protected with a long rubber apron.

[7]*Organic Syntheses;* Adams, Roger, Ed.; John Wiley and Sons: New York, 1921; Vol. 1. Marvel reported his synthesis of *p*-bromophenol in the first volume of this series.

to the young graduate student who loved to make chemicals. Speed Marvel became the key member of the group perfecting the methods for making the now unavailable German chemicals.

He stayed on at Illinois as an assistant; he then became an instructor after he earned his Ph.D. He and Adams decided to publish a formal record of the best of the chemical preparations carried out during the war. They were collaborating to repeat the experiments, to have them checked by an independent research group, and to publish the results.[8]

Marvel noticed Wallace Carothers. He called Carothers "Doc" from the beginning. He was clearly the finest student in the organic class; he seemed to have an intuitive feel for the principles of the course. Carothers was only mildly interested in the preparations—the shifting of atoms that signaled the synthesis of one material from another. He wanted to understand the nature of the cloud of electrons surrounding the atoms, even if the atoms were not in chemical transition. At times, Carothers dominated Marvel's class. He was witty, bright, quick—an overachiever extending the known facts out beyond current knowledge. But he would appear at Marvel's laboratory at night and sit off to the side, looking straight ahead, quiet, mute.[9]

■ ■

As Wallace Carothers left Tarkio for Illinois in the fall of 1920, Arthur Pardee left Washington and Jefferson to become chairman of the small chemistry department at the University of South Dakota in Vermillion. Over the winter, he asked Carothers to join him on the faculty in the fall of 1921. Carothers' stipend at Illinois failed to meet his expenses, and the opportunity to be with Pardee as an instructor in chemistry attracted him.

Carothers knew he must complete the four units required each semester at Illinois for his master's degree to qualify for the South

---

[8]C. S. Marvel interviews with J. Mulvaney (Chemical Heritage Foundation, Philadelphia, PA) and with David A. Hounshell and John K. Smith, May 2, 1983. (Hagley Museum and Library Collection, Wilmington, DE, 1878.)

[9]"Doc was a very interesting person. When he was in good humor, there was no one funnier than he was, clever, witty. If he was low, everybody got low. He had a personality that affected everybody around him. When he was low, he was low, lower than a snake's belly." C. S. Marvel interview with David A. Hounshell and John K. Smith, May 2, 1983, p 9. (Hagley Museum and Library Collection, Wilmington, DE, 1878.)

Dakota position. He delayed answering Pardee, explaining he had been forced to drop his thermodynamics course because of a conflict with teaching. He "wanted to have some assurance of being able of accepting the position if elected," he finally wrote Pardee in an awkward and fawning letter. "...I shall attempt to satisfy the demands of the position in so far as my abilities and experience coupled with hard work may permit. With best wishes, believe me. Sincerely your friend, Wallace H. Carothers"[10]

On April 14, 1921, Wallace Carothers wrote to President Robert Slagle of the University of South Dakota and accepted a position as instructor at a salary of $2000 for the 1921–1922 academic year.[11] And on August 12, 1921, Wallace Hume Carothers, along with six others, received his master of science degree in chemistry from the University of Illinois.[12]

■ ■

Vermillion, South Dakota, was a small town, 200 miles up the Missouri from Tarkio and on the eastern side of the river. The University of South Dakota enrolled only a few hundred students. Its entire faculty was no more numerous than the staff of the Illinois chemistry department.[13]

Carothers shuttled between his high-ceilinged, drafty chemistry laboratory and his rooms with Dean Stockton of arts and sciences. The Stocktons encouraged their boarder to get out onto the land that fall, out in the bracing air to hunt or fish, or at least, to watch the quail and pheasant and deer silhouetted against the rolling, ocher fields. But Carothers attended to his teaching and to the beginning of his first, independent research investigations.[14]

Carothers prepared his lecture assignments well. But the dazzling intellect was "not brilliant" in the classroom. He was "methodical,"

---

[10]Wallace Carothers to Arthur Pardee, undated, 1921. (University of South Dakota Chemistry Department records.)

[11]Wallace Carothers to Robert Slagle, April 14, 1921. (University of South Dakota Chemistry Department records.)

[12]University of Illinois Board of Trustees report. (University of Illinois Archives, Urbana, IL, folio RS 1/2/802.)

[13]*Encyclopedia Britannica*, 14th ed.; Chicago, IL, 1939; Vol. 21, p 89. The 1934–1935 enrollment at the university was 785, with a faculty of 90.

[14]Arthur Pardee to Betty Jo Travis, February 12, 1947. (University of South Dakota Chemistry Department records.)

"careful," and "systematic," wrote Pardee.[15] "Neither was he interested in people."[16] Pardee was looking for more of a showman, an exhibitionist, and soon realized that Carothers was not suited to teach.

But Wallace Carothers was thinking about organic chemistry. Chemists have developed an abstraction—a chemical alphabet, a scheme of representation in which a letter on a page is understood as an atom of material: C for carbon, H for hydrogen, N for nitrogen, and O for oxygen.

Chemical inquiry for the academic scientist—his basic task, his virginal science—is the task of holding in one hand 'stuff': a shiny, crystalline gallstone or silky white needles obtained from coffee extracts or an orange-yellow solid obtained from living cells, which fluoresces green in water, and naming them and then representing them with a precise, unique, written structure, drafted in the chemist's unequivocal alphabetic notation. This is the first step of pure science—describing reality in abstract form.

Wallace Carothers' first scientific research dealt with precisely the problem of identifying the structure of a material by a proper notation. His interest was a substance he called diazobenzene-imide (*see* **2** in Appendix A). He knew the oily liquid was slightly more dense and had a much higher boiling point than water. Chemists before him determined that each of its molecules possessed six carbon atoms, five hydrogen atoms, and three nitrogens. The carbon and hydrogen structure was well known, and he ignored its precise details and expressed that portion of the molecule as C_6H_5-. The question he chose to answer was, are the three nitrogens, which he knew were attached to one of the carbons, all in a row or in a circle of three?

Carothers worked late in the cold laboratory, first using Emil Fisher's 1878 method of preparing diazobenzene-imide. He carved a small slice of fiery sodium metal and immersed it in the diazobenzene-imide to scavenge from it the last traces of moisture. He built a simple manometer to measure vapor pressure, and a specialized tube to determine viscous flow.

During his preparations he waited for a commercial sample of phenyl isocyanate (*see* **3** in Appendix A) to arrive from Germany. He

---

[15]Pardee, Arthur. *Contribution to the Biographical Memoir of Wallace Carothers*, February 19, 1938. (Adams Papers, University of Illinois Archives, Urbana, IL.)

[16]Arthur Pardee to Betty Jo Travis, February 12, 1947. (University of South Dakota Chemistry Department records.)

knew this compound has the linear structure $C_6H_5-N=C=O$. The single line between carbon and nitrogen and the double line between the nitrogen, carbon, and oxygen atoms build on the strictly alphabetic abstraction of structure to express further information about the molecule— information about its subatomic, electronic structure.

The lines implied the presence, between the atoms, of the mysterious particles called electrons. A single dash represented two electrons; the parallel double dash implied four. Carothers was fascinated with the electron accounting system developed by Professor Gilbert Newton Lewis of the University of California at Berkeley. In 1916, Lewis suggested chemists always assumed each of the common atoms (except hydrogen) was surrounded by eight electrons in its chemical compounds. These electrons participated in bonding atoms to each other, generally two at a time, but some bonds contained four and even six electrons. All eight electrons surrounding oxygen and nitrogen atoms seldom participated in bonding. Some existed as unbonded pairs. If the single, double, and triple dashes did not account for all eight electrons around an atom, the remaining electrons required to total eight must be these unbonded, unshared pairs.[17] Thus, in the $-N=C=O$ group of phenyl isocyanate, the tally of electrons required nitrogen to have one, and oxygen two, unshared electron pairs. Carothers compared the vapor pressure and density and liquid viscosity of the two oily liquids, diazobenzene-imide and phenyl isocyanate. Their properties were virtually identical. Carothers concluded on the basis of this indirect evidence that diazobenzene-imide was linear, the nitrogens are in a row, and the structure of diazobenzene-imide should be written $C_6H_5-N=N=N$. In this structure the first nitrogen has one unshared pair and the third has two—the same positional electron pattern as phenyl isocyanate.

Carothers understood the esoteric elements of this structural determination but was not overcome with its importance. The next year when the *Journal of the American Chemical Society* published his work, he observed wryly to his grandmother McMullen in Des Moines, "We will talk over (my work on Diazobenzene-imide) on the occasion of my next visit, and while you may not agree with all my opinions, I hope that our differences may not lead to any physical violence as they often do. And by way of warning I may add that I am keeping in training."[18, 19]

---

[17]Lewis, G. N. *Journal of the American Chemical Society* **1916**, *38*, 762.

[18]Carothers, W. H. *Journal of the American Chemical Society* **1923**, *45*, 1734.

[19]Wallace Carothers to Mrs. J. L. McMullen, March 7, 1923. (Terrill.)

This was Wallace Carothers' first scientific publication. When it appeared in the journal he mailed a reprint to Pardee, adding, "grateful appreciation of your generous interest in this little work—."[20]

■ ■

Wallace Carothers stayed only that one year at South Dakota. He noted on two personnel forms he filed later at the University of Illinois that after he left Vermillion he spent the summer of 1922 at the University of Chicago. Frances Gelvin Spencer wrote in 1990:

> The last time I saw W(allace) was in the second year of my marriage. I was walking on the Chicago University campus when I ran into this man with his felt hat all pushed up into a choc(olate) drop soiled & dusty clothes & what looked like a weeks growth of whiskers on his face. I never saw a railroad vagrant look more pathetic. I finally recognized W. accompanied by another man. I spoke to him, and W. started gasping, opening and closing his mouth. I thought he was having a heart attack. He kept gasping, the companion supported him. I jabbered on as if I didn't notice his discomfiture. I said I was in Chicago. My husband was getting his doctorate in Classics at Chicago U. & that I was just going to hear him sing in the Leon Mandel Hall choir at the Chicago U. chapel. Finally W. recovered, we chatted a bit & I asked him if he would like to come to dinner with us. He agreed & I said I'd call, which I did several days later. He came to dinner all polished up & we all had a pleasant time. Several days later, my husband came home with a box of candy and said he had run into W. on the campus & W. gave him the candy to give to me. I'll never know why he was coming to see me in the afternoon nor what he wanted to say. Nor will I ever know why he reacted as he did on the campus.[21]

What is this mysterious apparition of the bright young man whose strange Chicago season is held in Frances Spencer's memory for nearly 70 years? Carothers had to be somewhere, of course, in the summer of 1922 between Vermillion and his reentry at Illinois. But his chance encounter with Frances gave her a sudden and unexpected glance beneath Wallace's carpet of respectability. Choose whatever explanation you will. The look of a vagrant, unshaven and dirty, may come with sufficient cause in a camp, in the country, but in the city, in summertime, it is just the look of a vagrant.

---

[20]Reprint held in the University of South Dakota Chemistry Department records.)

[21]Frances Gelvin Spencer to Matthew E. Hermes, August 13, 1990.

Whatever the secret of Wallace Carothers' summer in Chicago was, he never revealed it to Frances. Not at their chance meeting on the street, not when he cleaned up and came to dinner, and not years later, when Wallace revealed to Frances tortured details of a fading spirit.

Most men harbor secrets. Life, it seems, is a journey to maturity—a place where those secrets no longer matter. A mature man, wherever he goes, has no need to gasp in dismay, regardless of whom he meets.

■ ■

By the time Wallace Carothers reenrolled at Illinois to chase his doctorate, Roger Adams' department of chemistry had become the most prolific in the country. Adams himself trained 4% of all the nation's new chemistry Ph.D.s between 1920 and 1924. Marvel was now promoted to associate. Adams directed nearly 20 graduate students. He placed them four or more to a lab, mixing new students beginning research with the more experienced students. Adams visited the laboratories day or night, greeting each student with a Bostonian, "Waall what's new"? Some were eager to respond; often Adams and his students generated new directions for the research work. Roger Adams seldom forgot the conversations and he would be back in a week. "Did you try...," he would ask. Given a free moment, Adams talked expansively—on politics or business conditions or sports. Or other chemists. Adams comments were "shrewd and usually pungently expressed." Adams made himself a generous advisor to his students, but he contained his familiarity so that there was never any question who was the boss. By the time Wallace Carothers joined Roger Adams' research group, everyone called Adams, "The Chief."[22]

Adams asked Carothers to extend an important discovery Adams had just made. The surface of finely divided platinum or palladium or rhodium serves as a template for the addition of elemental hydrogen to electron-rich, unsaturated organic compounds. The equation summarizing the reaction of hydrogen with this four-electron bond is

$$-CH=CH- + H_2 \rightarrow -CH_2CH_2-$$

Even in the 1920s, hydrogenation of unsaturated fats to prepare hardened lipids for shortening and soaps was a major industry. But the powdered metal catalysts acted erratically. In one preparation, hydrogen would be absorbed rapidly; in the next, the chemical transformation would take place sluggishly if at all. Adams' invention was largely

---

[22] *Organic Syntheses;* Adams, Roger, Ed.; John Wiley and Sons: New York, 1921; Vol. 1, pp 59–75.

accidental. In setting out to make the traditional platinum catalyst, one of Adams' students spilled a solution of the valuable platinum salt. The recovery of all the metal from the bench top turned into an arduous task. The recovered platinum salt was contaminated with debris. Adams burned the residue with sodium nitrate to purify the platinum. But instead of metallic platinum he obtained a brown powder, platinum oxide. Hydrogen converted the oxide smoothly and quickly to "Adams' platinum," a reactive, reliable, and reusable form of platinum metal used ever since in hydrogenation.[23]

Adams instructed Carothers to extend the catalyst work to the reduction of electron-rich aldehydes to alcohols:

$$-CH=O + H_2 \rightarrow -CH_2OH$$

Carothers found Adams' platinum failed to catalyze the hydrogenation. He said the aldehyde had "poisoned" the catalyst but found that a small amount of iron salt restored the activity of the catalyst, and hydrogen cleanly converted the aldehydes to alcohols. When he tried the modified catalyst on a compound containing both groups, an olefinic aldehyde, he recorded a surprising result. In the presence of iron, only the aldehyde group reacted with hydrogen; the olefinic carbon-to-carbon double bond remained unreacted.

Carothers developed an experimental routine for one part of the work. He was determined to explore as many metallic co-catalysts besides iron as he could find. Each morning he added platinum and the test metal to a solution of benzaldehyde. He pressured the small glass vessel with hydrogen. As Carothers worked in the laboratory, preparing the co-catalyst for the next day's experiment, he watched the pressure of the hydrogen fall as it was absorbed and he graphed the results. At the end, he published a dictionary of more than 50 metal salts and their effect on reduction.[24] The work was a daily grind.

With this chemistry, Carothers added a detailed and useful extension to Adams' platinum studies (*see* reaction 2 in Appendix A). He completed the work in less than six months, and by late February 1923 he reported to his mother that he had sent his first manuscript describing the catalytic effects of the iron salts to the *Journal of the American Chemical Society*. He told his mother, too, that he had taken

---

[23]Ibid., p 71.

[24]Carothers, W. H.; with Adams, R. A. *Journal of the American Chemical Society* **1925**, *47*, 1047.

additional course work that winter and received all A grades; he completed four departmental examinations in the subdisciplines of chemistry and received two A's and two B's. He passed his French and German language requirements in routine fashion.[25]

But Carothers' successful laboratory and course work serves simply as a grace note on the full score of his elusive but emerging presence at Urbana. For Carothers made no further discoveries in his work for Adams. The aldehyde reduction work he sent to the *Journal* in February 1923 became the framework for two more papers and his Ph.D. thesis defense in May 1924, more than a year later.

Here is the puzzle. Near the end of his first year back at Urbana, Wallace Carothers looked at the steppingstones toward his Ph.D. as "a form of slavery." In April 1923 he was in the middle of his assembly-line studies of the salt co-catalysts. The requirements of the university held for him "all the elements of adventure and enterprise which a nut screwer in a Ford Factory must feel in setting out for work each morning." Illinois would give him his doctorate provided he wait out the required three years and continued to "show intelligence slightly above that of the pathologically subnormal." And he complained of the lack of true scholarship at Illinois; he saw none of the "cultural appendages of education."[26] "No time for critical comments or fine writing," he wrote his mother.[27] He was tired from years of schooling, ready to move on, but he could bring into focus only vague and unrealistic goals. "I meditate complete escape," he wrote Wilko Machetanz, his Tarkio roommate, as he congratulated Mach for passing his bar examinations in California.

Carothers dreamed of a private laboratory—in New York, Berlin, Vienna, or Paris—where he "could test out some ideas of vast commercial importance." But he passively accepted classical destiny in deciding the timing and direction of his future. In the spring of 1923, the university awarded Carothers its Carr Fellowship for the upcoming year.[28] The fellowship, with its stipend of $750 per year, required nothing in return beyond Carothers' presence at the university. "No

---

[25]Wallace Carothers to Mary Carothers. Undated but probably written in late February 1923. Wallace asks for his grandmother's address in the letter. He wrote a related letter to his grandmother on March 7, 1923. (Terrill.)

[26]Wallace Carothers to Wilko Machetanz, April 22, 1923. (Hagley Museum and Library Collection, Wilmington, DE, 1850.)

[27]Wallace Carothers to Mary McMullen Carothers, February 1923. (Terrill.)

[28]University of Illinois Board of Trustees reports, April 11, 1923. (University of Illinois Archives, Urbana, IL, folio RS 1/1/802.)

other inducement than the Carr," Carothers wrote, "would cause me to continue the struggle for a Ph.D. for another year, and the Ph.D. has so long been in sight as a goal that it seemed improbable that the fates would deprive me of it, especially since my desire for it continuously diminishes...all this rests as yet on the knees of the Gods."[29]

Somewhere from his background, Carothers drew upon the word "abulia" to characterize his mental state that spring.[30] Abulia is a psychiatric term defining the inability to come to choose, to make a decision, to act independently.

Carothers was broke, too, with expenses outrunning income by $20 per month. The expected financial cushion from his year at South Dakota had disappeared.

Yet in many ways Carothers seemed to be leading a life in full dimension. Spalding, the violinist, visited Urbana. Carothers heard him play. Sinclair Lewis published *Babbitt*. Carothers worked at the book, struggling to get beyond Lewis' malicious veneer and found "that the realism is not only literal but dramatically true..."

He reflected on the value of his education at Illinois. "They are attempting to raise the standards of the graduate school," he wrote,

> and of course they want to eliminate the ones they consider unfit as early as possible, hence the departmentals. Incidentally they bumped about fifty percent of the graduate students who came up this time. In some cases it is rather tragic, for of course in graduate work everyone is in earnest about the matter of getting by. It isn't like the undergraduate where students flunk just because they don't care. A majority of the graduate students are undergoing some sacrifice to stay in school for three or four years after graduation from college and during the best years of life, and to deny oneself everything but study and work in the laboratory only to discover that one is unfit for graduate work must be very disappointing. But when one gets through I suppose that ultimately it will be some satisfaction to have it known that one came from Illinois. There appear not to be more than two other schools in the country with standards as high as those here.[31]

---

[29] Wallace Carothers to Wilko Machetanz, April 22, 1923. (Hagley Museum and Library Collection, Wilmington, DE, 1850.)

[30] Ibid. "A brief escape from the abulia which has been enfolding me is to be devoted to replying to your letters."

[31] Wallace Carothers to Mary McMullen Carothers, February 1923. (Terrill.)

He developed a new passion for billiards and coffee at four in the afternoon in the company of a couple of "enlightened spirits."[32]

Carothers' younger sister, Isobel, now a freshman at Northwestern, came to Urbana to visit Wallace. She drove to Urbana from Evanston, through Chicago, Kankakee, and the series of smaller towns, Del Ray, Loda, Ludlow, on the narrow road along the Illinois Central right of way. It was April 1923 and Isobel motored slowly along the gravel-surfaced roads, which were improved with a covering of concrete only in short sections near the towns.

Isobel was spending a long weekend of relief from her studies. She was immersed in a frenzy of activity at Northwestern. She planned to stage a satirical one-act play—dialogues, character sketches, lighting, costumes, scenery, and rehearsal schedule—and then switch to a children's drama. She visited Jane Addams' Hull House to tell a Japanese fairy tale, "The Mirror of Matsuyama," to the children living there. Isobel was witty and wordy, and at 23 she was beginning to see herself as an entertainer.[33] Her stage voice was high-pitched and fast-paced, and she laughed with a birdlike cackle.[34]

Wallace walked her through the empty halls of the chemistry department that April Saturday. They returned late because Carothers was working. Isobel was dazzled. She took the stub of a pencil and tore open an envelope bearing the return address of the university and wrote on the inside of her improvised stationary:

Saturday nite

Dear Family:

Ain't this fun?

I'm over in Wally's lab. He's working on something—don't ask me what. You know he has a room in this great immense university all by himself to work in the importance of which you would realize if you could see the hugeness of the place and the hundreds and hundreds of people who attend this place.... I came over here again with

---

[32]Wallace Carothers to Wilko Machetanz, April 22, 1923. (Hagley Museum and Library Collection, Wilmington, DE, 1850.)

[33]Isobel Carothers to Mary McMullen Carothers, undated, probably 1923. (Terrill.)

[34]Isobel Carothers appeared on the radio sketch, *Clara, Lu, 'n Em,* from 1930 until her death in 1937. Audiotapes of four of these performances are held in the Museum of Broadcast Communications, 800 South Wells St., Chicago, IL.

him as he had left something here in process of — procedure. It seems so unreal that I should actually be hear [*sic*]. Tho't it would be fun to write you a line with him actually here.[35]

Wallace Carothers delighted in Isobel's visit. He typed a neat note immediately after she left:

Sunday 4 p m

Dear Sister:

I have to write you a few lines to tell you how glad I was that you were able to come down to Urbana, and how much your visit meant to me. I didn't realize myself how much it meant until you had piled into the car and pulled out. Then lightly touched with the sadness of the thought that you had gone was the graciousness of the memory that you had been here, and that the monotonous quotidianism had been illumined for a few hours with your presence....

Wallace[36]

Two days later Isobel wrote home and sent along Wallace's letter. "Thanks for the dictionary dad!" she wrote. "You see I had almost immediate need for it. <u>Only</u> that word wasn't in it."[37]

■ ■

Adams and Marvel saw beyond the modest experimental research contribution Wallace Carothers was making. They contrasted Carothers' desultory performance in the laboratory with his emerging understanding of chemical bonding and structure. Among all of them, both faculty and students, Carothers read the scientific literature most assiduously, and among all of them he had the most retentive memory.

Carothers clarified for them G. N. Lewis' octet theory, which he had used at South Dakota to define the structure of diazobenzeneimide. His year on the faculty at South Dakota and the 1923 publication of the paper in the *Journal of the American Chemical Society* under

---

[35]Isobel Carothers to Ira and Mary Carothers, April 1923. Years later, Carothers' sister Elizabeth wrote on the improvised stationery, "This envelope smells like the lab in a very much diminished degree—." (Terrill.)

[36]Wallace Carothers to Isobel Carothers, April 1923. (Terrill.)

[37]Isobel Carothers to Ira and Mary Carothers, April 1923. (Terrill.)

his own name and apart from the tutelage of a senior faculty member began for its author an early transition from student to faculty. He was counsel to both groups: His completed and independent research represented for the students a target; his depth of analysis represented for the faculty a resource.

Lewis' 1916 interpretation of the chemical bond between carbon and nitrogen, hydrogen, or oxygen as a sharing of paired electrons solved the final puzzle of the atomic theory, accounting for the bonds between the atoms we know exist in organic compounds. Irving Langmuir, the General Electric chemist who effectively promoted Lewis' work to the extent that the octet rule came to be known as the Lewis–Langmuir theory, said in 1921:

> These things mark the beginning of a new chemistry, a deductive chemistry, one in which we can reason out chemical relationships without falling back on chemical intuition.... I think that within a few years we will be able to deduce 90 percent of everything that is in every textbook on chemistry, deduce it as you need it, from simple ordinary principles, knowing definite facts in regard to the structure of the atoms.[38]

Carothers' reflections on the Lewis–Langmuir concepts ran deeper than the synthetic empiricism of Adams and Marvel and the Illinois group. In the months after he completed the core of his thesis, catalytic reduction work, doctoral candidate Carothers manipulated the octet theory to explain chemical reactivity of all the carbon-containing double-bond systems. Carothers deciphered nearly a dozen types of known chemical reactions of the four-electron bond in a paper called "The Double Bond" he submitted to the *Journal of the American Chemical Society* early in February 1924. Lewis imagined the nuclear atom surrounded by electrons disported to the eight corners of a cube.[39] Because the core of the Lewis theory was that electrons tended to pair, Carothers saw no need for his three-dimensional representations of electron disposition. Carothers suggested a planar

---

[38]Jensen, W. B. "Abegg, Lewis, Langmuir and the Octet Rule," *Journal of Chemical Education* **1984**, *61*, 196.

[39]Lewis, G. N. *Journal of the American Chemical Society* **1916**, *38*, 762. Lewis had drawn these cubical structures as early as 1902. *See* Lewis, G. N. *Valence and the Structure of Atoms and Molecules;* American Chemical Society Monograph; American Chemical Society: Washington, DC, 1923; p 29. Available from University Microfilms International, 300 North Zeeb Road, Ann Arbor, MI 48106.

illustration with two electrons at the corners of a square surrounding the atom. Carothers' new representation was just one step away from the shorthand Lewis was then adopting to represent valence electron structure: placing four colons around the atom to represent a completely filled valence shell.[40]

In the paper, Carothers visualized the movements of the electrons in the individual molecules as chemical reagents attacking the carbon bonds. He described a detailed and predictive mechanism of the catalytic hydrogenation afforded by the metal surface of the Adams' catalyst.

His mechanistic thinking differentiated him from his experimentally skilled senior colleagues now. In fact the general run of organic chemists distrusted the introduction of the Lewis–Langmuir electron manipulation into synthetic chemistry. Carothers was the sole author of the work and gave no credit to Adams in the manuscript.[41] Perhaps this omission was Roger Adams' choice. Was Roger Adams troubled with the paper and with its ramifications? After all, Adams knew any position he took would carry great weight among his academic peers. But Adams also knew these men saw little of use in the new speculations. He may have chosen an easier, softer way by letting Carothers alone take the heat for his theoretical work. A *Journal* reviewer took a critical stance against Carothers' paper, but the *Journal* published it late in 1924.[42]

The Illinois department was more than just its senior faculty. It was undergraduates and graduate students and assistants and the newer staff. They all began to come to Carothers as they saw his mechanistic vision as a complement to their focus on synthesis and structure determination of new organic compounds.[43]

"The Double Bond" fails the test of time as an accurate description of the electronic structure of organic molecules. Carothers saw for ethylene, the simplest of the olefins, a mixture of electronic struc-

---

[40]Lewis, G. N. *Valence and the Structure of Atoms and Molecules;* American Chemical Society Monograph; American Chemical Society: Washington, DC, 1923; p 88. Available from University Microfilms International, 300 North Zeeb Road, Ann Arbor, MI 48106.

[41]Carothers, W. H. "The Double Bond," *Journal of the American Chemical Society* **1924**, *46*, 2226.

[42]Ibid.

[43]A knowledgable and anonymous reviewer of this book pointed out that "American organic chemists in general rather resented the intrusion of the physical concepts of sub-atomic structure into organic chemistry." The reviewer cited Kohler, Robert E. "The Lewis–Langmuir Theory of Valence and the Chemical Community," *Historical Studies in the Physical Sciences* **1975**, *6*, 431.

The reviewer asks whether Carothers' "The Double Bond," judged in this light, truly made his Illinois reputation. Perhaps not. But the impact on the department of a sudden, independent, innovative paper from a brand new Ph.D. cannot, in my thinking, be underestimated.

tures. He pictured electrons rocking back and forth from a stable form into two equivalent, electrically charged structures with unsymmetrical electron distribution:

$$:CH_2-CH_2 \rightleftharpoons CH_2=CH_2 \rightleftharpoons CH_2-CH_2:$$
$$(-)\ (+) \qquad \rightleftharpoons \qquad \rightleftharpoons (+) \qquad (-)$$

Carothers viewed this electron polarization as the core element in chemical reactivity of the double bond. The belief in the existence of discreet, charge separated forms of electron-rich bonds ended (except as a common misunderstanding by entering students of organic chemistry) with Linus Pauling's adaptation of quantum theory to the chemical bond.[44] We now know the double bond has an average low-energy electronic structure as shown by the doubly bound construct. The bond is not polarized but is polarizable and can assume the extreme charged structures only under the influence of approaching, truly electronically charged species.

But for those of the synthetic faculty at Illinois, who saw past the liquids and crystals in their new Pyrex flasks to the accounting of arrangements and shapes of their molecules, came their student Carothers' new vision, a revelation of the electronic presence of those atoms, and the ability to predict how to manage them.

Two weeks after he submitted "The Double Bond" to the *Journal*, on February 6, 1924, Illinois named Carothers an assistant in the chemistry department. He gave up the second half of the Carr Fellowship, with its $75 a month stipend, trading it for the $1800 per year salary as the most junior member of the department.[45]

On the afternoon of Monday, May 19, 1924, Wallace Carothers defended his Ph.D. thesis in front of a committee of Professor Adams and four other senior faculty members. At this symbolic, formal affair he reviewed the experimental work he had done in 1922–1923 under Professor Adams' direction. The invitation to the examination suggests Carothers did not plan to discuss his independent thinking on structural chemistry—the stuff of which his growing reputation in the department may well have been made.[46]

Carothers stayed quietly at Illinois for two years as an assistant. He worked at times in a small laboratory opposite the entrance to the

---

[44]Pauling, Linus. *The Nature of the Chemical Bond*, 2nd ed.; Cornell University Press: Ithaca, NY, 1942.

[45]University of Illinois Board of Trustees records, March 11, 1924. (University of Illinois Archives, Urbana, IL, folio RS 1/1/802.)

[46]The announcement of Wallace Carothers' final Ph.D. examination shows it took place on Monday, May 19, 1924, at 4 p.m. in the chemistry building in front of Professors Adams, Rodebush, Beal, Hopkins, and Townsend. (University of Illinois Archives, Urbana, IL, folio RS 2/5/15.)

chemical library. He wrote two additional papers for the *Journal of the American Chemical Society*, completing publication of his thesis work.[47] He toyed with the chemical reaction of succinic acid with glycols, but the experiments "had not proceeded very far at the time he left."[48] He published one short paper.[49] Speed Marvel, Jack Johnson, and "Doc" Carothers along with two graduate students, Paul Salzberg and Merlin Brubaker, met regularly at Marvel's apartment for a seminar on "problems that none of us knew anything about but wanted to learn about. ...we would debate these things over a can of beer...."[50] Carothers taught Paul Salzberg in the course on identification of organic compounds. He would hand the student an "unknown," a material whose identity was known only to the instructor. Salzberg climbed through the laboratory window one Thanksgiving day to put in extra hours on the course. "Everybody else was there already," he said. "It was a great place to be."[51] These five men and their science were associated with each other and with DuPont until the end of Carothers' life.

Adams called Carothers an "unusual man," a man who could excel at physics, mathematics, and physical chemistry as easily as he handled organic chemistry. He was modest and quiet with an interest in music, philosophy, and labor relations.[52] Marvel found that Carothers' moodiness grew with time.[53] Brubaker reacted angrily in 1978 when he was asked about Carothers. "What do you want to work over that manure pit for? The guy was a manic depressive. He knew

---

[47]Carothers, W. H.; with Adams, R. A. *Journal of the American Chemical Society* **1924**, *46*, 1675; **1925**, *47*, 1047. Both papers list Carothers with Adams, rather than the more common Carothers and Adams. Both Adams and Marvel published this way on occasion. The modified conjunction expresses the collegiality of the Illinois department.

[48]John R. Johnson to Roger Adams, November 3, 1938. (Adams Papers, University of Illinois Archives, Urbana, IL.)

[49]Carothers, W. H.; Jones, G. A. *Journal of the American Chemical Society* **1925**, *47*, 3052.

[50]C. S. Marvel interview with David A. Hounshell and John K. Smith, May 2, 1983, (Hagley Museum and Library Collection, Wilmington, DE, 1878, p 8.)

[51]Paul L. Salzburg interview with David A. Hounshell, September 29, 1982, (Hagley Museum and Library Collection, Wilmington, DE, 1878, p 4.)

[52]Audiotape of interview with Roger Adams at the Chemist's Club, New York, 1964. (Adams Collection, University of Illinois Archives, Urbana, IL, Box 9.)

[53]C. S. Marvel interview with David A. Hounshell and John K. Smith, May 2, 1983. (Hagley Museum and Library Collection, Wilmington, DE, 1878, p 10.)

more about the subject than any psychiatrist in the country. He carried cyanide pills in his pocket—YES—even in the University of Illinois days."[54]

Once Carothers slipped his hand around the cyanide, Adams, Marvel, and Johnson began a task they would continue until Carothers died. They regularly measured Carothers' mood, hoping each day to find him "up," functioning, and happy. They began to talk among themselves about their friend, attempting to analyze his temperament, trying to arrange his environment, trying to fix him.

Wallace Carothers' chemical reputation took on mythic proportion at Illinois, assembled as it was from his study of the growing body of American chemistry, his currency with German science, *and* his interpretation of modern theory from Berkeley and the European capitals. He spread his knowledge and advice for the benefit of his colleagues, the pragmatic midwesterners making new chemicals at Urbana. But this prominence came from fragile stuff. Carothers stayed at Urbana, stuck as an instructor, leading the "aboulic life," repeating each day the previous day's activities, coming awake each morning, his first conscious awareness the tight pain that lies across the gut, the pain of reality.[55] Another day beginning for Carothers with the dreadful combination of yearning for success, for real achievement as a reward for his chemical intuition and measured by invention and metamorphosis. But, fixated as he was on his own depression to the exclusion of the laboratory, the department, the chemistry, he found an increasing inability to raise his arms, to write the letter, to choose a specific research objective, to assemble the equipment, to do the experiment, to make the critical decision to move on. The teaching rooms to which he was committed animated him. Immersed in the repetitive teaching, he could forget in daylight,

---

[54]Merlin Brubaker interview with Mrs. Adeline C. Strange, 1978. (Hagley Museum and Library Collection, Wilmington, DE, 1985.) In 1982, in an interview with John K. Smith, whom he knew was writing a history of DuPont research, Brubaker took a more diplomatic approach. (Hagley Museum and Library Collection, Wilmington, DE, 1878, p 14.) "He was a manic depressive, and he knew it. I suspect he knew pretty near as much about that ailment as the psychiatrist he consulted with. He had read all the books on it and knew all the details of it. He was reported, and I don't know how true that is, to have threatened suicide in Illinois, when he and I were both there."

Both Merlin Brubaker and Paul Salzburg worked later with Carothers at DuPont: Both spent their entire careers at the company.

[55]Sartre, Jean-Paul *Baudelaire;* Turnell, M., Trans.; New Directions: New York, 1950; p 15.

the shadows of the night. The depressed man, thrust to his duty, responds in a workmanlike fashion.

Experts think and talk. In the simplest grammatical terms, they give advice. In contrast, revolutionaries are men of action, standing with a new set of principles, an order of values they have themselves invented, and they make change. Verb, object. Give advice. Make change. Two very different tasks.

Carothers saw no value in his role as expert presence at Urbana. His dream of a laboratory on the continent, of work in the magical world of the postwar expatriate; Vienna, Paris flashed on occasion, through his cloud of melancholy. Marvel got Carothers to catch his first Northern Pike on Squaw Lake in Wisconsin in the summer of 1925.[56] Those two years, from the spring of 1924 through September 1926, were the only time Carothers and Marvel were colleagues, and with all Marvel said and wrote about his friend after his death, he had little specific to note about their time in the Illinois department. "Since early manhood he suffered from periods of depression." He was, "A complex man; shy, inquisitive, intellectually bright and quick; self-demanding and an over-achiever; pleasant but at times despondent; had the ability to extend (chemical) reactions and known information; a motivator."[57]

Harvard came fishing for Carothers early in 1926.

In the club of elite American organic chemists, James B. Conant joined Roger Adams. Conant's career as the young president of Harvard and as high commissioner in Europe after World War II was far ahead of him when he met Adams while he pursued his doctorate in Cambridge. Adams left Harvard for Illinois in 1916; Conant replaced him on the Harvard faculty. Adams ran one of three sections of the Chemical Warfare Service during World War I; Conant directed another. Adams planned to give a summer course at G. N. Lewis' innovative University of California at Berkeley department in 1924. At the last minute, Adams canceled; Conant substituted again. He took his wife and new baby on the westbound Union Pacific express. He found a universal collegiality there, in contrast with his Harvard provincialism. Berkeley's international seminar chuckled with Conant as

---

[56]Carothers and the fish are seen in a photograph from C. S. Marvel to C. E. Carraher, November 30, 1982. (Marvel Papers, Chemical Heritage Foundation, Philadelphia, PA.)

[57]Marvel, C. S.; Carraher, C. "Wallace Hume Carothers, Innovator, Motivator, Pioneer;" Draft for *Pioneers of Polymers*, 1982. (Marvel Papers, Chemical Heritage Foundation, Philadelphia, PA.)

he apologized for his bachelor's and Ph.D., and now his faculty position, all realized in the same halls of Harvard University.[58] But after studying the German system under which the great professors gained world status on the backs of isolated groups of new Ph.D.s who came to them as assistants, Conant chose to bring the influence of this competitive European system of basic academic research to Harvard. His ideal was as far from California's as Cambridge is from Berkeley. There, Lewis insisted all his professors teach laboratories, perhaps at the expense of research progress; Conant envisioned his Harvard as a select department with the major professors responsible only for research. Adams' Illinois department took a middle ground. The professors would depend on graduate students for laboratory teaching; they would count on these same doctoral candidates as the hands performing their research.

Conant, on Adams' advice, recommended Harvard hire Wallace Carothers to fill an open instructor position in 1926. The university needed Carothers as a teacher to handle elementary organic chemistry and its laboratory, and Harvard offered a chance at the advanced courses in structural organic chemistry. Carothers welcomed the opportunity and hoped to move to Harvard in the fall of that year. But first he would go to Paris.

[58]Conant, James B. *My Several Lives;* Harper & Row: New York, 1970; p 64.

Chapter 3

Paris and Harvard

> You can't get away from yourself by moving from one place to another. There's nothing to that....Why don't you start living your life in Paris?
>
> Jake to Robert Cohn
> *The Sun Also Rises*[1]

Carothers' friend, Jack Johnson left Illinois in the fall of 1922, lured by the spell of the continent, of Paris. Across from the Sorbonne, Johnson pulled the strings of his black rubber laboratory apron tight around his waist each day and toiled in his laboratory at the Collège de France. He became fluent in French. He returned to Urbana in 1924 and worked on Adams' staff with Marvel and Carothers for the next two years. Carothers and Johnson roomed together in an apartment owned by a professor of engineering. They met most mornings at 10 and walked off campus for coffee and a cigarette. They were not allowed to smoke on the Illinois campus.[2]

In the spring of 1926 Johnson persuaded his friend Wallace Carothers to return to Paris with him, to cash in on his anticipated position at Harvard, to bring reality to his vision of the continent. They used an excuse that they would attend a chemical meeting at Oxford in August. Carothers and Johnson left New York on June 13 on the French Line's *Paris*.

---

[1] Hemingway, Ernest. *The Sun Also Rises;* Scribners: New York, 1926; p 11.

[2] J. R. Johnson to Speed Marvel, July 4, 1974. Johnson's footnote to this chatty letter reads, "50 years ago I was just returning from France and getting set to start teaching at U. of I. You and I and Wallace were together the following two years." (Marvel Papers, Chemical Heritage Foundation, Philadelphia, PA.)

The French Line called itself "The Longest Gangplank in the World." As you stepped aboard the *Paris*, you were in France, surrounded by music, engulfed within the ship's majestic, double-tiered dining salon, the smart set in first class rubbing shoulders with the Greenwich Village artists and writers and the intellectuals who traveled in the always crowded cabin and third class. Carothers and Johnson shared a third-class cabin. But Carothers found the trip "a little boring at times." The problem was their companions: "mostly miscellaneous foreigners—Italians, Poles, of the partly Americanized class."

"If it is reported," he wrote his mother, "that any articles of my clothing have been washed up on the Atlantic coast you need not fear for my safety on that account. Monday evening my cap blew off quite suddenly and dropped gently into the green & silver foam at the side of the boat."[3]

If celebration and delight was Carothers' and Johnson's goal, their timing was excellent. They arrived in Paris as spring turned to summer, as the students and the exiles and the artists and the writers, anticipating languid August days in the countryside, prepared their masquerade costumery to the theme of the great masked ball.

Paris was a city of fantasy, punctuated at the end of June by the Four Arts Ball, a "full throttle orgy," attended by 3000 or 4000 students, celebrating with models and prostitutes and rich ladies from America the closing of school. The Parisians and their guests reveled in costumes of little more than body paint and snaked through the streets and hotels, rejoicing with splendid drink and greeting each other with public fornication.[4]

By night the Americans in Paris, the Main Street exiles, cast about at the Dingo Bar in the rue Delambre for a glimpse of the startlingly young Hemingway whom they now knew stood for them all: aggressive, pugnacious, and heroic, scraping his living with long cold days writing in his flat above the Luxembourg Gardens. They all talked and drank and made love like Hemingway's new characters. "You are

---

[3]Wallace Carothers to Mary Carothers, June 16, 1926. (Terrill.)

[4]Wolff, Geoffrey. *Black Sun;* Vintage Books: New York, 1977; p 167. Wolff's biography of poet Harry Crosby rests on the detailed record of Crosby's Paris debauch. Crosby killed himself in 1929 at the age of 31. He planned his suicide and the record that preceded it. Unfortunately, we have little of this same detail from the quiet Carothers.

all a lost generation," Gertrude Stein observed as Paris, city of drugs and sex and alcohol, filled with America's young writers and artists.[5]

Scott Fitzgerald, with Zelda, was in Paris, Ezra Pound, Archibald MacLeish, Gerald Murphy and Sara, Hemingway and Hadley, Harry Crosby and Caresse, all of them at Sylvia Beach's bookstore and drinking coffee and inexpensive red wine at the outdoor cafes and all of them writing critically about each other with minds impaired by the impact of their own alcoholic excesses.

Carothers in France was drawn to the flame, enticed by the charm and vigor and the mysterious sensuality of Paris; lured by the promise of its music and literary trysts. When Carothers arrived at the American University Union on the Boulevard St. Germain, he found a telegram. The Harvard Corporation had appointed him as instructor for the 1926–1927 year at a salary of $2250.[6] Now, with Harvard in hand, Carothers was no longer shopping for a laboratory or a scientific opportunity on the continent. After all, when he realistically could have accepted such a posting during his two years as an assistant at Urbana, he never sailed to Europe, never took his reputation and fluent German to Berlin or Vienna—or Paris—to fulfill his long-expressed yearning. In Paris, he could now be safely wistful for the continent, secure that he need do nothing to convert his melancholy dreams.

Gerard Berchet, a young French chemist at the Collége de France, who later played a pivotal role in the invention of nylon, says that Johnson's nights were taken with touring Carothers around his now familiar City of Light.[7] Carothers and Johnson wandered among the book sellers on the Left Bank. Carothers stopped one day and purchased a strange, small volume. It was called *21 Ways to Commit Suicide.* The book was illustrated.[8]

Through July and August, Carothers found a home at the laboratory once shared by Johnson and Berchet. He spoke no French and Berchet spoke no English, but Johnson fluently translated for them.

[5]*See* Cowley, Malcolm. *Exile's Return;* Viking: New York, 1956, for the flavor of Paris in 1926.

[6]The recommendation by the dean of the faculty of arts and sciences for the division of chemistry, approved by C. H. Moore, is dated June 15, 1926, and lists the American University Union, 173 Blvd. St Germain, Paris, as Carothers' address. (Harvard University Archives, Cambridge, MA, UA III 5.55.)

[7]Gerard Berchet interview with Adeline C. Strange, 1978. (Hagley Museum and Library Collection, Wilmington, DE.)

[8]John R. Johnson interview with Adeline C. Strange, July 1978. (Hagley Museum and Library Collection, Wilmington, DE.)

Carothers tagged along with the two former laboratory mates as
Johnson repeatedly visited the American Consulate to help Berchet
gain his visa to the United States. Carothers, Johnson, and Berchet
sailed together, third class, for the States on the *Paris* in early September. During the voyage they were entertained nightly by five Brown University students who drew even the first-class passengers to dance to
their "diabolical rhythm." Facing their landing in New York, with the
country's continuing Prohibition laws, the three young scientists purchased a modest cache of six bottles of champagne at Le Havre and carried it onto the boat. They ceremoniously opened and drank one each
day, celebrating with the final bottle within the long shadow of the
Statue of Liberty.[9] They arrived in New York on September 14.
Carothers successfully smuggled a copy of James Joyce's still-banned
Ulysses past customs; Carothers and Johnson stayed together at the
McAlpin Hotel, waiting for Berchet to clear immigration at Ellis
Island.[10]

■■

At Boylston Hall in Cambridge, Harvard men taught chemistry to
Harvard men. The 1912 appointment of E. P. Kohler, who earned his
Ph.D. at Johns Hopkins in 1892, was a noteworthy exception. But the
quiet Kohler became the symbol of the Harvard department.
Inwardly directed, a lifelong bachelor, he was an avid reader of chemical literature so he became totally familiar with the organic chemistry
of his decades of service. Kohler developed into a polished lecturer
who remade his entire organic chemistry course, Chemistry 5, each
year, transcribing on the back of used examination books the first
sentence of each lecture paragraph.
    Kohler loved the laboratory, lived for the simple thrill of watch-
ing a newly isolated oily material, a new chemical compound, coming
into being of his own hand, form its first solid crystal, then develop
strings of starry crystalline forms along the path where Kohler would
pass his glass stirring rod.
    Kohler was a classroom artist and a laboratory craftsman. Profes-
sor Conant called him "a great tactician, but no strategist." The
introverted Kohler would never leave Cambridge, never attend a scientific meeting, never chat with other chemists around the mahogany

---

[9]Gerard Berchet interview with Adeline C. Strange, 1978. (Hagley Museum
and Library Collection, Wilmington, DE.) Although I met Berchet during my
days at DuPont, by the time I tried to contact him to discuss Carothers, he was
ill and unable to respond. Dr. Berchet died in 1992.

[10]Berchet, Gerard. *Ellis Island and Other Reminiscences*, 1982. (Dennis Berchet.)

table at the Chemist's Club in New York. Although his Chemistry 5 course was a classic of preparation and presentation, in which Kohler became its star performer, Professor Kohler refused to talk publicly outside his classroom. He would offer nothing of his expertise and insight beyond the stage of his Harvard.[11] Julian Hill, a young Ph.D. student from the Midwest, who was studying at the Massachusetts Institute of Technology (MIT) went to a joint MIT–Harvard symposium one night and met Kohler. "But after I met him," he said, "we sat around listening, waiting for the words of wisdom. He was a bachelor, and he had an old New England house down in some town south of Boston. He had a picket fence around his house, and he talked at great length about what a great thing it was to paint a picket fence. He got enormous joy and satisfaction out of this, and I thought this was awfully stupid. Finally he said, 'The reason this is so wonderful; it's the only thing I do that has a beginning, has an end, and at any time I know exactly where I stand.' Twenty years later it finally dawned on me that I'd heard some words of wisdom."[12] Kohler, now 62, was in his sixteenth year at Harvard when James Conant and Roger Adams presented to Kohler—"the King" he was called—the convincing case for Wallace Carothers as a new Harvard instructor.[13]

Kohler and Carothers, faculty senior and junior, shared much. Both were quiet, introverted bachelors with encyclopedic knowledge of the literature. Kohler had developed an astute ability to judge other men. He would have expected much from the similarity of their natures. But Kohler and Carothers diverged scientifically and in the critical internal fires that governed each of their activities.

Kohler drew joy from the maneuvers of the laboratory; Carothers preferred thought and consideration—tactics versus strategy, perhaps. Kohler taught with a dramatic flair. He became famous for his lectures and for his laboratory acumen through focused, hard work, without any effort on his part to seek or capitalize on his reputation. Conflicting reports of Carothers' ability in the classroom leave some questions. Pardee, the master teacher at South Dakota, found him "methodical," "not brilliant." Salzburg, his Illinois student, remembered the excitement of his laboratories. Elmer Bolton, Carothers'

---

[11] Conant, James B. *My Several Lives;* Harper & Row: New York, 1970; pp 31–37.

[12] Julian Hill interview with David A. Hounshell and John K. Smith, December 1, 1982. (Hagley Museum and Library Collection, Wilmington, DE, 1878, p 7.)

[13] R. C. Fuson, autobiographical notes, September 6, 1966. (University of Illinois Archives, Urbana, IL.)

boss throughout much of his Dupont career said, "He was not a particularly good lecturer. There was too much tension. It apparently was quite an ordeal for him and he was perspiring you know, go(ing) through a terrible time.... I think he felt that he was never going to be a top professor but he was a wonderful research man."[14] But Conant, in a memorial after Carothers' death, wrote, "Although he was always loath to speak in public even at scientific meetings, his diffidence seemed to disappear in the classroom. His lectures were well ordered, interesting, and enthusiastically received...."[15] Conant described Kohler similarly 30 years later: "One of the strange characteristics of Kohler's character was his success as a lecturer and his life long diffidence as a public speaker. His delight...was clear when he presented a brilliant lecture.... His anxiety...was equally real."[16] Carothers yearned for the prestige accrued to the senior faculty but could not muster the drive, the obsession with the toil, to take him there.

Reynold C. Fuson, Kohler's assistant in Chemistry 5, moved on to Illinois the next year, and Kohler wrote Adams, just after he learned of Carothers' decision to move to DuPont, as if Carothers were already gone, "I have had a letter from Fuson which delighted me because he evidently is very happy with you and likes the way in which you do things, thinks much more highly of your ways than Carothers ever thought of ours...."[17] But Conant's remembrance of Carothers' three semesters in Cambridge found no fault. He showed, "that high degree of originality which marked his later work. He was never content to follow the beaten track or to accept the usual interpretations of organic reactions. His first thinking about polymerization and the structure of substances of high molecular weight began while he was at Harvard."[18]

---

[14]E. K. Bolton interview with Alfred D. Chandler, Richmond D. Williams, and Norman Wilkinson, 1961. (Hagley Museum and Library Collection, Wilmington, DE, 1689, p 20.)

[15]James B. Conant in Roger Adams' biographical sketch of Carothers in *Collected Papers of Wallace Hume Carothers;* Mark, H.; Whitby, G., Eds.; Interscience: New York, 1940; p XVIII.

[16]Conant, James B. *My Several Lives;* Harper & Row: New York, 1970, p 35.

[17]E. P. Kohler to R. Adams, December 7, 1927. Quoted by Tarbell in Tarbell, D. Stanley; Tarbell, Ann Tracy. *Roger Adams: Scientist and Statesman;* American Chemical Society: Washington, DC, 1981; p 57.

[18]James B. Conant in Roger Adams' biographical sketch of Carothers in *Collected Papers of Wallace Hume Carothers;* Mark, H.; Whitby, G., Eds.; Interscience: New York, 1940; p XVIII.

In the fall of 1926, Carothers found Johnson a handy foil for his quick discontent with Harvard and with his own performance there. He replied playfully to a letter from his Illinois friend by applauding "that the Humane Society is keeping watch over the sadistic impulses of the Illini sophomores. This is quite in line with New England ideals," he wrote.

> The students here do not use decrepit horses, being provided for the most part with Rolls-Royces and Hispanosas, if anything, but some of the citizens do, & two men were recently sent up for six months for driving a horse which was somewhat ill & besides had a bad sore under his collar. Every cultural person will applaud such manifestations of justice towards these cruel & debased monsters.

He continued in his letter that night by telling Johnson a story he felt developed an important comparison between Harvard and Illinois:

> An interesting sidelight developed the other day on the question of efficient laboratory administration. Dr. Fuson, who assists in Chem. 5, went to the office of the director to get thumb tacks for use in posting notices in connection with that course. It was made evident that the directors office is not in any sense a stationary supply house, but is willing to go to any lengths to assist in procuring necessary supplies.
> So Fuson after establishing that 10 thumb tacks would be, if not necessary for the proper administration of the course, at least of considerable service, was permitted to put in a requisition for thumb tacks to the number of ten (10), said thumb tacks to be purchased by the University and given to the assistant for the use in connection with Chemistry 5, and the cost (purchase price & the expenses incurred in their purchase) to be pro rated among the students of the course & charged to their individual accounts. This scheme while involving some slight delay seems much more equable than that in force at Illinois where thumb tacks, paper clips etc are distributed in a wasteful fashion without any question as to the propriety of their eventual use or as to whether certain individuals may receive more or less than their just share of thumb tacks.[19]

---

[19]Wallace Carothers to John R. Johnson, November 3, 1926. (Hagley Museum and Library Collection, Wilmington, DE, 1842.)

Carothers presented a research seminar to Kohler soon after he arrived. "It didn't go so well," he wrote Johnson. Carothers began a laboratory program to measure the basicity of a series of compounds called amides. "I can apply myself in a fairly calm fashion to the business of the laboratory...," he wrote. He was discouraged by a common problem in chemical research. "...it will take quite a while to get the details of the method worked out & there may not be time to work on more than a very few substances." Six months later the situation had not improved. "My own work goes pretty slowly," he wrote Johnson on May 30, 1927. "...For one thing I have too many things cooking at once. I hope however that June will see the completion of the measurements on the amides."[20] Carothers sent this brief work off to the *Journal* in August 1927.[21]

Carothers published just this one paper while at Harvard, but he began a scientific dialogue with the management of the chemical department of E. I. duPont de Nemours & Co. in Wilmington, Delaware. For the chemical movers and shakers of the mid-1920s were once again talking and pointing to Carothers as a potential star performer. Now DuPont wanted him in Wilmington to lead a new group of fundamental researchers in organic chemistry. But for the moment we will discuss the science, the chemistry, that began to flow between Carothers and DuPont in that late-Harvard period. Let us leave the maneuvers of Roger Adams and the ambiguous role played by James Conant, which led directly to the hiring of Carothers by DuPont, with its alarming revelations of the inventor's uncertainties, to a later chapter.

In September 1927, at the beginning of his second academic year at Harvard, Carothers interviewed DuPont managers at Wilmington. In a series of letters subsequent to the trip, and after a quick job offer from DuPont, Carothers set out a research scenario for DuPont's consideration:

> I understand that it is quite possible that I might continue working on purely theoretical problems already underway. Suppose that this problem were so theoretical as the following:[22]

---

[20]Wallace Carothers to John R. Johnson, May 30, 1927. (Hagley Museum and Library Collection, Wilmington, DE, 1842.)

[21]Carothers, W. H.; Bickford, C. F.; Hurwitz, G. J. *Journal of the American Chemical Society* **1927**, *49*, 2908.

[22]Wallace Carothers to Charles M. A. Stine, September 23, 1927. (Hagley Museum and Library Collection, Wilmington, DE, 1896.)

He proposed a study of ethylsodium and ethyllithium, searching for confirmation of the presence and existence of the negatively charged ethyl group:

$$CH_3CH_2^-$$

which many believed was an intermediate in the pathways of certain chemical transformations from reactants to products. Later he refined the idea. He asked DuPont's Hamilton Bradshaw to promise equipment to pass an electric current through a magnesium compound called a Grignard reagent.

$$CH_3CH_2MgBr$$

If this compound was dissociated in the solutions he proposed to study, if it were broken apart giving the negative ethyl anion, this ion would travel to the electron-poor anode of his electrolytic cell. The electron would be stripped from the ethyl anion, and Carothers expected to identify products, perhaps ethylene by loss of a hydrogen atom, perhaps butane by a coupling of two ethyl units at the electrode. These findings would prove the presence of the charged anion.[23]

Carothers then introduced a highly practical proposal to DuPont. He suggested synthesis of succinaldehyde (*see* 4 in Appendix A), noting "a case in which a Ph.D. thesis problem was abandoned after about a years work because of the inability to get any of this aldehyde." Carothers suggested succinaldehyde was an intermediate for compounds of pharmacological interest for DuPont. Cocaine (*see* 5 in Appendix A) was an example he cited.[24] Throughout the century, cocaine's reputation has cycled between that of a damaging and addictive substance and a useful and versatile drug. Freud believed in it. Healers prescribed cocaine topically for its anesthetic properties. They read that moderate doses of cocaine taken internally, "increase the bodily and mental power and give a sense of calmness and happiness; fatigue is abolished."[25] Cocaine had recently been synthesized by the German Richard Willstätter, assisted by an American studying with him, Wilfred Bode. The preparation of synthetic cocaine, identical to the natural product, proved the skeletal structure of the mole-

---

[23]Wallace Carothers to Hamilton Bradshaw, November 9, 1927. (Hagley Museum and Library Collection, Wilmington, DE, 1927.)

[24]Wallace Carothers to Charles M. A. Stine, September 23, 1927. (Hagley Museum and Library Collection, Wilmington, DE, 1896.)

[25]*Encyclopedia Britannica*, 14th ed.; Chicago, IL, 1939; Vol. 5, p 937.

cule, proved how the carbon framework was assembled. Total synthesis was accepted by chemists as proof of structure because the chemical reactions used to assemble the structural elements of the object compound give unequivocal results. The chemist knows beforehand, from previous studies, how the units combine.

Carothers indicated to his potential DuPont managers that, "I know that my own tendency is to work on too many unrelated things at once, and a good deal of the time can be wasted hopping from one idea to another continuously; on the other hand, if more than one pair of hands is available there is a good deal of advantage in having several things going."[26]

Perhaps Carothers' admission explains his lack of research progress at Harvard, his absence of laboratory results. Carothers substituted his thinking and consideration for hours at the bench. He gained expert status at Illinois in five years at Urbana; at Cambridge he was the newcomer, an unknown. He found it difficult to reestablish a cachet as chemical advisor in Harvard's private structure, where the insular chemistry department, confronted with the lack of space in Boylston Hall, had taken the path of constructing small, individual, isolated laboratory buildings for its top-producing faculty.[27]

But Carothers began to codify—perhaps for himself as much as for DuPont—his maturing attitude about the conduct of research, about Harvard, and about his own prospects there. He wrote DuPont's Arthur Tanberg:

> Some years ago I became convinced that the important contributions of the future to chemistry would be made almost entirely by powerful and compact research organizations; but I have since decided that discovery is too elusive an art for this to be necessarily true. Nevertheless it must be admitted that, other things being equal the amount of research which one can accomplish will depend directly up to certain point on the number of hands one can command.

Carothers seemed to decide this slippery task of discovery would work best for him at Harvard. For the moment, with a large, new building in the offing and the prospect of greatly increased funding for research throughout the department, Carothers looked at Harvard as a place of "real freedom and independence and relative sta-

---

[26]Wallace Carothers to Arthur P. Tanberg, November 20, 1927. (Hagley Museum and Library Collection, Wilmington, DE, 1927.)

[27]Conant, James B. *My Several Lives;* Harper & Row: New York, 1970, p 29.

bility." He predicted, "My teaching will be practically confined to one semester, and will not require an excessive amount of time, and will be of an interesting character."[28] He wrote to Speed Marvel with optimism about his future at Harvard.[29] "So far as teaching goes Harvard is the academic paradise," he wrote later.[30]

■ ■

And then, out of nowhere, Carothers launched his inventive, fluttering arrow toward an unknown target which would become nylon, that fiber with its strength and toughness, its durability and flexibility, and its final, universal appeal. Carothers sent that arrow from Harvard on eight years of wobbly flight with a letter to Hamilton Bradshaw of DuPont. He wrote Bradshaw immediately after accepting a revised job offer from the company, one which would take him in February 1928 to DuPont's Experimental Station in Wilmington. Carothers wrote on November 9, 1927:

> When you were at Cambridge this summer you referred to your interest in polymerization. I have some appreciation of the commercial importance of this subject because rubber, cellulose and its derivatives, resins and gums, and proteins may all be classified as large or polymerized molecules. This is a class of substances about which relatively little is known in terms of structure. None of these substances is very amenable to the classical tools of the organic chemist, and no doubt some of the most important contributions in this field will be made by experts in colloidal chemistry. From the standpoint of organic chemistry one of the first problems is to find out what is the size of these molecules and whether the forces involved in holding together the different units are of the same kind as those which operate in holding the atoms in ethyl alcohol together, or whether some other kind of valence is involved—more or less peculiar to highly polymerized substances.

---

[28]Wallace Carothers to Arthur P. Tanberg, October 13, 1927. (Hagley Museum and Library Collection, Wilmington, DE, 1784, Box 18.)

[29]As related in letters from Roger Adams to James Conant, October 17, 1927, and from C. S. Marvel to James Conant, October 24, 1927. (Conant Papers, Harvard University Archives, Cambridge, MA, Box 5.)

[30] Wallace Carothers to Wilko Machetanz, January 15, 1928. (Hagley Museum and Library Collection, Wilmington, DE, 1850.)

I have been reading with a good deal of interest lately some of the fundamental work which is being done by Staudinger and by Pummerer on rubber. Rubber may be regarded as a comparatively simple example of a highly polymerized substance—it contains only carbon and hydrogen, and it has only one kind of functional or reactive group, and this group in most respects shows the reactions which are characteristic of that group in a simple olefine like ethylene. Staudinger has demonstrated rather convincingly during the past few years that the molecule is practically an endless chain, and that only ordinary valence forces are involved in holding the atoms together. But in a journal which has only appeared quite lately Pummerer has succeeded in casting serious doubts on Staudinger's conclusions.

For some time I have been hoping that it might be possible to tackle this problem from the synthetic side. The idea would be to build up some very large molecules by simple and definite reactions in such a way that there could be no doubt as to their structures. This idea is no doubt a little fantastic and one might run up against insuperable difficulties The point is that if it were possible to build up a molecule containing 300 or 400 carbon atoms and having a definitely known structure, one could study its properties and find out to what extent they compare with polymeric substances. The bearing of this isn't restricted to rubber but is common to it and such other materials as cellulose, etc.

I don't know what you will think of such a problem as this or whether you want me to offer voluntary suggestions out of the depths of my ignorance of such subjects as rubber and cellulose; but I plan to talk it over especially with Dr. Conant.[31]

Carothers brought together parallel lines of thinking for this proposal—a proposal that reads as clearly today as it did for his new DuPont managers in 1927. He drew from years of interest in chemistry and from Duncan's invitation to the chemists of the first decade of the century to work on practical problems. He extracted from the electronic structure of molecules—the valence forces—the inference that the bonding in small molecules was repeated multifold in polymers. The synthetic scheme he suggested would simply repeat again and

---

[31]Wallace Carothers to Hamilton Bradshaw, November 9, 1927. (Hagley Museum and Library Collection, Wilmington, DE, 1896.)

again the molecular bonding from those simple materials. And he applied the critical concept that synthesis provided unequivocal proof of structure, for the first time, to polymers.

Carothers fluency in German served him well. He understood Staudinger's difficult work as few of his American counterparts could; he fathomed the debate raging among the Germans over the structure of these macromolecules. If he were correct in siding with Staudinger, synthesis would offer a decisive answer to the argument.

As he promised, Carothers talked to Professor Conant. Carothers' next letter to Bradshaw thinly conceals Conant's self-centered response to Carothers' proposals. As a result, Carothers begins to excuse himself again, as he did when making the original suggestion to Bradshaw. Conant's reaction had taken some of the enthusiasm from Carothers:

> He thinks that the attempt to synthesize large molecules is very much worth while and indeed says that he has had in mind work of that kind himself for some time. Of course there is nothing very original in the idea, and a good many chemists have made attempts along this line or have planned to make such attempts.[32]

Years later, Dr. Arthur Tanberg of DuPont urged Carothers to prepare a history of his fibers work. In his 1936 summary, Carothers hesitantly described the history of a research project, which at this point was just moving from the laboratory. "Dates of some events that, in retrospect, seem important, cannot be established with certainty," he wrote. "As a matter of fact many of these events had no very clearly defined dates: they were simply ideas first grasped as possibilities, which by slow growth became firm convictions." He then apologized to Tanberg for the "tentative outline," which pinned his first thinking about the long-chain, polymeric materials destined to birth nylon, to his last six months at Harvard:

> I first became interested in what I have since called bifunctional condensations during the autumn of 1927 in connection with some work that one of my students was doing on acetylene (di-) magnesium bromide. It seemed to me inevitable that such condensations would, in many cases, lead to long chains. In a rather vague way I planned some

---

[32]Wallace Carothers to Arthur P. Tanberg, November 20, 1927. (Hagley Museum and Library Collection, Wilmington, DE, 1896.)

experimental work. Then in the latter part of 1927, during a visit I made to the Experimental Station, you told me of the glyptal reaction. This, it seemed at once, must be a polyfunctional intermolecular esterification. For purposes of theoretical exploration, a purely bifunctional type seemed more interesting and appropriate and esterifications more suitable than Grignard reactions.[33]

And so, Carothers' most productive work at Harvard was not academic research, but careful consideration of a fresh problem, initiated by Bradshaw's and catalyzed by Wallace Carothers' own anticipation and consideration of a new opportunity with the chemical giant.

---

[33]Wallace Carothers to Arthur P. Tanberg, February 19, 1936. (Hagley Museum and Library Collection, Wilmington, DE, 1784, Box 18.)

Chapter 4

The Road to Pure Science

North of Wilmington, Delaware, on Montchanin Road, a gabled country church, with stucco siding and slate roof and plain glass windows still stands on open land near the crest of a gentle rise. The duPont's built this Catholic church in the mid-nineteenth century. They raised the building to provide for the spirit of the men who worked in the powder mills just below the church, in the stone caissons built with a single, fragile wall designed to burst toward Brandywine River when the powder, as it must on occasion, ignited.

Eighty years later, in 1927, a visitor gazing from the steps of the simple country church could not venture that these duPont's of Montchanin Road, these duPont's of Delaware, now "controlled the greatest industrial empire then known to history."[1]

The duPont's built their own church too, in 1856, the elegant stone Christ Church Christiana Hundred, north up Montchanin Road, hidden, off the road, off to the east on Buck Road, nestled in the sycamores and oaks above the creek. Montchanin Road now served as a duPont thoroughfare. From Christ Church north, they drove to their manors—to Winterthur, to Guyencourt, to Granogue. Their patriarch, Pierre Samuel duPont would be driven to the grandest of the estates, Longwood, with its six exotic, theater-sized greenhouses and two square miles of gardens and waterfalls and fountains.[2]

[1]Burk, Robert F. *The Corporate State and the Broker State, The DuPonts and American National Politics, 1925–1940;* Harvard University Press: Cambridge, MA, 1990; p 3.

[2]*Delaware, A Guide to the First State;* Compiled by the Federal Writer's Project of the Works Progress Administration for the State of Delaware; Viking: New York, 1938; pp 423–445.

For in the first quarter of the century, the duPonts mastered American industrial growth. First, Pierre Samuel, the most aggressive of the three cousins who wrenched the family powder company from his older relatives in 1903, adopted a new management theory, management by return on invested capital.[3] By contrast to 1900s industrial practice, P. S. duPont's strategies, his sales and acquisitions of businesses, took on an unmatched fiscal precision. With exquisite timing he monopolized the powder business in this country. He split off the obsolete black powder and dynamite segments, while keeping the military-bound smokeless powder franchise, when the courts convicted DuPont for violation of the Sherman Antitrust Act in 1913.[4] He slid cousins T. Coleman and Alfred I. out of control of the company. The Allies fought the World War with DuPont powder. The duPonts, with Pierre in command, achieved immense wealth.

Pierre duPont and his unlikely financial mastermind, John Jakob Raskob, became shrewd bankers for William Crapo Durant's General Motors. They achieved ironclad control of the motor company, and with Pierre as its new president, General Motors surpassed Ford as the largest of the world's automotive manufacturers.[5]

The DuPont Company grew, too. With Pierre's brother Irénée as president from 1919 to 1926, the company moved away from munitions into the role of supplier to the grand array of consumer-oriented firms beginning to serve the newly aroused mass markets of the 1920s. DuPont's chemical experts would first tackle dyestuffs.

Fine textiles required bright dyestuffs for color and shading. German chemical companies had become so adept at the manipulation and development of these vividly colored organic molecules and so masterful at hiding the secrets of their structure and manufacture, that virtually no domestic dyestuff industry existed when the British wartime embargo shut down imports of the German chemicals. In July and again in November 1916, the German submarine *Deutschland* crossed the Atlantic and docked at Baltimore and New London, unloading drums of dyes in exchange for American gold.[6]

[3]Chandler, Alfred D. *The Visible Hand, The Managerial Revolution in American Business;* Belknap: Cambridge, MA, 1977; p 446.

[4]Hounshell, David; Smith, John Kenly. *Science and Corporate Strategy, DuPont R&D 1902–1980;* Cambridge University Press: Cambridge, England, 1988; p 272.

[5]Chandler, Alfred D.; Salsbury, Stephen. *Pierre S. duPont and the Making of the Modern Corporation;* Harper & Row: New York, 1971.

[6]Haynes, William. *This Chemical Age, The Miracle of Man-Made Materials;* Alfred A. Knopf: New York, 1942; p 168. Morawetz, Gerbert. *Polymers, The Origin and Growth of a Science;* John Wiley and Sons: New York, 1985; p 65.

DuPont moved to utilize its chemical talents, even before the war ended, to enter the dyestuffs market. They took the direction of acquisition as the fastest route and in 1916 purchased the technology of the British firm, Levinstein. With the end of the war, however, DuPont's accumulated dyestuff experience was insufficient to challenge the Germans, should Badische and Hoechst and the other German firms be allowed to reintroduce their 900 different dye chemicals back into North America. The company elected to lure experienced German chemists to the United States, with pay packages of up to $25,000 per year—10 to 15 times their German salaries. Five Ph.D.s from Bayer came to DuPont, hired by the director of dye research, Elmer K. Bolton.[7]

Bolton was a Philadelphian, educated to his Ph.D. at Harvard, with Adams and Conant. He went to Europe, into the German university system for his obligatory postdoctoral training, and worked under Willstätter in Berlin, isolating and characterizing the colored molecules from scarlet sage and dark red chrysanthemums.

He joined DuPont in 1915 and quickly established himself as DuPont's dye expert. Bolton maneuvered the German scientists out of Europe and past U. S. Immigration. "I was in the group that went over to make the contacts with the representative of this group of Germans and that was done in Lucerne at the Grand National Hotel," said Dr. Bolton. "We made a contract with those Germans there. Then the problem was to get the Germans out of Germany and into the United States. The army was occupying some of that territory along the Rhine and they may have been quite helpful. Two of them came on the Dutch ship *Ryndam*, and then the question was to get them through customs and Ellis Island."

DuPont's "Major" Sylvester, who had headed internal plant security for the company's important munitions plants during the war, met the ship in Hoboken, New Jersey, but failed to get the scientists off the incoming steamer. Bolton was then sent to Ellis Island. "They had three or four people back of the railing that served something of the jury. So I went up and I of course told them we were trying to start a new industry in the United States and these men were very valuable—we've tried it but we are not making any progress without the assistance of experts. It was a long line of talk."

---

[7]Hounshell, David; Smith, John Kenly. *Science and Corporate Strategy, DuPont R&D 1902–1980;* Cambridge University Press: Cambridge, England, 1988; p 76.

The German scientists and their wives and children and their maids were transported to Wilmington and formed the core of the new DuPont dye research effort. DuPont's own dye research chemists were surprised at the arrival of their new co-workers, particularly because their own ranks had been thinned substantially by the post-war recession. "Why did they bring over the Germans?" they asked, and Bolton answered, "Just to help you fellows, that's all. And you can take it or leave it. But if you leave something good, then you've got to account for it." Despite the commotion in Germany and at DuPont over this clear case of industrial raiding, Bolton was able to integrate his new, high-priced scientists into the DuPont laboratory at Deepwater, New Jersey.[8]

Bolton's group directly applied the technology brought across the Atlantic to the production of the colored chemicals. He sought the efficient development of quality dyestuffs—anything else was superfluous. And as time passed, Bolton assembled a clear, personal vision of how to conduct research.

He organized the workers into compact groups of five to eight scientists under the direction of an experienced "division leader." He referred to his German training. "When Emil Fischer was most productive, he had only eight to ten chemists working at any one time. When Willstätter was carrying out his work on chlorophyll and plant pigments, he complained because he had thirteen chemists..."

Bolton looked at untried DuPont scientists as a raw material to be fashioned into talented experimenters. "Before the output of a chemist can be increased it is necessary that he first knows how to work rapidly, cleanly and with accuracy. As we all appreciate, the chemist in the beginning of a new problem must have a thorough knowledge of the literature bearing on the subject.... The most important thing that we should remember is that the highest type of research work is carried out in the mind and confirmed by laboratory experiments." Bolton reminded his staff that they "must realize there is no part of the day that will pay higher dividends in the long run than a few hours of quiet study in the evenings."[9]

DuPont entered the fibers market for the first time in 1920. The company formed a joint venture with Comptoir des Textiles Artificiels, a manufacturer of the "artificial silk" called rayon. Rayon was

[8]E. K. Bolton interview with Alfred D. Chandler, Richmond D. Williams, and Norman Wilkinson, 1961. (Hagley Museum and Library Collection, Wilmington, DE, 1689, pp 5 and 6.)

[9]Bolton, E. K. *Certain Phases of Research Work;* Oct. 27, 1928. (Hagley Museum and Library Collection, Wilmington, DE, 1662, Box 17.)

cellulose, the stuff of the structural fiber of plants. The Englishmen Charles F. Cross and Edward F. Bevan learned to make viscous solutions of cellulose, then regenerate a solid in the form of filaments by extruding the 'viscose' into an acid bath. Rayon was commercial in England by 1908. DuPont duplicated a French rayon plant at Buffalo, New York, and soon earned explosive profits as the fiber competed with high-priced silk. Rayon dressed "Judy O'Grady" and silk the "Colonel's lady," and you could no longer tell them apart.[10]

The company expanded into paint and varnish, cellophane film, ammonia manufacturing, and photographic film. In 1922, President Pierre S. duPont of General Motors informed his brother, President Irénée duPont of DuPont of the invention by the General Motors scientist, Thomas Midgley, of an effective combustion moderator for gasoline, tetraethyl lead. DuPont went ahead to supply that ubiquitous, toxic component of "ethyl" fuel.

In only one of these new markets was there a dramatic presence of DuPont's own research. DuPont's scientists took the task of adapting German dyestuff technology, French fiber and ammonia processes, General Motors' invention of tetraethyl lead. But DuPont could point with pride to its own revolution in automotive finishes, Duco lacquer, invented in 1920 by Edmund Flaherty and J. D. Shiels. DuPont's high-solids, quick-drying, and durable automotive finishes rapidly replaced the tortuously slow-drying varnishes then currently used at General Motors.[11]

■ ■

"The man who puts into words some of the things he believes chemistry will do would seem a fit subject for an insane asylum," said Irénée duPont to a reporter's question in 1923. But the 49-year-old president of DuPont anticipated an elixir to "maintain the vigor of youth far beyond three-score-years-and-ten." This 'reagent' would enable a person to work continuously, without tiring. In fact, duPont suggested, "it is likely that material will be found which, taken in the human system, will accomplish the results of eight hours sleep." He suggested the physicists, not the chemists, would achieve atomic power. And, with his knowledge of the DuPont–General Motors program on tetraethyl lead, he hinted that, as a practical matter, "Chem-

---

[10]Hounshell, David; Smith, John Kenly. *Science and Corporate Strategy, DuPont R&D 1902–1980;* Cambridge University Press: Cambridge, England, 1988; pp 162–164.

[11]Ibid., pp 138–189.

istry is on the eve of doubling the efficiency of internal combustion engines using the present fuel."

The newsman asked duPont, "Will chemistry produce a material as warm as wool, yet cheaper"?

"Research, whether it be chemical or mechanical, surely will. In fact, I applied for a patent on changing cotton fiber in a manner calculated to make it 'as warm as wool'. The application was not granted because I did not demonstrate that it would work. I have never taken the time to try experiments required to get conditions which will regularly produce the required results."[12]

Through the end of the war, research and development in Irénée duPont's company was centered in the chemical department, whose numbers reached more than 650 by the end of 1919. Director Charles Reese, operating with a large staff from the duPont building facing Rodney Square in Wilmington, ran the Experimental Station out along the Brandywine, and the Jackson Laboratory and the Eastern Laboratory at Deepwater and Repuano, New Jersey.[13] Postwar depression in 1919–1921, the rapid expansion of the company toward the acquired new businesses, along with the decision to run these units as separate, decentralized sections and Irénée duPont's dissatisfaction with his research director, Reese, led to a decision by the executive committee of the company in 1921 to allow each new business to establish its own research unit, independent of central control.[14, 15]

The decentralization of research in 1921 seriously undermined the franchise of the chemical department. Transfers to the new departments eviscerated the section, leaving a unit Irénée duPont described as capable of only "filing reports and doing some general research work."[16] Charles Reese remained on as titular head of the department until his retirement in 1924, but the mantle of leadership of the research unit passed to his assistant, Dr. Charles M. A. Stine.

Years earlier, in 1909, DuPont asked young Stine, 27 years old and two years from his Ph.D. at Johns Hopkins, to develop the process

---

[12]Excerpts of an interview of duPont published in the *St. Louis Globe Democrat* appears in the *Literary Digest*, November 3, 1923, pp 23, 24.

[13]Hounshell, David; Smith, John Kenly. *Science and Corporate Strategy, DuPont R&D 1902–1980;* Cambridge University Press: Cambridge, England, 1988; p 100.

[14]Ibid., p 75. "I have misgivings for the progress of research work under Reese, as I do not think he is a deep thinker on research matters; his chemical knowledge is not as broad as another available man..."

[15]Ibid., p 109.

[16]Ibid., p 107.

for the new European explosive, trinitrotoluene (TNT). Stine carried out the dangerous laboratory work, but he was reluctant to recommend a full-scale plant for his nitration process. Stine constructed an intermediate-scale manufacturing unit—a semiworks. Stine's successful, small-scale expansion of his laboratory work became a model for process development. Stine contended that intermediate-scale process development, before full plant manufacture, provided a margin of safety and assured efficient expansion to the full-scale plant.[17]

Later, as assistant chemical director in 1921, with the chemical department at "semiworks" size itself, Stine struggled to find an appropriate mission for the department. He proposed the chemical department act as a contract laboratory to do research for noncompeting companies. The executive committee of the company reacted coolly to this proposal, but Stine had narrowed his sights. He would negotiate a contract with General Motors, planning to supply the DuPont-owned automaker with all its chemical research.

Stine wanted to study fuels; he wanted to know exactly how gasoline burned in an automobile engine; he hoped to solve the engine knock problems caused by inefficient combustion that plagued the automakers. But the empirical work of Thomas Midgley at General Motors in Dayton revealed the antiknock properties of tetraethyl lead before Stine could sell his more basic program to his DuPont managers. DuPont contracted with General Motors to develop the additive in a separate research and development contract, leaving Stine uninvolved.[18]

Nevertheless, over time, Stine persuaded the general managers of the operating units of the growing company, now blessed with their own research groups, to support "general investigations" in his chemical department. He promised to conduct research closely aligned with current business interests, but he teased the operating departments with the hope of new opportunity from his "pioneering applied" program. He revived the chemical department and in a broad extension of his mission, in 1926, Stine proposed a program of "Pure Science Work."[19]

■ ■

In 1900, General Electric built the first industrial research laboratory in America at Schenectady. A spirit of innovation ran through the young company. Elihu Thomson, the company's co-founder and scientific sage, along with technical director E. Wilbur Rice, inventor

---

[17]Ibid., pp 51, 52.
[18]Ibid., pp 126–132.
[19]Ibid., pp 132–137.

Charles Steinmetz, and patent director Albert G. Davis, moved to expand the company's laboratories to start a program of "original research."

Willis R. Whitney came from MIT to organize the laboratory. Whitney recalled, "Mr. Rice's idea, from the very first, was to develop a laboratory for research in pure science. He wished it set sufficiently apart in the company organization to be free from the responsibilities of the current problems of the company." By 1926, Whitney led a staff of nearly 500, but his fundamental research mandate was always diluted by the priority to serve "calls for assistance...if they involve ...possible loss to the company or unsatisfactorily meeting customer's needs."

At AT&T, Frank B. Jewett joined with Edwin H. Colpitts to establish the core of Bell Labs in 1911. The laboratory would go beyond haphazard invention and begin fundamental research in the field of telephony. As Stine pondered a small fundamental research approach for DuPont in 1926, Bell Labs could point to more than 3500 employees.[20] But AT&T's long-time chief engineer, J. J. Carty, warned, "Unless the work promises practical results it is not undertaken, and unless as a whole the work yields practical results it cannot and should not be continued. The practical question is, 'Does this kind of scientific research pay?' "[21]

Chemical researchers could identify three disparate definitions for their work. Most industrial chemists did applied work, technology: the practical application of their science to a particular issue. DuPont's Duco automotive finish was successful because it overcame two problems of conventional paints. These paint resins gave viscous solutions that required extreme dilution before application in order to give a smooth coating. But dilution meant painters applied many coats to build up the required film thickness on the car. And each coat dried slowly.

Flaherty and Shiels' invention came from an accidental observation that a small amount of a salt, sodium acetate, added to a nitrocellulose solution used as a film base caused the thick gel to become syrupy and flowable at high-solids content. Flaherty and Shiels realized

---

[20]Noble, David F. *America By Design;* Alfred A. Knopf: New York, 1977; pp 112–116.

[21]Carty, J. J. *Science and Business;* NRC Reprint, No. 55; p 4. (Address to the Chamber of Commerce of the United States, Cleveland, May 8, 1924.)

the salt trick would yield an efficient automotive lacquer. They applied a scientific observation to an existing problem.[22]

A few scientists looked for domains to master beyond their current experience. Carothers' boyhood model, Robert Kennedy Duncan, had listed such conceptual vision, cataloging for young Carothers and for the rest of his large audience, the advances he expected to see. In his essay, "The Prizes of Chemistry," Duncan anticipates prolific, product-focused fundamental research bringing dreams to reality. Synthetic medicinals, fibers, fuels, catalysis —it is all there, in a list of clearcut objectives.[23] Duncan was not an inventor, but he was an imaginative chronicler and organizer. He called for a German model, for close cooperation between emerging technology and the expert scientists of the universities:

> The Germany of the days prior to the Prussian conquest has passed away and the new Germany is a Germany of workshops; and workshops, too, in which, in the intelligent application of the means to the ends, which constitutes the scientific method, in the eagerness to harness new knowledge to their service, and in a willingness to spend money in intelligent experimentation, there is demonstrated a condition of almost perfect functioning.
>
> ...indeed typical, is the case of a German university professor who discovered a new process. His first step was to present it to the experts of one of the factories concerned; his second was to present it to the Deutsche Bank which employed its own experts to report on the validity and practicality of the process. As a result, the professor with his discovery, the Deutsche Bank with its funds, and the company with its immense facilities for investigating the discovery on a large scale, formed a little company of three for the exploitation of the process. How impossible would be such an arrangement in this country![24]

In 1907, Duncan proposed a system of industrial fellowships, housed at the universities and focused on practical discovery. Industrial sponsors would specify and pay for the academic work and gain the benefit

[22]Hounshell, David; Smith, John Kenly. *Science and Corporate Strategy, DuPont R&D 1902-1980;* Cambridge University Press: Cambridge, England, 1988; p 140.

[23]Duncan, Robert Kennedy. *Some Chemical Problems of Today;* Harpers: New York, 1911; pp 1–18.

[24]Duncan, Robert Kennedy. *The Chemistry of Commerce;* Harper and Brothers: New York, 1907; pp 242, 243.

of the close association with the best university faculty. In return, faculty and their students gained the right to publish after the industrial sponsor had nailed down patent coverage of important developments.[25] Four years later Duncan reported on a long list of these fellowships funded at the University of Kansas and pointed to remediating the "amateurishness of American manufacture."[26] The National Association of Master Bakers supported discovery of a bacillus for salt-rising bread. A. J. Weith and F. P. Brock developed a superior enamel for lining steel tanks. Kansas scientists chemically treated wood; extracted the ductless glands of whales; and studied borax, gilsonite, cement, and glass. Duncan extended his fellowship scheme to the University of Pittsburgh where a major portion of the industrial support began petroleum and natural gas studies.[27] But it fell to the new industrial laboratories to convert the visionary thinking of Duncan and others to practical reality.

Willis Whitney modeled his General Electric laboratory after his own experience. Whitney was born in 1868; his scientific interests came too soon for advanced schooling in the United States, and he left MIT where he was an instructor and took his Ph.D. at Leipzig under Wilhelm Ostwald. There he worked on problems of industrial interest. He became a successful businessman after his return to MIT, operating a small solvent-recovery operation fashioned after a fragment of research he and Professor Noyes of MIT completed together.

As Edison's basic patents on lamps expired, General Electric's influential immigrant maverick, Charles Steinmetz, propelled the company toward a separate research operation based on his own German model. Whitney, after a trial period in which he worked part-time at MIT and part-time at Schenectady fell enthusiastically into the task of "...accomplish(ing) some great thing for the 'General Electric'."

The basic research goal for the new laboratory was a more efficient incandescent light bulb, replacing carbon fiber as the high-resistance filament. Professor Nernst at the University of Göttingen had perfected a metal oxide filament and had sold the rights to Allgemeine Elektrizitäts-Gesellschaft. Westinghouse had purchased the U.S. patent rights to the Nernst lamp. The European development threatened General Electric's core franchise. Whitney stumbled badly

---

[25]Noble, David F. *America By Design;* Alfred A. Knopf: New York, 1977; pp 112–116.

[26]Duncan, Robert Kennedy. *Some Chemical Problems of Today;* Harpers: New York, 1911; p 225.

[27]Ibid., 231–238.

until he hired William D. Coolidge, who had also received his Ph.D. from Ostwald at Leipzig, and Irving Langmuir, who had a Ph.D. under Nernst and was now teaching at a small college in New Jersey and had anxiously asked Whitney for a trial at an industrial position. By 1909, Coolidge had learned to extrude a ductile, high-resistance tungsten wire; soon Langmuir developed the inert-gas-filled lamp General Electric called the "Mazda." By 1928, General Electric held 98% of the entire U.S. electric lamp business.[28]

AT&T took the specific challenge of its chief engineer, J. J. Carty, to have coast-to-coast telephony in operation by 1914. Frank Jewett, who had defined the electron with Robert Millikan at the University of Chicago by measuring its charge, urged his boss Carty to purchase the rights to Lee deForest's triode. His AT&T engineers converted it to give the necessary signal amplification required for cross-country voice transmission.[29]

Three of these men are familiar names to Carothers. Carothers read Duncan as a youth, then benefitted from Duncan's prototype of industrial support for academic work as the Carr Fellow in his final year at Illinois. Nernst, who wrote Carothers' Illinois physical chemistry text, was the practical inventor of a lamp system. Langmuir who extended G. N. Lewis' electron theory, which so attracted Carothers, made the critical invention for the incandescent bulb, then went on to win the 1932 Nobel Prize in chemistry for his fundamental work on surfaces, all carried out under Whitney at Schenectady.

■ ■

It was a third type of research, not technology in its most practical form, nor the studied search for new methods to meet specific, ambitious research goals, that Charles A. Stine proposed in 1926 to the president of DuPont, Lammot duPont, who had that year succeeded his brother Irénée.

On December 18, 1926, Stine highlighted for DuPont's executive committee a "radical departure from previous policy," a $20,000 item in his proposed 1927 budget that he earmarked for "pure science or fundamental research work:"

> The prosecution of fundamental or pioneer research work
> by industrial laboratories is not an untried experiment.
> Not only is it fostered to a considerable extent by foreign

---

[28]Hughes, Thomas P. *American Genesis;* Viking: New York, 1989; pp 159–169.
[29]Ibid., pp 156–159.

industries, particularly in Germany, but also by certain concerns in this country, notably General Electric. The sort of work we refer to is work undertaken with the object of establishing or discovering scientific facts.

Stine argued from DuPont's position as one of the world's largest chemical manufacturers that his program to search for new scientific facts would benefit DuPont four ways:

1. The advertising value of staff publications in the world's scientific literature.
2. The ability of DuPont to attract and hire top scientists to do this academic-style research. The prestige of the laboratory staff would improve training and morale of the younger chemists.
3. DuPont could use its new science as a tactical ploy, as a talking point, in the informal, personal network of chemists. DuPonters would have something to say and perhaps would benefit by early disclosure of "the work of the other men."
4. The potential for practical application of the work.

Stine promised no practical result; he even anticipated the program would be worthwhile should they achieve no pragmatic conclusion. But Stine referenced General Electric's experience. Whitney had indicated, Stine wrote, "that although they have tried from the beginning to confine their activities to fundamental or pure science work, the rapidity with which they find practical applications staring them in the face is so great that for some time not more than 10% of their whole effort has been devoted to pure science work."

Stine asked for a long-term commitment to pure science, and he indicated the program would require additional money to expand an already-anticipated new building at his Experimental Station.[30]

DuPont's executive committee wasn't buying—yet. After all, research directors have the job to stand with a research program that they expect will drive the business. The top researcher must have the vision, defining for the company where science can take it. AT&T knew worldwide telephone service was its goal. Whitney's General Electric was now lighting everything; its RCA division boasted of bringing sound to the movies and of radiotelephones to trolley cars. But Stine's program seemed to whisper only of tactics.

Advertising. Morale. Networking. Stine's first proposal failed to focus, failed to go beyond the seeming wish to be like General Electric, and anticipated only that new science would, somehow, lead to

---

[30]Stine, C. M. A. *Pure Science Work*, December 18, 1926. (Hagley Museum and Library Collection, Wilmington, DE, 1784.)

useful ends. Irénée duPont left Dr. Stine a long list of chemical dreams in his 1923 interview in *Literary Digest*: conversion of the sun's radiant energy; a balanced, synthetic food ration; heat from atomic disintegration; cures for cancer, tuberculosis, and hardening of the arteries and, of course; his own unfinished hope to make a synthetic fiber, warmer than wool.[31] But Stine ignored them.

DuPont's executive committee sat on Stine's proposal for a time, then asked him for more details. They may have suggested that their chemical director take the time to visit Dr. Whitney at Schenectady and to evaluate the activities of Bell Labs in New York. In any event, Charles Stine wrote Willis Whitney on March 7, 1927, asking for a meeting. He knew Whitney would probably attend a research directors dinner at the American Institute of Chemists in New York on March 15; Stine offered to travel to Schenectady on the fourteenth and take the opportunity to ride the train to New York City with Dr. Whitney the next day.

The two men sat side-by-side as the afternoon train rolled down the Hudson Valley that Tuesday. Stine wanted advice on a series of questions about his fundamental research program: the hiring and handling of personnel, exchange of information with other laboratories, and publication.[32] These were procedural questions, in line with the tactical nature of Stine's original research concept. But Whitney apparently seized a moment that afternoon to expose a larger view of his research charter. Whitney clarified his views two weeks later, after Stine had inquired about the "pure science" work emanating from General Electric's Nela Park laboratory:[33]

> A great deal of the work at the Nela Park in the past has certainly been of an ultra scientific nature. This was due to the policy of the first director, who avoided problems which might have industrial application. It was his idea that there were a number of problems connected with the general subject of illumination, such as standardization of

[31] duPont, Irénée. *Literary Digest*, November 3, 1923.

[32] Charles M. A. Stine to Willis R. Whitney, March 7, 1927. (Hagley Museum and Library Collection, Wilmington, DE, 1784.) Stine's proposal that he and Whitney travel together was accepted by Mary Christie of General Electric who reserved two chairs on the 3:25 from Schenectady. Mary Christie to Charles M. A. Stine, March 10, 1927. (Hagley Museum and Library Collection, Wilmington, DE, 1784.)

[33] Charles M. A. Stine to Willis R. Whitney, March 24, 1927. (Hagley Museum and Library Collection, Wilmington, DE, 1784.)

a high temperature scale and the physiological effects of light, which ought to be studied apart from any practical issues. The results obtained are undoubtedly of scientific interest and of some value in the determination of accurate standards for high temperature investigations.

This policy has been gradually altered under more recent management, and the tendency now is to work on such problems as may lead to results of practical importance.

It is, of course, difficult to assign a value to the results of purely scientific work. This is always relative. I feel, however, as you know, that since we are part of an industrial organization our investigations should be directed along those lines which are not only of scientific interest, but also promise to yield knowledge of importance to the industry.[34]

The result of Stine's visit with Whitney, and a similar trip he and Elmer Bolton made on Saturday, March 26, to the Bell Labs in New York, was his expanded proposal, retitled *Fundamental Research by the DuPont Company*, sent to the executive committee on March 31, 1927.

Stine stuck to his principles. He differentiated his fundamental program from "pioneering applied research," such as the work at Badische Anilin in Germany on the reaction of hydrogen with carbon monoxide. He wrote that pioneering applied work could succeed or fail, but he made the curious distinction that his fundamental program must succeed, because it would of necessity meet its own limited goals—the identification of new knowledge. Stine proposed hiring one top researcher for each of five diffuse "lines of work": catalysis, colloids, polymerization, physical and chemical data, and synthetic organic work. Stine wrote:

We have been at some pains to discuss this matter of fundamental research work from every aspect with men like Dr. Whitney, of the General Electric Laboratories at Schenectady, and Mr. Edward B. Craft, Vice-President of the Bell telephone Company, and with men of long experience in fundamental research in university work, and we believe that we have a correct idea of how to carry on this work.[35]

---

[34]Willis R. Whitney to Charles M. A. Stine, March 28, 1927. (Hagley Museum and Library Collection, Wilmington, DE, 1784, Box 21.)

[35]Charles M. A. Stine to DuPont Executive Committee, March 31, 1927. (Hagley Museum and Library Collection, Wilmington, DE, 1784, Box 21.)

DuPont's executive committee, no longer led by Irénée duPont, but by his brother Lammot, approved Stine's proposal within a week.[36] Stine rushed to publicize the new program with a note in the industry weekly, *Industrial and Engineering Chemistry*. But Lammot duPont preferred caution, immediately asking Stine to " 'saw the wood' and let publicity take care of itself."[37]

[36]Hounshell, David; Smith, John Kenly. *Science and Corporate Strategy, DuPont R&D 1902–1980;* Cambridge University Press: Cambridge, England, 1988; pp 225, 226.

[37]Charles M. A. Stine to Lammot duPont, April 19, 1927; Lammot duPont to Charles M. A. Stine, April 19, 1927. (Hagley Museum and Library Collection, Wilmington, DE, 1784, Box 21.)

Chapter 5

DuPont Hires Carothers

Stine began to hire the "pure science" staff immediately in the spring of 1927. He quickly convinced Professor Elmer O. Kraemer from the University of Wisconsin to head the section on colloid research; Dr. Guy Taylor, from DuPont, started the catalysis group. Staffing for the organic chemistry section lagged behind.[1] University researchers were to be hired. Stine told the executive committee he wanted leading scientists, scientists willing to move from their university affiliations, away from the emotional ties to academic life and into a full commitment to the industrial program. And he indicated he would pay dearly for these leaders.[2] Elmer Bolton, through his friendships with Adams and with his classmate at Harvard, Conant, seemed to be the source for all the leading candidates, a series of university scientists—men who were to be the leaders of American chemistry through the 1960s.[3]

Adams says DuPont offered him a "very high position" in 1928. "I had some reputation in academic work and I decided if I left it and

[1]Hounshell, David; Smith, John Kenly. *Science and Corporate Strategy, Dupont R&D 1902–1980;* Cambridge University Press: Cambridge, England, 1988; pp 227, 228.

[2]Charles M. A. Stine to DuPont executive committee, March 31, 1927. (Hagley Museum and Library Collection, Wilmington, DE, 1784.)

[3]James Conant recommended Louis Fieser of Bryn Mawr College to Bolton just one month after Stines's program was approved. James Conant to Elmer Bolton, May 2, 1927. (Conant Papers, Harvard University Archives, Cambridge, MA.)

went into industry it would look like I was selling out for money."[4] Adams' standing in 1928 was preeminent in American chemistry. He earned a salary at Illinois of $8000 per year, but Stine had warned the DuPont executive committee it might take up to $15,000 per year to lure the people he wanted.[5] New Ph.D.s fresh from the university were receiving about $3000 per year at the time.[6]

There is no question that the step into DuPont would have been a backward move for the "Chief." His Illinois department trained many of the top scientists for industry. His highly placed students cemented a powerful bond between Adams and his university and these potential industrial clients. In the new position he would stand at a lower level than his friend Bolton, though not reporting directly to him, and be well removed from the politically active scientific stage he now held. Adams' image that working in industry represented an unsavory trade-off was shared by many in the academic world. Willis Whitney at General Electric and Frank Jewett at Bell Labs took great pains to counter the lower status of the industrial scientists and to establish their laboratories on a university model, yet maintain the drive for realistic, useful work.[7]

But Adams supported the DuPont program. And he respected most of the other chemists DuPont approached: Henry Gilman at Iowa State, who had been a student with Adams at Harvard; R. C. (Bob) Fuson, who was now a junior member of the staff at Illinois after leaving Harvard; and Speed Marvel, now head of the organic chemistry section in the Illinois department. Adams secured his industrial ties by recommending good candidates and minded his university responsibilities by giving the Illinois candidates implied promises of future status should they stay at Urbana. DuPont courted Fuson through 1928. Arthur Tanberg wrote the young scientist in March inviting his application after a chat with Adams. Fuson refused. Tanberg approached him again in November; this time Fuson noted Adams had "offered...certain inducements." Fuson

---

[4]Roger Adams, interview with John B. Melleker, 1964. (Adams Papers, University of Illinois Archives, Urbana, IL.)

[5]H. E. Cunningham to Roger Adams, July 3, 1928. (Adams Papers, University of Illinois Archives, Urbana, IL.)

[6]Hounshell, David; Smith, John Kenly. *Science and Corporate Strategy, Dupont R&D 1902–1980;* Cambridge University Press: Cambridge, England, 1988; p 657.

[7]Hughes, Thomas, P. *American Genesis;* Viking: New York, 1989; pp 169–171.

would stay at Illinois.[8] Gilman, Marvel, and Fieser of Bryn Mawr, who came with strong recommendations from Conant, refused to come to DuPont.[9] But Conant and Adams also recommended Wallace Carothers, who was now working in Harvard's department as an instructor.[10]

■■

Carothers was restless at Harvard. It appears he may have been talking to Adams about returning to Illinois early in 1927. Conant wrote Adams twice in March of that year. On March 8, he noted, "So far as Carruthers [*sic*] is concerned, I hope that he will turn you down and stay at Harvard, which I am personally in favor of..."[11] Then, two weeks later, Conant wrote, "I am glad to hear that Carruthers [*sic*] has decided to stay."[12] From the timing of these fragments, it is clear Conant and Adams were not discussing a DuPont move for Carothers, because Stine was still setting salary parameters for the pure science program at the end of March.

In mid-September 1927, Wallace Carothers accepted Stine's invitation to come down from Harvard and visit the chemical department to talk about heading up the organic chemistry portion of the pure science program soon to begin in Stine's new laboratory, which some were already calling Purity Hall.[13]

It is compelling to fashion the details of that trip, to take the afternoon train from Boston to Wilmington with Carothers, to share the short cab ride from Wilmington's Pennsylvania Station, up Market Street and past the statue of Caesar Rodney, then left to the entrance of the Hotel duPont on the east side of the DuPont building, through its dark, yet glowing travertine lobby under its crafted,

[8]Arthur P. Tanberg to R. C. Fuson, March 24 and November 20, 1928. R. C. Fuson to Arthur P. Tanberg, December 5, 1928. (Fuson Papers, University of Illinois Archives, Urbana, IL.)

[9]James B. Conant to Elmer K. Bolton, May 2, 1927. (Conant Papers, Box 5, Harvard University Archives, Cambridge, MA.)

[10]Conant, James B. *My Several Lives*; Harper & Row: New York, 1970; p 319. (Conant Papers, Box 5, Harvard University Archives, Cambridge, MA.)

[11]James B. Conant to Roger Adams, March 8, 1927. (Conant Papers, Box 3, Harvard University Archives, Cambridge, MA.)

[12]James B. Conant to Roger Adams, March 24, 1927. (Conant Papers, Box 3, Harvard University Archives, Cambridge, MA.)

[13]Hounshell, David; Smith, John Kenly. *Science and Corporate Strategy, Dupont R&D 1902–1980;* Cambridge University Press: Cambridge, England, 1988; p 227.

gold-leaf ceiling, to a small, high-ceilinged room, Carothers' lodging for the night.[14] For the candidate, the uncertainties of the event, the apprehensions over the interview, are calmed by the rooted stability of the hotel. Quiet, serene, solid, substantial, prosperous.

Candidates eat breakfast in the Green Room with their DuPont hosts. They meet beneath two-story, gothic windows and delicate chandeliers, on tables set with starched linens, eat from the Hotel duPont china and silverware, listen to violin music offered by the formally dressed musician bowing steadily from the balcony. Four at the table. Carothers with Stine, certainly, and his assistant, Hamilton Bradshaw. Certainly Arthur Tanberg who directed the recruiting for the chemical department.

Next, his hosts would take Carothers to the laboratory complex situated along the Brandywine River, just downstream from the first home of the giant company. They would ride out Pennsylvania Avenue past the manicured grass tennis courts of the Wilmington Country Club, then right on Rising Sun at St. Amour, the nineteenth century stone castle now home to DuPont president Lammot duPont, and with a sudden dip under the railroad, down to and across the Brandywine and into the main gate of DuPont's Experimental Station.[15]

Whatever happened that September day, whatever Wallace Carothers felt as he was courted by his DuPont hosts, as he looked out from the laboratory being constructed for his use on this quiet slope as the first yellowed and fallen sycamore leaves drifted along the languid, late summer Brandywine, has passed from us. There are no records. We cannot know which Carothers made that visit to Wilmington, ran his hand along the shelves of the Experimental Station library, peered into the darkened glass fabrication shop, lit only by the fire of the blower's torch. Was it the Wallace Carothers who lived paralyzed by his uncertainties, who needed to be moved rather than move himself, or was it the newly awakened scientist, filling with enthusiasm and mental verve as he considered this quiet industrial retreat? Did the opportunity with DuPont somehow ignite a latent creative fire, which poured from Carothers in the next few weeks, or did DuPont, with exquisite timing, chance upon Carothers at the precise moment when he would rouse himself in an insatiable active rush?

[14]*Delaware, A Guide to the First State;* Compiled by the Federal Writer's Project of the Works Progress Administration for the State of Delaware; Viking: New York, 1938; p 288.

[15]Ibid., pp 424, 425.

Carothers itemized his expenses of $34.41 and sent the bill to Hamilton Bradshaw, Stine's assistant chemical director on Monday, September 19. Bradshaw received the letter in Wilmington the next day, and by nightfall, DuPont's expense check was on its way back to Carothers. But Stine, not Bradshaw, wrote DuPont's response.

> I believe your qualifications fit you to undertake a part of our fundamental research work, and I hope that you will decide to cast in your lot with us. I suggest that an initial salary of $5,000 would appear to be suitable.[16]

Stine recognized from his discussions with Carothers during the Wilmington interview that Carothers could not leave Harvard in the middle of the semester. Carothers had suggested February 1, 1928, would be his starting date, should he accept an offer. With his letter, Stine agreed to the timing. Stine invited questions from Carothers, and offered a second, DuPont-paid visit.

Next-day, three-cent mail service, with twice-daily delivery was the norm up and down the East Coast in 1927; Carothers likely received his check and Stine's offer on the twenty-first. On Friday, September 23, from Harvard's Boylston Laboratory, Carothers typed his response. He asked for time to make a decision. He wanted Stine to purchase certain equipment "in advance" if he accepted so that it would be available when he arrived in February. Carothers wanted assurance of an indefinite extension of the fundamental work, and he began to outline, without benefit of any confidentiality agreement, the series of researches he would initiate. Finally, he noted:

> I don't want to make the change unless it appears that there will be some real gain in making it. My present position is not altogether ideal, but it promises to develop into something which will be very nearly that providing I stay with it for two or three years and make good.[17]

Stine responded immediately on Monday, September 26, assuring Carothers that DuPont was hiring him to exercise his own judgment on research direction; that Carothers himself would be the judge of how long the fundamental work would last. Carothers' own

---

[16]Charles M. A. Stine to Wallace Carothers, September 20, 1927. (Hagley Museum and Library Collection, Wilmington, DE, 1784, Box 18.)

[17]Wallace Carothers to Charles M. A. Stine, September 23, 1927. (Hagley Museum and Library Collection, Wilmington, DE, 1896.)

scientific discretion would rule. Stine went on to site a specific case in which J. B. Nichols was to join DuPont in December, coming from a postdoctoral appointment in Sweden. Nichols had been fronted $1500 already to purchase equipment in Europe to bring with him to the Experimental Station. There would be no problem with funding Carothers' needs.[18]

For the next two weeks, Wallace Carothers attended to his Harvard duties: taught class, worked in his small laboratory, and thought about his new opportunity. On Sunday, October 9, Carothers typed a brief note to Dr. Stine. He rejected DuPont's offer. "In view of the complete balance in my mind between the advantages of my present and possible positions, it seems unwise to make the change," he wrote Stine.[19]

Had Carothers talked over the DuPont proposal with anyone? Did he seek help with this decision, did he talk with Professor Conant or Adams or Jack Johnson? Probably not. Carothers believed Professor Conant set the standards at Harvard, but the young instructor stayed his distance. During the spring of 1927, Conant finally refused an offer to go to the California Institute of Technology, after Harvard acceded to Conant's desires to establish his program on the German university model and agreed to supply funds for postdoctoral assistants.[20] Carothers learned of this new money, not from Conant, but through one of Conant's current assistants. He wrote Johnson it was "fortunate for Harvard" that Conant would stay on.[21] Carothers' early relationship with Conant lacked the close bond he held with Roger Adams, as is indicated by Conant's inability to spell Carothers' name correctly.

Carothers wrote Johnson some time in late October or early November, inviting him to come over from Cornell and tour the Boylston Laboratory. He invited him for Thanksgiving dinner, for a drink in honor of the occasion, a last look, "if you are ever to see this illustrious and gorgeous place really functioning." But it was not his leaving Harvard that occasioned the invitation, but the final days for Boyl-

---

[18]Charles M. A. Stine to Wallace Carothers, September 26, 1927. (Hagley Museum and Library Collection, Wilmington, DE, 1784.)

[19]Wallace Carothers to Charles M. A. Stine, October 9, 1927. (Hagley Museum and Library Collection, Wilmington, DE, 1784, Box 18.)

[20]Conant, James B. *My Several Lives;* Harper & Row: New York, 1970; pp 73–75.

[21]Wallace Carothers to John R. Johnson, May 30, 1927. (Hagley Museum and Library Collection, Wilmington, DE, 1842.)

ston as a chemistry laboratory. He related to Johnson that the laboratory would be replaced by a new facility on Cambridge Street within a few months. Carothers had made his decision to go to DuPont by November 1, as we will see, but he was not actively discussing the plan with his friend Johnson.[22]

The absence of a record showing that Carothers revealed his DuPont opportunity to Conant, Adams, Marvel, Johnson, or even his family does not, of course, mean he avoided talking to them. But the old Harvard connection of Adams, Bolton and perhaps, Conant, was working to bring Carothers to DuPont, yet none of these three document any influence on Carothers' decision at this point.

Carothers wrote Speed Marvel once or twice after he rejected the DuPont offer. Adams talked to Speed, then made Conant aware on October 17, 1927, that "[Carothers has] shown an entirely changed attitude toward his present position. He seems to be pleased with it and very optimistic about the future. It apparently came after he decided to turn down the DuPont offer. I feel certain with this change of attitude you will find him very contented from now on."[23] Speed wrote Conant, too, echoing Adams' report. "You may be interested to know that recently I had a letter from Carothers in which he told me he was quite contented at Harvard this fall."[24]

Carothers' refusal letter of October 9 crossed a letter from Arthur Tanberg that was mailed on October 7 but, for some reason, was delayed.[25] Carothers received Tanberg's encouraging note, which enclosed a copy of the *DuPont Magazine*, on Thursday, October 13. He replied by hand that night in a long, revealing, and ultimately prophetic message.

Carothers wrote of his reluctance in rejecting the DuPont offer in the face of his "kind reception" in Wilmington. His visit left him with an obligation to his DuPont hosts. "I am not altogether sure, yet", he wrote, "that my prospects are better at Harvard than they would have been at Wilmington." He inserted the word "yet" after completing the sentence.

---

[22]Wallace Carothers to John R. Johnson, undated, placed by context to about November 1, 1927. (Hagley Museum and Library Collection, Wilmington, DE, 1842.)

[23]Roger Adams to James B. Conant, October 17, 1927. (Conant Papers, Box 5, Harvard University Archives, Cambridge, MA.)

[24]Carl S. Marvel to James B. Conant, October 24, 1927. (Conant Papers, Box 5, Harvard University Archives, Cambridge, MA.)

[25]Arthur P. Tanberg to Wallace Carothers, October 7, 1927. (Hagley Museum and Library Collection, Wilmington, DE, 1784.)

He looked forward to the new laboratory at Harvard, and with it, new students and opportunity in the framework of "real freedom and independence and relative stability of a university position..." Scientific fulfillment seemed to depend, at Harvard, on fewer "contingencies." After all, Carothers anticipated, Stine might well have to modify or curtail his fundamental research program.

Then there was the money. Carothers suggested he would be paid almost $3200 for teaching in the 1927–1928 year. That amount represented a substantial increase over the $2250 Harvard offered him to come in 1926. It may have been an exaggeration on Carothers' part, including an ambitious estimate of added summer earnings. At his interview, Stine, in the presence of Tanberg, had asked Carothers what he anticipated for a salary. Carothers' reply is lost to us. But he now admitted to Tanberg, "The salary which I suggested in reply to Dr. Stine's question was as I realized at the time somewhat fantastic...."

Carothers' "somewhat fantastic" requirement came from a new understanding. Wallace Carothers had taken the time before his Wilmington visit to consider what Harvard meant to him. And it meant a lot. Harvard was important not just because of its name and reputation, not just because he saw the opportunity for chemical achievement and recognition, but also because of a difficult and winding emotional trail which, as he wrote Jack Johnson, left his mind "a perfect chaos...for most of the time since my stay here at Cambridge."[26] Carothers' admitted, "That to leave Harvard now would involve the relinquishing of some attachments and some plans, and that these had become considerations of real value." Carothers spelled out for Tanberg just a bit of his history:

> I had considerable difficulty in adjusting myself to the change from Urbana to Cambridge, and now that the adjustment has been pretty well made do not like to make another change unless there is a clear cut and indisputable advantage to be gained in so doing.

And then anticipating the worst case at DuPont:

> I suppose it more or less inevitable that circumstances might arise there making it necessary to suppress the development of an investigation; and there is the question

---

[26]Wallace Carothers to Arthur P. Tanberg, October 13, 1927. (Hagley Museum and Library Collection, Wilmington, DE, 1784, Box 18.)

of how my ability to fit in and adapt myself to the quite different situation might affect the possibility of doing productive work.

■ ■

"I suffer," Carothers informed Tanberg, in a clear hand, as a final justification for his refusal of DuPont's offer, "from neurotic spells of diminished capacity which might constitute a much more serious handicap there than here."[27]

■ ■

Carothers' explanation of his refusal to come to DuPont, along with his admission of mental difficulties, landed on Arthur Tanberg's desk at the Experimental Station on Monday morning, October 17.

Dr. Tanberg hired the chemical department's chemists. He knew how personnel departments at General Electric and AT&T and Westinghouse had come to lean on new forms of personnel testing to identify top engineering prospects. He knew that these test methods were thought to characterize quantitative ability, along with emotional and psychological fitness.[28] He ignored all that. Tanberg was evolving for Stine's chemical department a selective, private approach to recruiting. He avoided DuPont's personnel department entirely. He hired the staff through the individual recommendations from men, like Roger Adams, with whom he had established close ties. Stine and Tanberg supported DuPont's interests at a list of universities, including Illinois, with fellowships after Duncan's model. "The matter of getting a reliable opinion about a prospect is very largely a personal matter, and not one which can be reduced to a formula," Tanberg wrote.[29] So Carothers' admission of his "spells" was unlikely to be completely new to Tanberg; Adams certainly had disclosed what he had observed over the years.

Dr. Tanberg immediately presented Carothers' letter to Hamilton Bradshaw. Both men saw Carothers' text as an opportunity to work an improved arrangement, to turn his "no" to a "yes." That evening, Bradshaw mailed an innocent-appearing letter back to Carothers at Harvard:

---

[27]Ibid.

[28]Ibid.

[29]Hounshell, David; Smith, John Kenly. *Science and Corporate Strategy, Dupont R&D 1902–1980;* Cambridge University Press: Cambridge, England, 1988; pp 294, 295.

According to my present plans, I shall be in Cambridge on Friday of this week, namely October 21, and should like very much to have an opportunity of talking with you a little further about fundamental research at the Experimental Station. Your letter to Dr. Tanberg arrived this morning, and we appreciate very much your going into detail about the reasons for not accepting our offer. I wish you would write me whether you will be in Cambridge on Friday, and I shall appreciate it very much if you will also ask Dr. Conant whether he will be in Cambridge on that date. There are two or three things I should like to talk with him about.[30]

Bradshaw's visit to Cambridge was an almost immediate success. On October 31, 1927, Carothers wrote, "I have decided to accept your offer if it still holds."[31] In retrospect it took very little from DuPont to change Wallace Carothers' decision. Carothers responded to the last person to whom he talked. In a period of three weeks he had refused DuPont's offer, written Tanberg with a long explanation, given Speed every sense that his Harvard position was secure and represented his future, and then changed his mind and accepted DuPont's offer.

The evidence is Conant changed his mind, too, since spring, as he got to know Carothers. He now helped DuPont push the industrial opportunity. Elmer Bolton remembered Conant's role in ending Carothers' career as a teacher at Harvard. "Conant and Adams were the ones who persuaded him to accept DuPont's offer. He was not a particularly good lecturer. There was too much tension. It apparently was quite an ordeal for him...he was perspiring...a terrible time."[32]

Bradshaw had carried financial authority with him to Cambridge. He used little of it. He offered Carothers $6000 per year—20% more than DuPont's initial offer but far less than Stine had been willing to go.

Hamilton Bradshaw wrote Carothers on November 4, 1927, confirming the deal. For confidentiality purposes, Bradshaw himself inked the salary, $500 per month, into the typed letter. Bradshaw

---

[30]Hamilton Bradshaw to Wallace Carothers, October 17, 1927. (Hagley Museum and Library Collection, Wilmington, DE, 1784, Box 18.)

[31]Wallace Carothers to Hamilton Bradshaw, October 31, 1927. (Hagley Museum and Library Collection, Wilmington, DE, 1784, Box 18.)

[32]Elmer K. Bolton interview with Alfred D. Chandler, Richmond D. Williams, and Norman Wilkinson, 1961, p 20. (Hagley Museum and Library Collection, Wilmington, DE, 1689.)

wrote to Conant that same day telling him, "...how much we appreciate your cooperation in our efforts to persuade Dr. Carothers to join our staff."[33]

So a small amount of money and Carothers' inability to hold to a constancy of values pushed him across the barrier between Harvard and DuPont. I believe the increasingly ambitious Conant weighed Carothers' shaky start with Kohler and his potential in a department Conant himself hoped to shape against his future at DuPont. Perhaps Wallace Carothers' move from Harvard would benefit both organizations by removing a perplexing personality from Cambridge and establishing a small new obligation, owed to the Harvard department, by Conant's friends at DuPont.

■ ■

Wallace Carothers worked through the late fall at Harvard, through gray days, lengthening nights, Thanksgiving, then Christmas and New Year's Day passing in Cambridge. Carothers in his room on Kirkland Street, alone for the most part, anticipated his new chemical frontier now set to begin in February.

Just after the first of the year, three weeks before Wallace Carothers would leave Cambridge for Wilmington, Wilko Machetanz, Carothers' Tarkio roommate wrote to his friend. Mach was a Yale law graduate now, married to Tort Gelvin and beginning his practice in Visalia, California. Mach reported he had recovered from a lingering typhoid infection and was well with his wife, his work, and a new son. But Mach's letter brought additional news, information of greater import to Carothers. Tort's half-sister, Frances Gelvin, now Frances Spencer, was separated from her husband. Since Carothers had last seen his college companion at that strange summer meeting in Chicago in 1922, she had borne a son, but now she was back in her hometown of Maitland, Missouri, teaching school while awaiting her divorce.

Mach's letter found his friend struggling once again. Carothers got the letter at Boylston on the morning of January 16, 1928. He was tempted to go directly home and respond, but words failed him. The

---

[33]Hamilton Bradshaw to Wallace Carothers, November 4, 1927. (Hagley Museum and Library Collection, Wilmington, DE, 1784.) Hamilton Bradshaw to James B. Conant, November 4, 1927. (Conant Papers, Box 4, Harvard University Archives, Cambridge, MA.)

composite of joy at Mach's letter—"one of the few genuine events of the year" he called it—blended with the melancholy of his reflections on Frances Spencer.

Carothers replied to Mach that cold night, late, from his room. "You know," he began, "that I am not a very sociable person and so life gets rather lonesome and worthless at times; and your hearty words are like drink to the thirsty."

He turned to Frances; he faced, on paper, 13 years of memories of the woman he had been so attentive to at Tarkio; the woman he had seen but once in the past eight years. He hadn't corresponded with her. He hadn't known of the end of her marriage. He fantasized a different life if he had been aware Fran Spencer was free. "I really believe that if I had known about that prospect before it would have [been] the straw to tip the balance in favor of Urbana rather than Cambridge." But he feared his own "clumsy" behavior would have left him in a situation, "even worse than...now." He wrote:

> I can't deny that your tale of Fran's affairs aroused considerable interest. If you were here we might talk over this matter among others in great detail. We might discuss for example the question of the probable upshot of matrimonial partnership between me and Fran, going back say to the year of 1922 and erasing subsequent events as they have actually happened.

Carothers suggested marriage would have been a persistent battle:

> Fran has several qualities that I detest; and I have several which she detests with equal or better grounds. The war might have been interesting; and life would never probably have sunk to the level of common boredom, but I have a feeling that things would have ended up in a big crash one way or another.

Carothers now used his vision of a stormy marriage with Frances Spencer as a vehicle to immolate his own character. He could neither cure nor excuse his "asinine and treacherous behavior." He wrote that night

> I find myself, even now, accepting incalculable benefits proffered out of sheer magnanimity and good will and failing to make even such trivial return as circumstances permit and human feeling and decency demand, out of obtuseness or fear or selfishness or mere indifference and complete lack of feeling.

That's it for Carothers. Thankless, stubborn, fearful, selfish, indifferent, apathetic. But in his mental minefield Wallace Carothers confused cause and effect. For Carothers, all those thoughts, all the negative adjectives he could muster to describe his character, dulled the ability to take action. He was paralyzed to inaction by the mental accumulation and powerless to be rid of it. But just by the action of writing Wilko Machetanz, by making the "trivial return" of which he said he was incapable, Carothers took a small step away from indifference and toward his own emotions. He wrote, "I don't suppose you will be able to make anything out of these vague wanderings; but I pour them out to you because I happen to be in a confiding mood, and there is no one else to confide to."

Halfway through his long letter, Carothers glanced at the clock. At the late hour he feared the noise of his typing might now disturb others in the house. He switched to longhand. He wrote of his "completely arrested development of personality on a certain side." He described how he feared his father, yet he could not excuse the defects in his father's character. These faults left Carothers seeing him as a transparent fraud: impenetrable from the outside but deficient in internal integrity. Carothers could not yet reconcile that, as an adult grown from the seed through childhood, he, himself, brought forward both strengths and weaknesses:

> I suddenly realize sometimes that people of my generation are really adults and not children...that Fin Brown is Pastor of a Congregation in Chicago and that Fran and her son are in Maitland and that she is suing for divorce. The years leave marks on me, but they teach me no wisdom.

Summing up for his friend, Wallace Carothers wrote, "The record of my tortuous wanderings during the past few years would make a rich case for the psychopathologists, but so far I have placed little of it in their hands." And in signing the letter, he remembered himself as the young man who had enrolled at Tarkio more than a decade ago; he signed the letter, "Prof."[34]

---

[34]Wallace Carothers to Wilko Machetanz, dated January 15, 1928. January 15, 1928, was a Sunday. The letter is postmarked January 17. Carothers wrote he received Machetanz's letter the morning he wrote the letter. I believe Carothers wrote the letter on January 16 but misdated it. (Hagley Museum and Library Collection, Wilmington, DE, 1850.)

Chapter 6

Arrow of Discovery

Wallace Carothers reported for work at the DuPont Experimental Station on Monday, February 6, 1928. Aside from the normal introductions and the tour of the station, no one presented him with guidelines for his new job. The Harvard scientist, the special hiring project of DuPont's Stine, Bradshaw, Bolton, and Tanberg, was now in place. Carothers was not the sole bearer of the fundamental research charter, of course. He was only one of the small cluster of chemists. Carothers headed organic chemistry, but no one as yet reported to him. Charles Stine kept his word. Wallace Carothers could begin research entirely of his own choosing.

What promise could Stine expect in giving that freedom to Carothers? As a researcher he had been invisible during his two years on the junior staff at Illinois. Carothers chose "abulia" to describe his mood.[1] With that psychiatric description, certainly picked up at a healing couch, Carothers leveled at himself the haunting charge of indecision and paralysis of resolution. He fumbled about for a year at Harvard, too, before DuPont came looking, suggesting his research progress was delayed by consideration of too many projects, each standing in the way of advancement on any of the others.[2] Now he had complete freedom, limitless funds, no academic distractions— and no instructions. After one week in Wilmington, Carothers wrote Jack Johnson at Cornell from his comfortable room in a Federal-style home on the north side of Wilmington along the city's fashionable

---

[1] Wallace Carothers to Wilko Machetanz, April 22, 1923. (Hagley Museum and Library Collection, Wilmington, DE, 1850.)

[2] Wallace Carothers to Arthur Tanberg, November 20, 1927. (Hagley Museum and Library Collection, Wilmington, DE, 1927.)

Delaware Avenue. He was optimistic. He joked that, "A week of industrial slavery has already elapsed without breaking my proud spirit." He was rising before the sun "like the child laborers in the spinning factories and the coal mines" and at work by eight, whereupon he spent the day quietly, with a cigarette, reading and talking and thinking about his open-ended task. The laboratories were new and unequipped. But that was of little concern because he could order what he wanted from the chemical supply houses and receive rapid delivery from Philadelphia or New York.

Within that first week, Wallace Carothers framed a specific goal for his work. He refined the proposal on the synthesis of polymers he had made to Bradshaw from his rooms at Harvard; he sharpened it to a precise target. Carothers would aim to synthesize a polymer with a molecular weight greater than 4200 in the laboratory. That was the world record, established in Emil Fischer's classic syntheses of proteinlike polypeptides at his laboratory in Berlin.[3] "It would be a satisfaction to do this," Carothers wrote, "and the facilities will soon be available here for studying such substances with the newest and most powerful tools."

Under the warm glow of Langmuir's new Mazda lamps that night, Carothers seemed different from the vulnerable, desolate, isolated man who had written to Wilko Machetanz from Cambridge in mid-January. Then he was resigned to self-loathing for his ingratitude. Now he proceeded without a backward glance to describe for Johnson a plan for the synthesis of the record-setting polymers. He would "study the action of substances xAx on yBy where A and B are divalent radicals and x and y are functional groups capable of reacting with each other." He anticipated that if A and B consisted of short chains of carbon atoms, rings would result. In this case, the first reaction of xAx with yBy would yield xABy, but the ends of the resulting molecule, the pendant x and y groups, would be spatially close to each other and would react to cyclize A with B. Carothers thought that A and B could be fashioned from longer carbon chains. The ends of xABy would now be spatially distant from each other; their tendency would be to react with another xABy to begin a process of building "endless chains":

xABABy   xABABABABy   xABABABABABABABABy

Dr. Stine's intent was a separate organic chemistry synthesis section and a separate polymer section for his fundamental research

---

[3]Carothers later cited Fischer, E. *Chemische Berichte* **1913**, *46*, 3287, for this goal.

program. He hired Wallace Carothers for the organic chemistry job. But from the very first moment of his arrival, Carothers set out on a polymer mission. Carothers never addressed organic synthesis and Stine never hired anyone else to lead a synthesis effort. "Nobody asks any questions as to how I am spending my time or what my plans are for the future. Apparently it is all up to me. So even though it was somewhat of a wrench to leave Harvard when the time finally came, the new job looks just as good from this side as it did from the other. According to any orthodox standards, making the move was certainly the correct thing," Carothers concluded.[4]

Carothers walked from his new laboratory, built on a gentle rise above the Brandywine. The laboratory was set amid rows of tilted racks of paint-test panels facing southward toward the promise of spring Delaware sunshine. Below him was the older, quiet library looking directly out on the river.[5] None of the other chemists spent the time Carothers did with the literature. On the chilly mornings, and on the few late February days when the librarian, Virginia Duncan, could push up the windows a crack to allow the tendril of a still-chilly breeze to clear the indoor, winter air, Carothers read and wrote. He sat in the third-floor library, looking out at eye level over the bare oaks and tulip poplars. If he tired of the work, he could pick up the binoculars Mrs. Duncan kept at the window and look through the bare trees toward the Brandywine River below him, searching for the winter birds. But most of the time he was at a long, yellowed wood table near the window, writing with a hard, pale pencil, which he held peculiarly between his first and second fingers.[6]

Within three weeks of his arrival, Wallace Carothers proposed a sweeping and detailed program of discovery. He addressed his letter to Dr. Stine, but he distributed his recommendations widely throughout the chemical department.[7]

---

[4]Wallace Carothers to John R. Johnson, February 14, 1928. (Hagley Museum and Library Collection, Wilmington, DE, 1842.)

[5]Hounshell, David; Smith, John Kenly. *Science and Corporate Strategy, Dupont R&D 1902–1980;* Cambridge University Press: Cambridge, England, 1988; p 320, had two photographs of the Experimental Station, setting Purity Hall in relationship to the older buildings.

[6]Virginia Duncan interview with Adeline C. Strange, June 1978. (Hagley Museum and Library Collection, Wilmington, DE.)

[7]Wallace Carothers to Charles M. A. Stine, March 1, 1928. (Hagley Museum and Library Collection, Wilmington, DE, 1784.) Carothers made 10 copies of the document *Proposed Research on Condensed or Polymerized Substances.*

He looked upon the synthesis of polyesters from dicarboxylic acids and glycols as the ideal example of the xAx + yBy type of chemical reaction he had described two weeks earlier to Jack Johnson. Chemists understood esterification, they knew esterification is not complicated by alternate chemical reactions, yet esterification is completely reversible. The esters produced in esterification can be converted completely back to acid and alcohol.

Diacids and glycols fitted Carothers' model for the synthesis of long-chain molecules, and he knew many examples of diacids and glycols were available for study. German chemists who had reported reactions of glycols and diacids in the nineteenth century had puzzled over their results. No two groups could achieve the same results, and in any one laboratory, the properties of these strange chemical reaction products would vary from day to day. Carothers suggested condensation to polymeric polyesters had already been achieved to some extent in these studies. Carothers now understood a series of equilibria allowed these mobile chemical systems to flow back and forth, from linear polymers back to the starting materials, and off to the ring systems he had first described for Johnson.

Carothers would begin to condense diacids and glycols to challenge Fischer's record synthesis of a molecule with a molecular weight of 4200. But he suggested greater importance lay in the study of the properties of the condensed polymers. DuPont already made resins called glyptals by controlled heating of diacids with alcohols, such as glycerol, which contained three –OH functional groups. Carothers hinted his precise studies would open new information on the relationship between the molecular structure of these complex systems and their useful properties.

In November 1927, Carothers had sketched for Hamilton Bradshaw the debate over the molecular structure of highly complex natural products, such as rubber, silk, and cellulose.[8] He now laid out for his new DuPont colleagues the details of the arguments. Carothers stood apart from those who believed these natural substances were merely aggregates of smaller molecules in which the small units retained some essential integrity. Natural polymers were not held together by valence bonds of the type G. N. Lewis and Langmuir—and Carothers—understood well, according to this colloidal theory of molecular structure. Carothers believed he could recognize the forces of nonbonded aggregation when he saw them, and that is not what he observed when considering rubber and cotton and silk.

---

[8]Wallace Carothers to Hamilton Bradshaw, November 9, 1927. (Hagley Museum and Library Collection, Wilmington, DE.)

Carothers believed synthesis of high molecular weight materials by unequivocal means would clarify the debate. Polymers would be seen as extensions of ordinary structural theory, theory "sufficient when used with some discrimination, to furnish a means for treating such substances in a rational fashion."

Wallace Carothers believed throughout the rest of his life he had managed this one good idea, a single useful concept. Perhaps. But with this internal DuPont document, he laid out the entire field of condensation polymers, an idea of broad scope and implication, amenable to immediate theoretical treatment and practical trial. Conceivably over the next few years, even he forgot one question he asked his readers:

> What is the influence of the nature of the structural unit on the physical properties of highly polymeric or condensed substances? I think that enough data already exists to state definitely that this influence will parallel that of the known relationship between the properties and constitution of the substances from which the structural unit is built up. Thus the esters in general have low melting points, and the ester resins are viscous liquids or resemble super-cooled liquids. On the other hand, amides have relatively high melting points; and those amides which have been prepared from diacid bases (such as benzidine) and diacids (such as carbonic) do not melt, but decompose at very high temperatures, and are insoluble in all solvents.[9]

In this short paragraph lay the critical key to nylon, spelled out by its inventor, right from the first, in his first blush of invention for DuPont. But nearly six years would pass before the decisive experiment would be run, not by Carothers, but by Dr. Donald Coffman.

■ ■

If we were to define an ideal job, especially for one employed somewhere in the middle of the work hierarchy, perhaps we would hope for a supportive and engaged management, cooperative peers, and a talented, eager staff. Wallace Carothers seemed to have a short ration of these assets at first. His management was supportive, of course, anxious to frame the fundamental research program into a trademark of the DuPont Company, to use the effort for its stated purposes: gaining prestige and high quality personnel and accessing

---

[9]Wallace Carothers to Charles M. A. Stine, March 1, 1928. (Hagley Museum and Library Collection, Wilmington, DE, 1784, Box 24, pp 2, 6.)

science. But from the first, Stine, Bradshaw, and Tanberg refrained from directing the effort. Not from these would Carothers or any of the other leaders find specific goals, either scientific or commercial. Stine believed to his core that this absence of management command was essential to his purposes. In time, this formless environment would lead to disaffection and clamor for change of administrative style. But in the meantime the new scientists, as they gathered at Purity Hall, might decry the absent leadership they might have desired, but as a result drew together with a seamless bond. They were a very special group among the chemists of DuPont's chemical department. In time, over a very long time, over the next 40 years, these men made fundamental research itself the trademark of the chemical department. This group, as it grew by adding just the type of outstanding chemists Stine had envisioned, became, simply, one of the most prestigious organizations for basic science in the United States.[10]

Dr. Stine built Purity Hall and Arthur Tanberg built the staff. But Carothers had no staff at the beginning, no chemists meeting in lingering discussions that would turn into gentle direction for new laboratory work. What little of his own time he spent in the laboratories he used collecting the equipment: the electrically heated *bloc Maquenne* from France for the determination of melting points, tubing and small, drilled spheres wooden for making models of his organic compounds, and the specially ordered Greiner and Friederichs glassware—"beautiful stuff, but the flasks are about as thin as paper."[11]

By the summer of 1928, Carothers had help. Donald Coffman came east from Illinois where he was still two years from his Ph.D. under Speed Marvel. He worked until his return to school in the fall on Carothers' old Harvard problem of the decomposition of ethyl sodium. Soon, Dr. James A. Arvin came to DuPont after graduating from Illinois with his Ph.D. from Roger Adams. Dr. Frank van Natta from Michigan began work with Carothers. Ralph Jacobson, another of Adams' Ph.D.s from Illinois joined the group in October. Dr. Gus Dorough came from Johns Hopkins.[12]

---

[10]Hounshell, David; Smith, John Kenly. *Science and Corporate Strategy, Dupont R&D 1902–1980;* Cambridge University Press: Cambridge, England, 1988; p 376.

[11]Wallace Carothers to John R. Johnson, July 28, 1928. (Hagley Museum and Library Collection, Wilmington, DE, 1842.)

[12]Arthur Tanberg to Charles M. A. Stine, October 24, 1928. (Hagley Museum and Library Collection, Wilmington, DE, 1784.)

But for Carothers the most direct and welcome additions to his work were two new part-time employees of DuPont. Within days after Carothers began his research, Dupont hired Roger Adams and Speed Marvel as consultants. DuPont wanted Adams, but he would not make the trip every month as Tanberg proposed. Adams suggested DuPont hire Marvel, too; they would share the travel and the visits to Wilmington. (Adams consulted for DuPont for more than 40 years; in 1978, Marvel received a silver and diamond belt buckle for 50 years with DuPont.)[13]

By the fall of 1928 Carothers had a group in place, and he set Arvin, van Natta, and Dorough on separate courses of synthesis. Arvin worked broadly, reacting with a range of glycols as many of the dicarboxylic acids he and Carothers could obtain. van Natta specialized in polyesters from the simplest diacid, carbonic acid (*see* reaction 3 in Appendix A). Dorough focused on the polymers from a single combination: the four-carbon succinic acid combined with the simple glycol, ethylene glycol (*see* reaction 4 in Appendix A). In the meantime, Wallace Carothers sat in his office at the Experimental Station. He was not alone. Purity Hall was a small place, and the chemists, working two to a lab, were nearby. They read their journals over lunch and wandered into the other laboratories or into Wallace Carothers' room when the boredom of standing in front of a chemical reaction or distillation got to them. Everyone called everyone else by their first names. That is, except for Dr. Tanberg, whom no one would call Arthur.[14] Carothers believed DuPont had provided him with men of "exceedingly great competence." Yet he complained of the slow progress of the work. He did so without indicting the chemists working for him. The slow pace seemed to be part of the industrial environment. "Our polymerization work hasn't been progressing very rapidly," he wrote Adams at the end of September.[15] "I rather doubt," he wrote to Jack Johnson two weeks later, "that one individual gets as much done here when he devotes all his time to research as in the University where he sandwiches it in as he is able." Carothers' former Illinois student, Merlin Brubaker, who was now

[13]Tarbell, D. Stanley; Tarbell, Ann Tracy. *Roger Adams: Scientist and Statesman;* American Chemical Society: Washington, DC, 1981; p 112.

[14]Julian Hill interview with David A. Hounshell and John K. Smith, December 1, 1982. (Hagley Museum and Library Collection, Wilmington, DE, 1878, p 17.)

[15]Wallace Carothers to Roger Adams, September 23, 1928. (Adams Papers, University of Illinois Archives, Urbana, IL.)

also at DuPont, told Carothers, "he hadn't done as much work in any week here as he had in the average day at Urbana."[16]

Wallace Carothers knew everything. And perhaps the work wasn't going as slowly as Carothers imagined. Perhaps it was Carothers who was heating up, driving faster, racing. In contrast everything else just seemed, to Carothers, to be moving in slow motion.

In the last few months at Harvard he had learned the literature of polymerization; now in his first few weeks at DuPont he became fluent with the company's own science. He was the leader of the fundamental effort, but each day his activities seemed to tie in, ever more tightly, to DuPont's unscientific, yet important polymer interests. "...that was his choice," one of the young chemists remembers.[17] Carothers' interests in the practical application of his science was broad. In August 1928, he advised Adams on the commercial prospects for organic silicon compounds and he described several synthetic routes to silicon-containing polymers and suggested their use in varnishes. He was 20 years ahead of the invention of industrial silicones and, chemically, just wide of the mark.[18]

Soon, Carothers began to write. He envisioned rapid publication of his synthetic work, but he would take the time to outline the new field of condensation polymers. He would prepare a tutorial, first, before he reported the experimental results. He would outline how the history of polymers, as he then knew it, fit into his general theories of condensation. He would describe how bifunctional materials could react to form small rings of five or six atoms, but if they could not form such small rings, they would instead form long-chain molecules—condensation polymers. By April 1929, his primer was ready and had passed inspection by DuPont. He was free to publish. He combined it with the results of Jim Arvin's extensive synthesis studies on the reactions of glycols with dicarboxylic acids, in which five- and six-membered rings could not be obtained (*see* **6** in Appendix A). Arvin had made a series of polyesters, similar in that each was a crystalline solid, each was a polymer with molecular weight about 3000, indicating linear assembly of up to 20 of the glycol–dicarboxylic acid

---

[16]Wallace Carothers to John R. Johnson, October 8, 1928. (Hagley Museum and Library Collection, Wilmington, DE, 1842.)

[17]Julian Hill interview with David A. Hounshell and John K. Smith, December 1, 1982. (Hagley Museum and Library Collection, Wilmington, DE, 1878, p 20.)

[18]Wallace Carothers to Roger Adams, August 18, 1928. (Adams Papers, University of Illinois Archives, Urbana, IL.)

units had occurred. Carothers sent the two papers off to the *Journal of the American Chemical Society*.[19] In May 1929, Roger Adams wrote Carothers and asked for carbon copies of the two papers Carothers had submitted. Adams was now following the work of his student closely and had in fact initiated related work at Illinois.[20]

■ ■

Despite his sense of a languid pace, by the fall of 1928, eight months into the new position, Carothers appeared comfortable with his lot. He attended a polymerization meeting at Swampscott, Massachusetts, and came back to Wilmington, confident he was on the right track.[21] He corresponded with Adams regularly, becoming more familiar with his mentor as time passed. His greeting was "Dear Dr. Adams" in August, "Dear Adams" by September, then finally "Dear Roger" in December. Adams replied to "Carothers," "Doc," or occasionally, "Wallace."

Carothers read the journals more assiduously than Adams and found an important reference for his mentor. Adams admitted he would "frequently miss things" and "skip a journal now and then by mistake."

Adams was a thrifty and practical man but was prone to speculate in the volatile stock market. Now in the summer of 1928, in the last year before the great crash, he wrote to Carothers of "another two weeks of bull weather" for his favorite rubber stocks. "Assuming everything goes well, I would expect Goodrich to sell well over a hundred within a year." Carothers had a toe in the market, too, and Adams recommended he stay for the long pull, "until you get a price which seems proper."[22] Adams sent Carothers his copy of the new publication, *Reader's Digest*, and recommended the magazine and an article titled "Nervous Liquidations." At the same time he passed on a broker's recommendation that once again advised purchase of Good-

---

[19]Carothers, W. H. *Journal of the American Chemical Society* **1929**, *51*, 2548. Carothers, W. H.; Arvin, J. A. *Journal of the American Chemical Society* **1929**, *51*, 2560.

[20]Roger Adams to Wallace Carothers, May 27, 1929. (Adams Papers, University of Illinois Archives, Urbana, IL.)

[21]Wallace Carothers to Roger Adams, undated, about September 20, 1928. (Adams Papers, University of Illinois Archives, Urbana, IL.)

[22]Roger Adams to Wallace Carothers, August 29, 1928. (Adams Papers, University of Illinois Archives, Urbana, IL.)

rich for a "long pull speculation."[23] Carothers wrote Adams that the
article "provided a diagnosis of my own case." He had bailed out of
Goodrich with a stop loss order at 80. He made some money with the
sale—but plowed it right back into 50 shares of Freeport Texas even
though he knew nothing about the company and the market was
"boiling violently." Then, having purchased the stock a few days
before, he asked Adams whether he had any "sad news" about the
company "before it is made manifest in the quotations."[24] The next
day, Carothers noted to Adams:

> Freeport Texas appears to be an excellent stock for get-
> ting quick action. I lose [*sic*] approximately $250 on it the
> first day. I suppose that means more loss to come. If it
> sinks very much further my speculating days will be over
> for a few months. It will be necessary to take time out for
> recuperation. As the colored man said when he was about
> to be hung—this will be a lesson to me.[25]

Adams chided Carothers for his compulsive purchase, then
reported information he had read about Freeport Texas in Barrons.
He refused to predict whether the stock would rise or fall. He told
Carothers he expected to eliminate many of his own holdings, soon.
"The general public," the self-styled expert, Adams, wrote, "will be
likely to want to get out...and when they do, of course, there will be a
considerable drop of prices down to proper investment value."[26]

Adams and Carothers had some serious business to conduct, too,
business about which they were much more expert. The chemical
department apparently asked Adams, as one of his first tasks as consultant, to critique the research management of the department.[27]
Adams quietly sent a draft of his report to Carothers, at his home,
before he sent it off to Stine and Tanberg. Carothers wrote back
immediately, confirming two of Adams' main points: "The two great
defects it seems to me are the two of those you mentioned: lack of

---

[23]Roger Adams to Wallace Carothers, September 17, 1928. (Adams Papers,
University of Illinois Archives, Urbana, IL.)

[24]Wallace Carothers to Roger Adams, September 23, 1928. (Adams Papers,
University of Illinois Archives, Urbana, IL.)

[25]Wallace Carothers to Roger Adams, September 24, 1928. (Adams Papers,
University of Illinois Archives, Urbana, IL.)

[26]Roger Adams to Wallace Carothers, September 28, 1928. (Adams Papers,
University of Illinois Archives, Urbana, IL.)

[27]Roger Adams to Wallace Carothers, September 22, 1928. (Adams Papers,
University of Illinois Archives, Urbana, IL.)

competent research directors provided with adequate authority, and the large amount of time spent in writing reports."[28]

But in fact, Carothers himself was making up unconsciously for Stine's conscious absence of direction. He drove the program forward, and with his self-effacing style drew the loyalty of his staff to him. He was their research director.

■ ■

In early 1929, Carothers' management asked him to take on a small practical problem. Carothers seemed more able to take on emergencies of this sort, since the charter of Purity Hall lacked an urgent timetable. Discovery, just this once, could wait.

The problem lay in the manufacture of the explosive, nitroglycerin. DuPont made the oily explosive by treating glycerol, the water-like component obtained by rendering fats, with nitric acid. After switching to glycerol made from fermented molasses, the explosive makers had a predicament. The nitroglycerin "fumed off" in the preparation vats. This was not a violent event, but the batch was ruined. Carothers was to find the cause of the irregular behavior.

A young Midwesterner, a casual friend of Carothers, Dr. Julian Hill, was asked to work on the problem in Carothers' group. He jumped at the chance. Hill sorted through the information, ran a few experimental studies and, within a few months, solved the problem. "We were quite lucky," he said.

Carothers immediately asked Hill to stay on in his group. "It was a very congenial place to be, believe me, and I said 'sure.'"[29]

Hill and Carothers talked. Carothers outlined his theory that if a bifunctional system could react chemically to form a five- or a six-membered ring, it would do so. But chained-out polymers would result from any system forced, as an alternative, to make a seven- or higher membered ring. Carothers' enthusiasm was infectious. Hill stopped up to the library and soon found reference to a compound made from adipic acid, called adipic anhydride (*see* 7 in Appendix A). Adipic anhydride has a seven-membered ring. "It seemed like an anomaly to the Carothers generalization, and we were offended," Hill said. "This was an insult to Carothers' principle."

---

[28]Wallace Carothers to Roger Adams, September 24, 1928. (Adams Papers, University of Illinois Archives, Urbana, IL.)

[29]Julian Hill interview with David A. Hounshell and John K. Smith, December 1, 1982. (Hagley Museum and Library Collection, Wilmington, DE, 1878, pp 12, 22.)

Julian Hill took to the laboratory and soon proved the substances identified in the literature as the seven-membered ring adipic anhydride were actually polymers. He did so by repeating the previous work and getting solid materials with the same properties as the earlier chemists. But then he heated his products vigorously. They broke down into an easily distillable liquid, dramatically different from the initially formed solids. He converted the liquid into well-known solid derivatives, irrefutably related to the cyclic anhydride. Hill published his adipic anhydride work in late 1930. It was one of few publications from the group not carrying Carothers' name.

So nature, here on the banks of the Brandywine, could not contest with the principles laid out by DuPont's Wallace Carothers. The offending literature reference was found incorrect. An appropriate explanation, matching the realities of nature to Carothers' thesis, was developed. At Purity Hall, in 1929, the chemists challenged nature to defy them. They would somehow prevail.

As early as December 1928, Carothers noted a curious phenomenon. He wrote Roger Adams the day after Christmas that work was going slowly, but favorably. And he said the polymers the group was making all seemed to be about the same molecular size, a molecular weight of about 3000 to 4000.[30] Carothers had set as a challenge the preparation of a material of higher molecular mass than Emil Fischer's 4200. But time and again, Carothers' new polyesters would fall just short of the mark. Getting accurate sizes for the new molecules was not easy, and some samples appeared to best Fischer's 4200, but Carothers could not be sure.

By the middle of 1929, failure to get well above 4000 was a personal, not a scientific, challenge to Hill. "We were annoyed..., this was an affront. We operated on defying things that we thought were affronts."[31] But for a time, nature stood its ground, and for the rest of the year, Carothers made little progress.

After the close of 1929, Charles Stine wrote to his executive committee. He had expanded Purity Hall during the year, adding a three-story section and a large auditorium. Stine pointed to the benefits of the fundamental research program:

> We have noted with satisfaction the increasing interest which these men are taking in our applied research, with

---

[30]Wallace Carothers to Roger Adams, December 26, 1928. (Adams Papers, University of Illinois Archives, Urbana, IL.)

[31]Julian Hill interview with David A. Hounshell and John K. Smith, December 1, 1982. (Hagley Museum and Library Collection, Wilmington, DE, 1878, p 23.)

the result that many of them are gradually assuming the functions of resident consultants in their special fields. Publication of results has occasioned favorable comment from numerous sources, and several of our men are earning increasing recognition in the scientific world.

Carothers' section: Carothers himself, Gus Dorough, Frank van Natta, Julian Hill, Ralph Jacobson, and newcomers G. A. Jones, J. E. Kirby, and Carothers' Paris companion Gerard Berchet had spent nearly $43,000 in 1929. Stine gave a patient summary. In essence, he wrote, they are studying polymerization over there, but nothing has come of it as yet.[32]

---

[32]Charles M. A. Stine to DuPont Executive Committee, January 15, 1930. (Hagley Museum and Library Collection, Wilmington, DE, 1784, Box 16a.)

Chapter 7

We Will Have Quite a Lot of Things in My Division To Show You[1]

Just before Christmas, 1929, Wallace Carothers had an unexpected surprise. For the first time since 1922 he heard from Frances Spencer. Carothers knew Frances was divorced, knew she was now teaching school in her home town of Maitland, Missouri, near Tarkio. Now, having heard from her directly, he supposed he might risk writing her and think about seeing her.

Wallace Carothers typed most of his letters. But he was a trained penman, schooled at his father's desk in Des Moines. Letters written in his own hand, his daytime hand, flow across the page. Precise one and one-half inch left-hand margins, deeply indented paragraphs, elegant lettering, fully formed, closed and connected, mark his distinctive script.

Deep at night, under the lamp, next to a glass, Carothers' margins and paragraphs are shaped as by day, but his words spread across the page, flatten out, scarcely lift from their base, sprawl as if, in his haste, Carothers' pen will not make an upward or backward stroke. The words flow, too, but sometimes they do not quite connect.

"Perhaps you know that I am now an industrial slave and clock-puncher, a circumstance which is mitigated only by the fact that I

---

[1]Wallace Carothers to Roger Adams, May 17, 1930. (Hagley Museum and Library Collection, Wilmington, DE, 1784.)

reside in rooms and get my meals at a boarding house (most blessed of privileges)," Carothers wrote Frances Spencer on December 20, 1929.

> It would in my presently slightly alcoholic mood be in the highest degree indiscreet if indeed not impossible to write you a letter, and although it seems desirable for me to abandon my intuitions in this respect, there is one thing which I cannot refrain from mentioning before proceeding to the customary concluding salutations: In case you should ever find yourself in the East and in no mood of definite repugnance toward my presence, I should appreciate it if you would let me know—including addresses, dates, etc. My idea of the East in this connection includes New York, Philadelphia, Washington, Baltimore and accessible, neighboring villages and even intends at times associated with alternate humor cycles to the western borders of Indiana. The value to me of this information would be that it might provide an opportunity of seeing you. Such an opportunity would, I assure you, be most welcome.
>
> Sincerely
Wallace[2]

What an invitation! If this single mother should awaken one day, two states, three states, east of her home, anywhere in Carothers' admittedly generously described neighborhood, the young and established scientist would like to see her. Seven years had passed since their last, awkward encounter in Chicago.

Carothers extended his range, his home territory, to western Indiana because his job now took him regularly to visit Notre Dame University. The itinerary for those trips was through Chicago: Wilmington to New York on the Pennsylvania railroad, then to Chicago on the New York Central, then to South Bend, and return. One month after his letter to Frances, Carothers made the trip: out to Notre Dame at the end of the week, the weekend in Chicago, probably visiting his sister Isobel, then starting home late on Sunday night. He did

---

[2]Wallace Carothers to Frances Spencer, December 20, 1929.

not invite Frances to meet him in Chicago. His terms were clear. If she were to travel, she should let him know when and where they could meet.[3]

■ ■

Now, in January 1930, there is little in the way of warning in the life of Wallace Carothers. He no longer writes of "neurotic spells." His group is assembled, his work pushes on. He writes one confusing and perhaps boozy letter to an old flame. We might look ahead. In just seven years, Wallace Carothers will kill himself as a conscious choosing. In six years, in 1936, he will marry a young co-worker. In four years, burdened with the end of a long, illicit, and public affair, his career of scientific inquiry will virtually end. And as of now he remains a scientific unknown. In just three months he will have in hand all the success most scientists ever imagine.

Carothers is DuPont's inventor of nylon. What of nylon? Nylon will come after all this. Nylon will appear at the end. It will develop as the product of the last gasp of Carothers' will, brought to reality by the loyal young men now united in their efforts to push back science, cooperating to establish that their leader had taken them to the core of chemical understanding.

■ ■

Elmer Bolton hovered over Stine's fundamental group, even as he still ran research at the dyestuffs division.[4] As early as 1914, when, as a student of Richard M. Willstätter's in Berlin, he asked Kaiser Wilhelm's chauffeur whether the tires on the Kaiser's limousine were made of synthetic rubber, Bolton looked for a future in man-made materials.[5] The Kaiser's synthetic tires, and a decade of rumors of German progress on synthetic rubber, proved inaccurate; in 1930, Bolton's opportunity still existed. Bolton looked to make the first syn-

---

[3]Wallace Carothers to John R. Johnson, January 30, 1930. (Hagley Museum and Library Collection, Wilmington, DE, 1842.) "I really do hope to pay you a visit before very long. As a matter of fact, I was in striking distance of Ithaca during the early hours of last Monday morning—on the NYC from Chicago to New York. I had just been making one of my semi-monthly visits to Notre Dame."

[4]Julian Hill interview with David A. Hounshell and John K. Smith, December 1, 1982. (Hagley Museum and Library Collection, Wilmington, DE, 1878, p 17.)

[5]Elmer K. Bolton interview with Alfred D. Chandler, Richmond D. Williams, and Norman Wilkinson, 1961. (Hagley Museum and Library Collection, Wilmington, DE, 1689, p 18.)

thetic rubber from the monomer butadiene by the process Carothers now called "A" or addition polymerization.

$$n\mathrm{CH_2{=}CH{-}CH{=}CH_2} \longrightarrow -(\mathrm{CH_2{-}CH{=}CH{-}CH_2})_n-$$

The basic structural element of rubber is isoprene,

$$-(\mathrm{CH_2{-}C(CH_3){=}CH{-}CH_2})_n-$$

and differs from butadiene only by the presence of an additional methyl group. Bolton told his DuPont management he planned to produce butadiene from two molecules of the available compound, acetylene. In late 1925, at an American Chemical Society meeting in Rochester, New York, Bolton heard Father Julius Nieuwland, professor of chemistry at Notre Dame, describe his use of a cuprous chloride catalyst that joined acetylene into chains. Bolton saw that Nieuwland's simplest potential product, vinylacetylene, would set up the four-carbon chain he needed to make butadiene (*see* reaction 5 in Appendix A). Bolton asked Father Nieuwland that day to drive with him on a tour of the Eastman Laboratories. They talked about the priest's experiments. Bolton worked with Nieuwland on an informal basis for more than two years, sending the cultured Virginian, Dr. William Calcott, out to South Bend to visit the shy priest. In 1928, Bolton signed a formal consulting agreement with the priest and with Notre Dame.[6] By 1929, Bolton still saw promise in Father Nieuwland's work. Bolton began to talk to Carothers about it, and by late December 1929, Carothers was writing down possible chemical transformations.[7] Bolton soon sent Carothers to visit Father Nieuwland.[8] On Saturday, January 18, 1930, Father Nieuwland described in detail to Wallace Carothers the experiments he was conducting for the preparation of divinylacetylene, including his latest formulations using cuprous chloride and the combination of cuprous chloride and

---

[6]Smith, John K. "The Ten-Year Invention: Neoprene and DuPont Research, 1930–1939;" *Technology and Culture* **1985**, *26*, 34.

[7]Wallace Carothers to Elmer K. Bolton, November 19, 1931. (Hagley Museum and Library Collection, Wilmington, DE, 1784.) Carothers described his recollection of the advent of the polychloroprene work in this memorandum.

[8]Wallace Carothers to John R. Johnson, January 23, 1930. (Hagley Museum and Library Collection, Wilmington, DE, 1842.)

hydrochloric acid.[9] The priest loved poetry and botany; Carothers brought to this visit his wide interests in literature and politics.[10]

Days later, on January 22, 1930, soon after Bolton moved from the dyestuffs division and became assistant chemical director at the chemical department, Carothers began intense laboratory work on divinylacetylene and the more attractive, but relatively inaccessible, monovinylacetylene, which was present as an impurity in Father Nieuwland's divinyl compound. Carothers assigned Gerard Berchet to isolate the vinylacetylene; Berchet worked quickly and had material for study by February 20. Berchet ran a series of inconclusive experiments that same week. Carothers himself went into the laboratory to make an identifying derivative of divinylacetylene. Carothers rarely visited the laboratory as an experimenter. But Julian Hill said Carothers got his results in two days. "Wallace knew just the right experiment to do, but that's the only experiment I recall he did."[11]

"On March 12," Carothers wrote, "we decided to treat M. V. A. (monovinyl acetylene) with a variety of reagents and (Dr. Berchet) put up in sealed containers about 25 samples. Included among these was M. V. A. with concentrated hydrochloric acid."[12] Berchet did not "work up" his samples immediately to determine what transformations might have occurred; he allowed them to stand on the laboratory bench for more than five weeks.

Arnold Collins was assigned by Bolton to Carothers' group in early February. Collins had worked with Father Nieuwland's divinylacetylene at the Jackson Laboratory, and he had worked for Bolton on the project. He had developed from divinylacetylene a modestly attractive, self-drying coating, which DuPont now called Synthetic Drying Oil.

■ ■

At the Experimental Station, Collins worked alone in a two-person laboratory. Julian Hill and Gus Dorough moved equipment onto the unoccupied benches. Dorough assembled a special still designed for careful liquid separations that he had brought from Johns Hopkins. Hill had fabricated a complex glass assembly for high

[9]Wallace Carothers to Elmer K. Bolton, November 19, 1931. (Hagley Museum and Library Collection, Wilmington, DE, 1896.)

[10]Haynes, William *This Chemical Age, The Miracle of Man-Made Materials;* Alfred A. Knopf: New York, 1942; p 212.

[11]Julian Hill interview with Matthew E. Hermes, May 29, 1990.

[12]Wallace Carothers to Elmer K. Bolton, November 19, 1931. (Hagley Museum and Library Collection, Wilmington, DE, 1850.)

vacuum work, a diffusion pump, powered by the condensation of mercury vapor. Hill used the pump to operate what he called a molecular still, an apparatus fashioned to remove the last traces of volatile gases from bulk samples at high temperatures, using low pressures and an extremely short path between the heated residue and the condensing surface. Carothers remembers he told Hill about this apparatus, which he had seen two years before at the 1928 Swampscott meeting at which its inventor, Washburn of the National Bureau of Standards, had shown it.[13] Hill remembers differently, claiming he had first seen the apparatus at an American Chemical Society meeting, probably in Buffalo.[14] Whatever the origin of the idea, both agree Hill built a redesigned version of the still.

Hill was often in Collins' laboratory, using the molecular still. Carothers and Julian Hill were frustrated now with their inability to move polyesters above a molecular weight of 4000. Hill dropped polyester work and was on a new version of an old problem. Carothers believed he would see an upper limit to the number of carbon atoms capable of extension in a linear chain. For relatively short chains, Carothers knew stability dropped dramatically with increasing chain length. He and Hill began to assemble the simplest carbon chains, $-(CH_2)_n-$, called linear alkanes, by using a molecule with 10 linear carbons as the building block. Hill sat patiently in Collins' laboratory and isolated linear alkanes of 40, 50, 60, and 70 carbon atoms by using his peculiar distillation apparatus.[15] Carothers found the drop in stability that he originally expected, in fact it decreased with chain length and became insignificant. He now speculated that a linear alkane or paraffin of a molecular weight of 200,000 or greater would be stable. Carothers noted the melting points of the paraffins increased with chain length. His C_{70} samples melted at about 105 °C.[16] As he had done with his comments to Roger Adams on silicon chemistry in 1928, Carothers' prediction of a stable, long-chain

---

[13]Washburn *Bureau of Standards Journal of Research* **1929**, *2*, 480. *Also see* Carothers, Wallace. *Early History of Polyamide Fibers.* (Hagley Museum and Library Collection, Wilmington, DE, 1784, p 3.)

[14]Julian Hill interview with David A. Hounshell and John K. Smith, December 1, 1982. (Hagley Museum and Library Collection, Wilmington, DE, 1878, p 23.)

[15]Carothers, W. H. "Paraffin Hydrocarbons of High Molecular Weight," *Fundamental Research, Summary Report;* January 1 to March 31, 1930. (Hagley Museum and Library Collection, Wilmington, DE, 1784.)

[16]Carothers, W. H.; Hill, J. W.; Kirby, G. E.; Jacobson, R. A. *Journal of the American Chemical Society* **1930**, *52*, 5279.

alkane foreshadowed the invention and universal commercial extension of the polymer, polyethylene (*see* reaction 6 in Appendix A). Polyethylene can have a molecular weight in the millions, melts at about 130 °C, and finds boundless usage. Carothers' synthetic process, which built up alkane chains by condensation, is useless compared with the addition polymerization of ethylene to polyethylene. But Carothers knew from his data on C_{70}, as early as 1930, that a linear polyalkane could be stable and might have useful properties.

Carothers was assembling an impressive life list of new thinking in polymers: condensation polymers to prove structure, polyesters, silicon-containing polymers, long-chain alkanes. However, at the end of March 1930 Carothers lacked even one experimental gem from his bank of ideas. He had nothing to show but activity, with little in the way of achievement.

■ ■

On April 17, 1930, Julian Hill was working with his molecular still in Collins' laboratory when he noticed Collins fumbling with a small glass container. It seemed as though Collins was trying to retrieve a small white mass of solid from the flask. But the material would not dislodge easily. Collins determinedly fished the mass out with a stiff wire. He felt it, rolled it in his hands—and bounced it across the laboratory bench.

Collins' "exceedingly elastic and extensible" sample was the compound polychloroprene. It became DuPont's Duprene, the world's first synthetic rubber.[17] Wallace Carothers and Arnold Collins determined immediately that the elastic material contained the element chlorine. They pulled down the shades in the laboratory to darken the room; they heated a copper wire in a Bunsen flame until it glowed red, then plunged the wire into the core of the rubbery sample. The rubber seared and smoked; they withdrew the wire and reinserted it into the flame. The flame turned a bright green. The hot metal had degraded the rubber, had combined with chlorine to make copper chloride, which gave the fleeting jet of green, clearly visible in the dimly lit room. They immediately theorized the polymer was derived from a new material, 2-chloro-1,3-butadiene (*see* reaction 7 in

---

[17]DuPont changed the name to neoprene in 1936. *See* Hounshell, David; Smith, John Kenly. *Science and Corporate Strategy, Dupont R&D 1902–1980;* Cambridge University Press: Cambridge, England, 1988; p 255. John K. Smith provides an extensive history of the rubber's development through the 1930s. *See* Smith, John K. "The Ten-Year Invention: Neoprene and DuPont Research, 1930–1939," *Technology and Culture* **1985,** *26,* 34.

Appendix A). The chloro compound had formed by the addition of traces of hydrochloric acid in the cuprous chloride-catalyzed system to vinylacetylene. Carothers reported the sequence of invention to Bolton the next day.

On March 27, Collins had distilled the liquid products from a cuprous chloride-catalyzed acetylene reaction modeled after Father Nieuwland's work. He used Gus Dorough's still and obtained a small fraction of a liquid, an impurity, boiling at 60–60.5 °C. On April 10, Collins tried to wash out contaminants in the new liquid with water, but some of the material had stayed in the water washings. That same day, he set aside two stoppered test tubes of the liquid. Seven days later, Collins noted the water emulsion had become a solid mass. Collins examined the two test tubes the next day: They, too, has become a "clear, tough, homogeneous, elastic mass."

Carothers and Collins suggested to Bolton that the chlorobutadiene could be made in significant quantities by altering Collins' procedures.[18] Three days later, Collins made a large sample, 42 g, of chloroprene.

Bolton passed Carothers' letter on to Dr. Stine, but not before writing out for Stine the structural comparison between chlorobutadiene and isoprene.

$$CH_2=C(Cl)-CH=CH_2 \qquad\qquad CH_2=C(CH_3)-CH=CH_2$$
2-chloro-1,3-butadiene (chloroprene)   isoprene

This molecule was similar to isoprene, the building block of natural rubber, and similar too, to butadiene—Bolton's long-held target for a synthetic rubber.

Carothers immediately realigned his group and began intensive study of chloroprene and the chemistry of a series of related compounds. Bolton moved the bulk of the practical work—the synthesis of chloroprene, the optimization of its polymerization, and the processing of the new rubber into useful articles—to two laboratories in Deepwater, New Jersey, the Jackson Laboratory and the Rubber Laboratory.[19]

Elmer Bolton prepared the strategic ground for the discovery of a synthetic rubber by patient pursuit of Notre Dame's Father Nieuwland. He assigned Carothers to stay in touch with Father Nieuwland,

---

[18]Wallace Carothers to Elmer K. Bolton, April 18, 1930. (Hagley Museum and Library Collection, Wilmington, DE, 1784.)

[19]Smith, John K. "The Ten-Year Invention: Neoprene and DuPont Research, 1930–1939," *Technology and Culture* **1985**, *26*, 34.

he brought Arnold Collins in to work for his masterly organic leader. Wallace Carothers did his job. He developed a program with Berchet and Collins. All were alert to the possibility of an unexpected finding. Collins' isolation of the unanticipated "impurity" is standard procedure; Collins' probing at the unusual white solid where once there had only been a liquid is the normal inquisitiveness of the organic chemist. Research is the dogged pursuit of the unexpected. For chemists, observation levels the playing field. Careful examination of results, diligent pursuit of the unforeseen, takes the average among us to dine with the distinguished theoreticians, the brilliant minds, the clever experimentalists.

What was this new invention for Carothers, this new addition to his developing catalog of polymer thinking, the first of his useful creations?

Polychloroprene lacked something in its discovery. That April morning, Collins observed something the gods had given him. A gift, a material accidentally but properly polymerized to a substance of obvious utility. Once Collins and Carothers bounced their sample to each other, the pathway to a synthetic rubber became clear. But for Carothers there had been no chase of an elusive intellectual goal, no painful series of unsuccessful trial and failure leading to success. Later Carothers claimed his parentage of chloroprene by reminding Bolton he had initiated Berchet's experiment in which Berchet mixed vinylacetylene and hydrochloric acid. He admitted to Bolton he and Berchet had not expected to make chloroprene from the reaction. But the young French chemist ignored the sealed glass tube, which certainly contained some chloroprene. He lacked diligence and let Collins discover chloroprene and its marvelous polymeric product.[20]

If, as Hill suggests, the behavior of the polyesters was an affront, the behavior of polychloroprene was a teasing harassment. Carothers' fundamental theories and their patient application was insufficient as yet to open all of nature's polymer secrets. Collins' observation and good luck had done it this time for Carothers.

In the scientific world, in the academy, scientific credit seems to go to the senior investigator. Carothers indeed saw the opportunity to expand his research, to start a new series of numbered papers for the journals. He recognized a long sequence of publications on the same topic gave a considerable advance to his chemical reputation over

---

[20]Wallace Carothers to Elmer K. Bolton, November 19, 1931. (Hagley Museum and Library Collection, Wilmington, DE, 1784.)

scattered publications on various subjects. Carothers had initiated the polyester series with the *Journal of the American Chemical Society*; now he recognized the opportunity to start a new series on the chemistry of monomers and polymers related to polychloroprene.[21]

But research took time. Eventually, Carothers' group prepared more than 260 new compounds and "many complex polymers" in the chloroprene work. Carothers would write and publish more than 20 papers from DuPont on his chloroprene research. However, Carothers seemed to regard the isolation of chloroprene and its polymerization into the synthetic rubber with some disdain. He called this scientific work "abundant in quantity but perhaps a little disappointing in quality."[22]

But only the first of these papers would come before the end of 1931.[23] By the time the *Journal* published Carothers' first paper, DuPont had plans to link publicly, the name Carothers, for all time, with its newest product, the first synthetic rubber. Wallace Carothers gained recognition for Collins' inventive observations that led to DuPont's neoprene, more by the subsequent actions of DuPont than by the deeds of the inventors.

■ ■

As the first azaleas budded in the warm, sunny gardens across the Brandywine, above the Experimental Station in Wilmington's Rockford Park, as May loomed on the calendar of 1930, Carothers was a busy man. Bolton wanted his old associates at the DuPont laboratories across the Delaware River at Deepwater to join in the chloroprene work immediately. But Bolton wasn't feeling well, he was suffering from intestinal difficulties that would hospitalize him within weeks for removal of his appendix, and Carothers would have to carry out the bulk of his work for a time. In addition to setting up the work on the new synthetic rubber, Bolton asked Carothers to assemble the quarterly report for his chemical department. So Wallace Carothers spent much of his time that April in his car, shuttling from the Experimental Station to Bolton's office in the DuPont Building in the center of Wilmington, then down Route 13 toward New Castle, where he

---

[21]Julian Hill interview with Matthew E. Hermes, May 29, 1990.

[22]Carothers, W. H. "Fundamental Research in Organic Chemistry at the Experimental Station, A Review," August 5, 1932. (Hagley Museum and Library Collection, Wilmington, DE, 1784, Box 21, p 9.)

[23]Carothers, W. H.; Williams, I.; Collins, A. M.; Kirby, J. E. "A New Synthetic Rubber: Chloroprene and Its Polymers," *Journal of the American Chemical Society* **1931**, *53*, 4203. This paper appeared November 5, 1931. Fifteen papers appeared in the *Journal of the American Chemical Society* in 1932.

took the ferry over to the Jackson Laboratory at Deepwater, New Jersey, where Ira Williams and the rest of Bolton's old rubber chemistry group were to begin serious scale-up of Arnold Collins' discovery of polychloroprene.[24]

Meanwhile, on the opposite bench in Arnold Collins' laboratory in the last week of April 1930, Julian Hill fired up his apparatus to carry out a new piece of work. He began boiling the mercury in his diffusion pump so that the metal vapor would entrain the last traces of gas from his molecular still. Hill was returning to the problem that had stumped him and the rest of Carothers' group: the apparent limitation of molecular weight of about 4000 in the preparation of polyesters.

Carothers and Hill knew Carothers' reaction of a dibasic acid with a glycol produced water. They projected that the troubling limitation on molecular weight was not fundamental, that it was instead caused by their practical inability to remove all the water. If traces of water remained in the boiling mixture of dibasic acid, glycol, and forming polymer, the water would simply reverse the polymer back to the lower molecular weight starting materials. Hill would work at very low pressures, 10^{-5} mm of mercury pressure, about 1 millionth of the normal atmospheric pressure, where virtually all traces of air are absent. Hill knew that in the absence of all gases, an individual water molecule, rifled off the surface of the roiling reactants, would fly straight across the small gap between his heated tray of reactants and freeze to the cold condenser he provided. The water molecules could not be diverted back by colliding with another molecule. They could not return to the hot reactants to break the forming chains. Carothers wrote of a "complicating circumstance." Each end of the growing polymer chain might be the glycol component. If so, Carothers envisioned the polymer could not extend by splitting off water. But Carothers saw the molecular still would draw off the glycol and force the chains to extend through a process he called transesterification (*see* reaction 8 in Appendix A).[25]

---

[24]Wallace Carothers to Roger Adams, May 17, 1930. (Adams Papers, University of Illinois Archives, Urbana, IL.) "I suppose you know Bolton is in the hospital now. He was operated on for appendicitis about a week ago and is doing pretty well now but he will probably be in the hospital for some weeks yet since the doctors think he needs a very strict diet for a while to remedy some of the troubles he has been having."

[25]Carothers, W. H.; Hill, J. W. *Journal of the American Chemical Society* **1932**, *54*, 1557.

Hill combined an unconventional, long-chain dicarboxylic acid of 16 carbon atoms with a 10% excess of propylene glycol, the dialcohol with three carbon atoms. The acid was part of Carothers' extensive collection of dicarboxylic acids assembled from all over the world. This one came from the Swiss chemist, Ruzicka, and was made by Rhone Poulenc, the specialty chemical producer, and was raw material for an investigation Carothers planned on large ring systems.

Hill was patient. First he made the polyester from the dibasic acid and a slight excess of the glycol. He got the usual results, a viscous liquid solidifying to a waxy solid. The molecular weight was 3300. He clipped alligator clamps to a heater beneath the tray of his molecular still and placed the polyester on the tray. Julian Hill began heating at about 200 °C under the high-vacuum conditions. Each morning he checked the cold condenser. After two days, about 5% of the material had condensed on the water-cooled surface above the tray. But for the next five days, no more material condensed.

Finally, on the morning of April 30, 1930, Hill disassembled the still and removed the tray. The polymer mass clung tenaciously to the

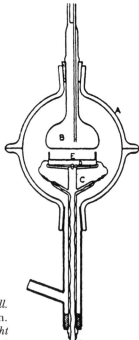

Wallace Carothers' molecular still.
(Reproduced from J. Am. Chem.
Soc. **1932**, 54, 1558. *Copyright*
1932 American Chemical Society.)

glass. Hill heated the tray and touched a glass rod to the tough, adherent mass. "...there was this *festoon* of fibers," Hill related.[26]

Within seconds, that morning, Julian Hill had observed three new and related phenomena, three inventions coming from the steady pursuit of longer chain synthetic polymers. First, the prolonged heating of the polyester in the molecular still had changed the material. Hill later showed the molecular weight far exceeded the offending barrier of about 4000—Carothers and Hill estimated from the data they could collect that they had reached 12,000. They had smashed Emil Fischer's synthesis record of 4200, finally and well. And with the increased molecular weight came fiber-forming properties. The molten mass of polymer could be pulled out into long strings of fiber. With Carothers downtown in Wilmington at the DuPont building, minding Bolton's business for a time, Hill assembled his young cohorts and they ran through Purity Hall, tweezers in hand, drawing long, lustrous, continuous filaments of their first artificial silk from the molten mass.

Hill's third observation came as a total surprise. As he elongated the mass of polymer into thin fibers, he saw at once a spontaneous ordering of the fiber, a playful, almost magical transformation. "The first thing you did was to pull on them and the cold drawing phenomenon was immediately evident, which was a lot of fun to do and show people," Hill said.[27] Carothers described Hill's observations for the *Journal of the American Chemical Society*:

> In connection with the formation of fibers, the polymers exhibit a rather spectacular phenomenon we call cold drawing. If stress is applied to a cylindrical sample of the opaque, unoriented 3-16 polyester at room temperature or at slightly elevated temperature, instead of breaking apart, it separates into two sections joined by a thinner section of the transparent, oriented fiber. As pulling is continued, this transparent section grows at the expense of the unoriented sections until the latter are completely exhausted. A remarkable feature of this phenomenon is the sharpness of the junction of the boundary between the transparent and the opaque sections of the filament. During the drawing, the shape of the boundary does not change; it merely advances through the opaque sections until the latter are exhausted. This operation can be car-

---

[26]Julian Hill interview with Matthew E. Hermes, May 29, 1990.
[27]Ibid.

*Cold-drawn fibers. (Reproduced from J. Am. Chem. Soc. **1932**, 54, 1580. Copyright 1932 American Chemical Society*

ried out very rapidly and smoothly and it leads to oriented fibers of uniform cross section.[28]

■ ■

Wallace Carothers wrote to Roger Adams in the middle of May, anticipating a summer visit from his mentor. He closed, "I am looking forward to your next visit. We will have quite a lot of things in my division to show you."[29]

■ ■

Examine, carefully, Carothers' critical pathway to fibers: his concept of condensation polymerization as a method to prove the structure of polymers by assembling long chains, his demonstration of polymerization with a wide variety of bifunctional monomers, and his final decision to wring the last trace of water and other volatile components from his incompletely condensed systems.

Without question, Carothers' designed synthesis of polymers from small molecules proved the case that these high molecular weight materials were normally bound and not the colloidal massing of small molecules preferred in the previous decade. But the success

---

[28]Carothers, W. H.; Hill, J. W. *Journal of the American Chemical Society* **1932**, *54*, 1579.

[29]Wallace Carothers to Roger Adams, May 17, 1930. (Hagley Museum and Library Collection, Wilmington, DE, 1784.)

of Wallace Carothers—as polyesters led to fibers and as fibers led to nylon—comes as it did for the synthesis of polychloroprene, from careful observation and laboratory technique, independent of Carothers' theoretical thinking.

Carothers' and Hill's thesis that the molecular still would end the "affront" of the upper limit of 4000 in the molecular weight of polyesters by squeezing off the final water and glycol molecules was correct. But throughout their early work, Carothers and all his co-workers showed curious inattention to one critical and seemingly obvious experimental characteristic in the theory they were to prove. Carothers and his group caused the limited growth of their polyesters by the reaction conditions they chose. If they wished to reach high molecular weight from a two-component system such as dicarboxylic acid and glycol, then reactants had to be in balance: one diacid molecule for each glycol. But in the laboratory, in experiment after experiment, Hill, Arvin, and Dorough combined their diacids with a 5 or 10% excess of glycol. The results were as expected. Carothers listed 13 polyesters in his first experimental paper. In every case Arvin used a 5% excess of glycol. Statistically, the average chain cannot be longer than 20 repeats of the diacid–glycol under these conditions, with a glycol at each end. And that is precisely what Carothers reports—13 polyesters, every one averaging between eight and 20 repeating units.[30]

Success came for Carothers only as Hill heated his new polyester in the molecular still—a polyester he had prepared with a 10% excess of propylene glycol. This polyester had approximately 10 repeat units and a molecular weight of 3300 with no chance to go higher. This chain length was right in line with what we can predict from the excess glycol Hill used.

It was the high temperature in the molecular still that caused a second reaction to occur, a reaction that Carothers noted after the fact.[31] Hill's molecular still drove off propylene glycol in a separate, high-temperature chemical reaction called transesterification, independent of the original, water-producing, polyester-forming reaction. Randomly the –OH group at the end of the polymer chain, forced by the energy in the hot melt, attacked other polymer chains in the fluid, driving off a segment. Some of these fragments were short, and

---

[30]Carothers, W. H.; Arvin, J. A. *Journal of the American Chemical Society* **1929**, *51*, 2560.

[31]Carothers, W. H.; Hill, J. W. *Journal of the Amerian Chemical Society* **1932**, *54*, 1559.

in the hot void, they escaped from the surface. As the short, volatile segments, notably $HO(CH_2)_3OH$, were stripped from the surface, the ratio of carboxylic groups and hydroxyl groups came more closely into even balance and the molecular weight of the residue soared.

Success came for Hill as he toyed curiously with the warm, amber residue he obtained, a residue somehow different, more viscous and promising, than the earlier syrups, a residue that yielded the tufts of silken fibers as Hill twisted and pulled away his glass rod.

■ ■

Wallace Carothers took full advantage of DuPont's generous program that allowed publication of all the results from his group. As the frustrating polyester work of 1929 opened into new promise in 1930, Carothers envisioned a major, personal, scientific contribution. He began a review paper, destined for the American Chemical Society's journal, *Chemical Reviews*, in which he would describe his polymer theories, introduce new polymer nomenclature, and summarize polyester synthesis. As he wrote the paper, he described many new potential polymers. He turned to the structure of cellulose and rubber, describing them as linear polymers with normal chemical bonds. He allowed that in the last few years, independent of his own work, new methods of analytical chemistry had largely resolved the issues of colloid versus large molecule. "No example is yet known in which a small molecule of known structure simulates a material of high molecular weight without undergoing any change in structure.... It appears that many naturally occurring macromolecular materials have a linear polymeric structure" he wrote.[32] Carothers spread his papers across the tables in the library above the Brandywine, opening the volumes containing the 177 references he would cite, stacking them, spine on spine, reaching for them as he needed. In the absence of any method to make copies, the preparation of a major manuscript was a public event, requiring an assembly of towers of books to be left in place overnight, so they need not be regathered the next day. Carothers would write this scientific manifesto, but as Bolton's new assistant, Ernest B. Benger, observed his writing, he became restless with Carothers' rush to publish. In the end, Carothers would win the battle to send off this review, but some of its disclosures had punishing aftereffects for DuPont.

■ ■

We cannot walk up behind Wallace Carothers in the summer of 1930, place our hand on his shoulder, turn him in his chair, and look

---

[32]Carothers, W. H. *Chemical Reviews* **1931**, *8*, 353.

into his eyes. If we could, would we see the blank stare of depression? Did Carothers stand at his office window in the afternoon, hands behind his back, fixed in his gaze on the imperceptible drift of the shadow of Purity Hall lengthening away from him? Where lay, in 1930, the mental and spiritual encounter that would physically kill him?

Did Julian Hill burst into Wallace Carothers' office, trailing the filaments of his new discovery, and see for a fleeting instant, before Carothers recognized his presence, that disquieting, vacant look, that gaze off into space that cannot be forgotten, that fixed look that is always remembered as an instant, uninvited revelation of the depths of another's soul.

We can know our own depression. Lacking any power to restore ourselves to normal behavior, we remain indifferent, confused. We look out on our world and see gray, muted colors. We awaken in our darkened rooms, paralyzed with fear of the unknown, even as the warming spring sunshine outside our window livens the tiny flecks of dew condensed in darkness on the grass and leaves and the shimmering webs of protein silk fiber, newly spun by nature's spiders.

Wallace Carothers promised DuPont "neurotic spells." He anticipated "diminished capacity." He stumbled at times, alone. But rounding through the summer of 1930, with DuPont friends in Wilmington—Julian Hill and Sam Lenher from Berkeley—and with visits from Adams and Marvel and regular correspondence with Jack Johnson, and with exciting new discovery and with action, with experiments lined up for the future, with paper after paper to write, there is no trace in Carothers' record of what William Styron recounts, a "struggle with the disorder of my mind—a struggle which...might have a fatal outcome."[33]

---

[33]Styron, William. *Darkness Visible, A Memoir of Madness;* Vintage Books: New York, 1992; p 3. Only a person who has suffered depression can know the accuracy of Styron's memoir.

Chapter 8

Elmer Bolton's Chemical Department

In that brief and troubled summertime of 1930, with the nation tilting toward the first economic uncertainties following the market crash the previous October, Carothers and his group had, indeed, something important to display. Collins' elastic synthetic rubber and Hill's shiny fibers were two unexpected, practical results from the fundamental research program: rubber and fiber, two enormous potential opportunities for DuPont realized in the first weeks of Elmer Bolton's newly established watch as head of the chemical department.

Bolton thought Carothers was a "wonderful research man."[1] He recognized how Carothers' continuous flow of ideas greatly benefitted DuPont. Carothers' intellectual curiosity and the questioning manner he developed within his group had suddenly generated these two obviously useful substances. Stine's program had succeeded foursquare in its mission. It brought good men to DuPont. Dupont would capitalize on the new research to get nationwide publicity, and the record of publications and the list of patents would provide the company with its trading stock of scientific knowledge. Rubber and fiber were the unexpected, fourth practical outcome of Charles Stine's fundamental research.

As he took charge of the chemical department, Elmer Bolton preserved a belief that his research unit must have a practical outlook. Chemical research at DuPont, for Bolton, could not look on its

---

[1] Elmer K. Bolton interview with Alfred D. Chandler, Richmond D. Williams, and Norman Wilkinson, 1961. (Hagley Museum & Library Collection, Wilmington, DE, 1689, p 20.)

industrial application as an option—it was the core of its meaning, the rationale for its existence. In late 1928, he composed a "white paper," attributing the success of his dyestuffs research to clear organizational focus, extensive training of the scientists, and a passion for providing them the services they needed "to conserve the chemist's time and avoid as far as possible, diversion from his laboratory work." Patent abstracts came from Berlin as soon as they were published, glassblowers fabricated new equipment, testers provided instant analysis of the new dye compounds. Bolton expected a strict discipline from his chemists; they would work only with purified starting materials in their chemical syntheses, so as not to be misled by chemical transformations of unwanted components, and they would study the literature, and they would do that at night, on their own time.[2] His chemists must "pay the price."

"Research work," he wrote, "has been, and will continue to be, the key to success in all of our industries." But he warned the results must be evaluated to make certain the "results obtained are commensurate with the investment."[3]

For years, Bolton had answered, as director of dyestuff division research, to Willis F. Harrington, a man who demanded, every day, that Bolton justify the existence of his research group in terms of immediate, useful results.[4]

Bolton had, in fact, from the start opposed the premise of the basic science program he now ran.[5] So in the spring of 1930, the thin veneer of lofty research purpose was rubbing from the walls and corridors of

---

[2]Robert Joyce to Barbara Pralle, June 14, 1995. "I think that Bolton's insistence on working with pure compounds is underplayed. The presence of an impurity *can* alter the course of a synthesis reaction. But more importantly an impurity can be devastating in a polymer-forming reaction; at the least, it can limit the molecular weight of the polymer; at worst, it can actually prevent a polymerization from proceeding at all. Bolton inculcated this tenet into all Chemical Department managers. When I came to the Station in 1938, the importance of working with pure starting materials was the First Commandment that was immediately imparted to new chemists."

[3]Bolton, E. K. "Certain Phases of Research Work," October 27, 1928. (Hagley Museum and Library Collection, Wilmington, DE, 1662, Box 17.)

[4]Hounshell, David; Smith, John Kenly. *Science and Corporate Strategy, Dupont R&D 1902–1980;* Cambridge University Press: Cambridge, England, 1988; p 149.

[5]Charles M. A. Stine to Roger Adams, December 2, 1938. (Adams Papers, University of Illinois Archives, Urbana, IL.)

Purity Hall, revealing a laboratory now under direct pressure from its leader, Elmer Bolton, to begin the "creation of new industries."[6]

What of Wallace Carothers and the practical side of chemical research? It is convenient to see Carothers in a simple dimension, to see him born and educated as a basic researcher matched to a fundamental research effort. Yet at Illinois, he once fantasized of a place to "test out some ideas of vast commercial importance."[7] He displayed clear interest in potential for practical applications of his pioneering polymer ideas as he negotiated for the DuPont job. Now he must attend to the pragmatic work necessary for the development of the synthetic rubber. Were his practical interests simply a thin but unmistakable gloss over his fundamental drive to do basic scientific work? If so, that mask, too, was rubbing thin. Was Carothers, as counterpoint to the emerging practical focus of Elmer Bolton, now developing a stiffened resolve to continue down the road of "pure science," now that that road had bred synthetic rubber and fiber?

A cautious reality was developing at the chemical department. Elmer Bolton, now tired of paying lip service to the fundamental objectives, moved toward a demand for practical application; his most notable scientist, Wallace Carothers, began to use his accomplishment as a lever to force decisions ratifying Stine's original fundamental research program.

■ ■

Another reality was that the chemical department suffered from an undercurrent of suspicion, skepticism, and doubtful purpose from the first days of Stine's program. One of Arthur Tanberg's earliest hires, Victor Cofman, leveled thinly concealed charges of antisemitic and sexist hiring practices at DuPont as the company readied to fire him.

Julian Hill described Cofman as "a charlatan," with what Hill recalls as an English background. "He sported a cane, was kind of a dude,...but somehow he had impressed Tanberg, who hired him. He was finally fired, the only person I ever knew that was fired from the DuPont Company. He was a real fraud." Hill said Cofman was fired for "moral turpitude." He kissed a mail girl in a stairwell. Hill remembers the firing as "sort of a frame." Cofman's story was he

---

[6]Hounshell, David; Smith, John Kenly. *Science and Corporate Strategy, Dupont R&D 1902–1980;* Cambridge University Press: Cambridge, England, 1988; p 238.

[7]Wallace Carothers to Wilko Machetanz, April 22, 1923. (Hagley Museum and Library Collection, Wilmington, DE, 1850.)

wanted something trivial done for him immediately; the girl said, "Well, what will you give me?" And so he kissed her.[8]

Cofman parted in the spring of 1929, not unhappily, for he shuttled immediately to France as the director of the Société Française du Lysol.[9] As he left he dropped a note to Stine, Hamilton Bradshaw, and Tanberg, reproaching DuPont for its hiring policies:

> Of course nothing is perfect in this world, at least people differ as to what perfection is. There are a few minor points that seem to mar unnecessarily the excellent picture which one can draw of the du Pont Research Organization. No chemist at the Experimental Station would care to be told that if Curie had been working there, radium would not have been discovered or that men like Steinmetz and Soddy, Michelson and Einstein would be considered undesirable among them. A harmonious organization is highly profitable and it could be maintained by choosing each individual on his merits, without excluding whole groups. An executive might well take greater pride in picking the right person from a group of doubtful human material rather than rely on "blanket" rules which are liable to exclude exceptional individuals because of sex, or political or religious views. These are not necessarily bound with his or her ability for scientific or for cooperative work.[10]

But that was the way Dr. Tanberg worked. He depended on university faculty for talent. And they sent him their best. But the new recruits had a uniformity to them, and the chemical department became "almost exclusively a man's world—a white, Anglo-Saxon, Protestant man's world."[11]

One of Tanberg's 1929 recruits was Sam Lenher. Tanberg, on his university rounds, visited Professor Farrington Daniels at Wisconsin.

---

[8]Julian Hill interview with David A. Hounshell and John K. Smith, December 1, 1982. (Hagley Museum and Library Collection, Wilmington, DE, 1878, p 13.)

[9]Hounshell, David; Smith, John Kenly. *Science and Corporate Strategy, Dupont R&D 1902–1980;* Cambridge University Press: Cambridge, England, 1988; p 295.

[10]Victor Cofman to Charles M. A. Stine, Hamilton Bradshaw, and Arthur P. Tanberg, May 9, 1929. (Hagley Museum and Library Collection, Wilmington, DE, 1784, Box 21.)

[11]Hounshell, David; Smith, John Kenly. *Science and Corporate Strategy, Dupont R&D 1902–1980;* Cambridge University Press: Cambridge, England, 1988; p 295.

Daniels told Tanberg of Lenher, a young physical chemist with a National Research Fellowship at the University of California. Daniels was connected with Lenher's family—Sam's father, who had just died, was on the Wisconsin faculty with him, and Daniels knew Lenher wanted an industrial job now that his widowed mother depended on him.

Tanberg quickly offered Lenher a position in the fundamental group; Lenher would work for Hugh Taylor in the physical chemistry unit. Lenher described his arrival at DuPont. "I was innocence," he said. But Lenher had a marvelous education—a doctorate from University College, London, a fellowship from the International Education Board in Paris to work in Berlin, then the award to go to California. "I was very, very, very well trained," he said. Through a list of publications in the English and German journals, Lenher already had a modest reputation. "I fell on my feet because I made a record...in the chemical literature. A very small, teensy record it's true, as I later found out with a full analysis of myself...."

Sam Lenher was unprepared for the industrial opportunity—he was an innocent because he came with almost no information about his new post. He didn't know the chemical industry, DuPont, or Delaware. But Lenher discovered immediately he had a unique opportunity—one of only a dozen, trained to the peak of their science, chosen for the fundamental work and welcomed to the Experimental Station. He was treated well, he said.

But Lenher noticed almost immediately a pervasive skepticism among the researchers. It wasn't simply the lack of direct guidance from Stine and Bradshaw and the undefined nature of their programs. There was a reaction to the lack of direction. Most of the men worked conservatively, sticking closely to narrow areas of research, rather than taking advantage of their research freedom. Their reaction to freedom was fear. "Most of the people played safe...so you were lost twice over," Lenher said.[12]

During the summer of 1929, Sam Lenher and Wallace Carothers, two bachelors, motored together on weekends, down the DuPont highway—the broad, four-lane concrete highway from Wilmington to

---

[12]Samuel Lenher interview with Matthew E. Hermes, August 24, 1990. Sam Lenher was affiliated with DuPont for 48 years. He eventually served as an influential member of the company's executive committee, was once thought to be a candidate for president of DuPont, and helped direct research policy through the 1960s. *See* Hounshell, David; Smith, John Kenly. *Science and Corporate Strategy, Dupont R&D 1902–1980;* Cambridge University Press: Cambridge, England, 1988; p 510.

Dover that T. Coleman duPont had built entirely with his own funds—to vacation at Rehoboth Beach. Back in Wilmington, Julian Hill joined Carothers and Lenher in a speculative try at the stock market, now only a few months away from its crash. "We are dealing in Goodrich," Carothers wrote Adams, "so if this stock shows any erratic behavior on the market you will know the reason." The three men owned a total of 30 shares.[13] Adams responded, recommending that the powerful buying group branch out to Allied Motors, a holding company on the Chicago exchange. Adams reported that he had "heard indirectly that those people in Chicago who know about the concern are very 'bullish' indeed..."[14]

Each of the three was still early in his DuPont career; Lenher and Hill in their first year, Carothers in his second. Hill was enthusiastic over his new opportunity to work with Carothers; Lenher was breaking ground with some new chemistry of his own choosing he thought might be of commercial value. For Carothers in the summer of 1929, "nothing of importance has happened at the lab."[15]

Lenher had a violent explosion in his laboratory that summer. He was treating propylene gas with oxygen in the gas phase. This mixture is extremely dangerous, and from somewhere, probably from a static spark, his container blew up. His summer assistant, George Kistiakowsky was gravely injured in the explosion. Lenher was not hurt badly. Lenher's boss, H. S. Taylor, treated Kistiakowsky "shabbily" after the explosion, Lenher claimed. ". . .he was inadequately treated for his injury," Lenher said. "George was badly treated; inadequately treated. Very seriously mistreated."[16]

Lenher continued his fundamental work on oxidation. He learned to convert propylene and ethylene to propylene oxide and

---

[13]Wallace Carothers to Roger Adams, July 1929. (Adams Papers, University of Illinois Archives, Urbana, IL.)

[14]Roger Adams to Wallace Carothers, August 3, 1929. (Adams Papers, University of Illinois Archives, Urbana, IL.)

[15]Wallace Carothers to Roger Adams, July 1929. (Adams Papers, University of Illinois Archives, Urbana, IL.)

[16]Samuel Lenher interview with Matthew E. Hermes, August 24, 1990. George Kistiakowsky was five years older than Lenher and had a Ph.D. from Berlin. He had worked for Taylor at Princeton for a time, before DuPont hired Taylor for the basic science group. Kistiakowsky became Harvard's chemistry department chairman after World War II and served as President Eisenhower's science advisor. *See Current Biography Yearbook, 1960;* H. W. Wilson: New York, 1960.

ethylene oxide, two reactive materials with three-membered rings that Lenher believed would be useful in themselves and be reactive chemical intermediates:

> I thought (these were) real opportunities for DuPont; new, but probably expensive. But what the hell, they had lots of money. I found that people were exercising judgement, limited judgement. (They) felt those were very good articles to publish in the journals of the chemical society. I thought they had admirable possibilities, pretty exciting. But I was encouraged to publish papers. This is what I didn't want, because I knew patents were stock. The DuPont executives were not interested in these subjects. It was beyond their cope.[17]

Within a few months, Sam Lenher realized he was not doing what he wanted to do. He went to Arthur Tanberg with his resignation in hand. Tanberg was aghast at the suggestion and asked Lenher to hold off on his plans while Tanberg found him an appropriate position in another part of the company.

"I was skeptical and thought I had made a mistake," Lenher said. Lenher claims Carothers shared his unrest. "He was very unsettled in his way of life," Lenher said. Only the fact that Carothers had learned patience with industrial practices through years of guidance from Adams kept Carothers in place at DuPont.

Lenher changed jobs in DuPont as Tanberg suggested, believing, in the long run, DuPont would hold promise for him. His interests turned slowly from the bachelor life. "I got married," he said.

> Married a girl who had a successful father in industry at General Electric and her parents were society figures in their own way. So I got engaged and got married successfully and my wife's family sized me up very carefully. I was viewed and not found wanting.

Sam Lenher drew slowly away from Carothers for another reason. "He told me he had an alcohol problem," Lenher said. "I was careful,

---

[17]Samuel Lenher interview with Matthew E. Hermes, August 24, 1990. Lenher said Union Carbide quickly converted the chemistry he revealed in his publications to plants for the manufacture of ethlene oxide, now ubiquitous in its use as as an antifreeze and in polyester fibers. "In a couple of years," Lenher said, "DuPont still bought ethylene oxide, their opportunity lost."

after his confession that alcohol was meaningful to him, not to encourage his drinking. And I continued to use alcohol, but not with Carothers."[18]

■ ■

Wallace Carothers and Julian Hill expanded their polyester work immediately after their discovery of superpolymers, synthetic fibers, and cold drawing. Hill began to make controlled, drawn fibers from the polymer they now called a superpolyester. Hill dissolved the superpolyester in chloroform solvent and squeezed the solution through a hole in a plate into a chamber of warm air. A similar process of extrusion was already used for the production of rayon. Hill's fibers were uniform, and to the hand, appeared to be strong, elastic, and resilient. The scientists took turns tying complex knots with samples of the new fiber. Carothers obtained X-ray diffraction patterns, measured the elastic recovery of the fibers, and they made a few similar, fibrous polyesters.

It soon became apparent that the new polyesters were simply laboratory curiosities. The fibers had some strength, approaching that of rayon. But they melted at low temperatures, so low that fibers in practical use could not be washed in hot water without returning the drawn fiber right back to the sticky mass from which it was drawn. And they were soluble in solvents, like Stoddard Solvent, a specially developed naphtha developed by the National Association of Dryers and Cleaners for the French cleaning process. DuPont would not commercialize these materials as Bolton had done with polychloroprene.

As Hill stayed in the laboratory in the summer of 1930, Carothers worked in the library on the *Chemical Reviews* article. He described in another paper he sent to the *Journal of the American Chemical Society* in early October how Gerard Berchet had failed to make high polymer from aminocaproic acid. The polyamide Berchet obtained was high melting but could not be dissolved in common solvents.[19] This work was completed before Carothers and Hill learned to make their superpolyesters, but Carothers directed Julian Hill to take one more try at aminocaproic acid. On July 21, 1930, Hill heated the low poly-

---

[18]Samuel Lenher interview with Matthew E. Hermes, August 24, 1990. Lenher was 85 years old when we had lunch together at his Martha's Vineyard home. His son George measured a single, precise, small glass of Scotch whiskey for Sam Lenher, a measure dictated by years of unchanging habit.

[19]Carothers, W. H.; Berchet, G. J. *Journal of the American Chemical Society* **1930**, *52*, 5289.

mer that Berchet had made in the molecular still. He got a hard, high-melting mass but couldn't make a fiber from it.[20]

Carothers wrote up Hill's work on the paraffins. Then he began to assemble the papers he would publish on the polyester work. First he dashed off a brief note on the use of the molecular still; he then described his methods for making the superpolyesters and followed that with a short description of some mixed polyester–polyamides that Hill had made to increase the melting points of the fibers.

Carothers was in high gear. He took two weeks off at the end of the summer to go to Rehoboth Beach but did not attend the American Chemical Society meeting in Cincinnati, which followed close on the heels of his vacation. He wasn't particularly eager to go, but he wrote Adams that, "...Dr. Tanberg, the other day, volunteered the suggestion that since I am going to take my vacation in the last two weeks of August, he could hardly afford to have me go to the meeting since this would take me away from the Station for nearly three weeks continuously. This goes to show how bang-up important I have become to the Company."[21]

Carothers was writing simultaneous patent applications for all these new inventions. He completed a patent draft on the superpolyesters in June 1930. He wrote Speed Marvel that he was "completely swamped with work."[22] As the mass of paper accumulated on the desks of Tanberg, Bolton, and Benger, Dr. Benger, who had come to the chemical department from the DuPont Rayon Company, took notice of something he didn't like.

In early November 1930, Benger told Carothers he objected to publishing the superpolyester work until both the paper and patent were thoroughly reviewed by the rayon department. His concern was the paper and patent might, "in any way be contrary to policies or actions being taken by the Rayon Company." Carothers responded to Benger after a few days, taking Benger's objection to a higher plane.

For Carothers, Benger's action reversed the publication policy in place from the start in Stine's program. Carothers wondered what more he would do on polymerization if his work could not get into the literature. Benger agreed with Carothers that no practical application could come from the current work, but he worried about future

---

[20]Wallace Carothers to Arthur P. Tanberg in *Early History of Polyamide Fibers*, February 19, 1936. (Hagley Museum and Library Collection, Wilmington, DE, 1784, p 6.)

[21]Wallace Carothers to Roger Adams, July 26, 1930. (Adams Papers, University of Illinois Archives, Urbana, IL.)

[22]Wallace Carothers to C. S. Marvel, November 15, 1930.

results. "It has been on the basis of the possible great importance of
the work if successful rather than on the likelihood of its being successful that I have taken the attitude that the work should not be published and that our position should be thoroughly protected by a well
planned patent program." Benger wavered after his discussions with
Carothers, and he asked Elmer Bolton for help. He wrote, "Dr.
Carothers is convinced that my earlier decision is wrong and I am not
at all sure it was right."[23]

■ ■

Carothers took a quick trip to New York on November 9 to Yankee Stadium. There, with 74,000 others he watched Illinois lose in an
inept football performance against the Army. Illinois made but one
first down and punted 14 times in a 13 to 0 loss. The day's major
attraction was the arrival of the entire corp of cadets and the music of
Illinois' 160-member band. Carothers scanned the stands during the
exhibition, unsuccessfully searching for Speed Marvel and his new
wife, Nell. He knew they were there, but he had made no prior
arrangements to meet up with them.[24]

Jack Johnson planned to visit Carothers at Wilmington in January
for the first time. Apparently Johnson asked how he would know he
was at the right place. Carothers wrote:

> I think that you will have no difficulty in recognizing the
> Main Grand Central Station of the Pennsy in Wilmington
> by the imposing majesty of its architectural conception
> and execution, the giant porphyry columns with archi-
> traves, etc. but in case there should be any doubt it is just
> as well to keep in mind that the first station you come to is
> the only one.

Carothers promised his friend an afternoon with "flagons of gloops
which if it is not of the most aristocratic lineage is at least not lacking
in authority."[25]

---

[23]Ernest B. Benger to Elmer K. Bolton, November 7, 1930. (Hagley Museum
and Library Collection, Wilmington, DE, 1784, Box 21.)

[24]Wallace Carothers to C. S. Marvel, November 15, 1930. (Hagley Museum
and Library Collection, Wilmington, DE, 1784.) *The New York Times,* November 10, 1930, reports the game. Across the Harlem River at the Polo Grounds
that same afternoon, 42,000 watched Georgia defeat New York University, 7
to 6.

[25]Wallace Carothers to John R. Johnson, about January 15, 1931. (Hagley
Museum and Library Collection, Wilmington, DE, 1842.)

■ ■

The Rayon Company tested some of Carothers' polyester fibers spun mechanically by Julian Hill, in January 1931, and liked them—a lot. Not that they represented a commercially viable product—they did not. But they had some strength, were soft and pliable, retained what strength they had when wet, and had a springiness that reminded Rayon's Hale Charch of wool. Charch particularly liked their elastic recovery, which was vastly better than rayon and better than silk.

Charch insisted the products should be rigorously patented and that the research work should not be abandoned. "We recognize," he wrote, "that very material obstacles appear to be in the way of considering the commercial application of the principles of the work at this time, but we feel it is entirely too early to be influenced by such considerations...." Charch called for the continuation of polyester research.[26]

Carothers' group proceeded to make fibers for testing but did little further experimental work on polyesters. Carothers kept writing as if all the work would soon be released. As winter turned to spring in 1931, the chemical department allowed Carothers to send off his *Chemical Reviews* paper, which did not disclose the superpolymers or their fiber-forming tendencies. The company held the papers on superpolyesters while Carothers developed, with the help of Hale Charch, a complete set of polymer and fiber specifications for the patent proposal.[27] The company filed the patent application with the U.S. Patent Office in late summer. They immediately learned they were in "interference;" an application claiming similar work had appeared in the office at about the same time. DuPont's legal department had Carothers chase around to find all the documentation he had on his original polyester ideas.[28]

Carothers finished his polyester papers and began composing a series of papers on the chloroprene work. By mid-summer of 1931, the dust-up over Carothers' superpolyester publications resulted in a meeting about the charter and the usefulness of the fundamental

---

[26]Charch, Hale. *Special Report, Examination of Properties of Threads and Filaments Spun from Experimental Station (Dr. Carothers) 3-16 Polymer,* January 19, 1931. (Hagley Museum and Library Collection, Wilmington, DE, PR records, 1931, File P-5.)

[27]Wallace Carothers to Hale Charch, June 8, 1931. (Hagley Museum and Library Collection, Wilmington, DE, 1784.)

[28]Wallace Carothers to Roger Adams, September 9, 1931. (Adams Papers, University of Illinois Archives, Urbana, IL.)

research, led by Benger and attended by Carothers and other group leaders, both in and out of the fundamental work.[29] Benger later cautioned his fundamental researchers they were to resist the "gambling spirit" that had led Carothers to his two basic findings. He was looking for "thoroughgoing cultivation of a field."[30]

Although Purity Hall still resounded with the summer's rhetoric over purpose and direction, Carothers seems to have won his point on publication. He traveled to Buffalo and on September 1, 1930, stood before a large proportion of the record 2057 chemists gathered for the American Chemical Society's semi-annual meeting and told his story of the high molecular weight polyesters and how they gave strong, lustrous fibers. Carothers' friend from Iowa State, Henry Gilman, wrote in his review of the meeting that Carothers' paper was received with "noted interest."[31]

In the fall of 1931, the DuPont Company thrust Wallace Carothers into the glare of nationwide publicity. It scheduled a scientific announcement of Carothers' invention of DuPrene for the November 3 meeting of the Akron group of the Rubber Division of the American Chemical Society. The talk coincided with publication of Carothers' first journal article on polychloroprene.[32] Carothers' talk to the local group, even at this center of America's rubber industry, would hardly draw notice, but on the day of the meeting DuPont sent a press release to 500 daily papers and to the Associated Press.[33] Wallace Carothers became a public figure. More than 400 newspapers responded to the release with editorials; the *New York Times* reported the story on its front page.[34] DuPont sent Ira Williams and F. B. Downing with Carothers to speak on aspects of the commercial development of the new rubber. Arnold Collins was given no notice in the company publicity.

---

[29]Kraemer, E. O. "Fundamental Research in an Industrial Laboratory," July 31, 1931. (Hagley Museum and Library Collection, Wilmington, DE, 1784.) No record exists of the meeting for which this memorandum is the agenda.

[30]Hounshell, David; Smith, John Kenly. *Science and Corporate Strategy, Dupont R&D 1902–1980;* Cambridge University Press: Cambridge, England, 1988; p 241.

[31]*Industrial and Engineering Chemistry, News Edition,* September 10, 1931.

[32]Carothers, W. H.; Williams, I.; Collins, A. M.; Kirby, J. E. *Journal of the American Chemical Society* **1931,** *53,* 4203.

[33]Weston, Charles F. *News Publication,* November 3, 1931.

[34]Smith, John K. "The Ten-Year Invention: Neoprene and DuPont Research, 1930–1939," *Technology and Culture* **1985,** *26,* 34.

Carothers returned immediately to Wilmington; he had another job to complete. He now had approval from DuPont for journal publication of all his polyester work. Within a week he sent off a packet of seven publications to the *Journal of the American Chemical Society*, telling the complete story of his superpolyesters and their conversion to fibers. And Carothers prepared one more presentation. He was invited, the week after Christmas, 1931, to talk to the fourth Organic Symposium of the American Chemical Society at the Sterling Chemical Laboratory on the Yale campus. There he would discuss the chemistry of chloroprene before the most prestigious audience American chemistry could muster.

■ ■

In retrospect, at the end of 1931, DuPont had gained little and lost much through its fundamental research program. Commercial introduction of DuPrene was years away and, with the price of competing natural rubber falling to three cents a pound during the Depression years, the prospects for the new material appeared grim at best. Carothers' fibers seemed no more than a curiosity. Carothers won the battle for publication of his work over the concerns of Benger and Bolton that by publishing, DuPont might be giving to the public valuable knowledge that might be better kept, as Sam Lenher would put it, as its stock in trade.

As it happens, Carothers had already placed in the literature, in late 1930, a single statement that would come back to haunt DuPont and cause the company to lose forever a significant portion of the nylon business. He and the so-far luckless Gerard Berchet, who had reacted vinylacetylene with hydrochloric acid weeks before Collins did but had not extracted the products from his glass tubes quickly enough, studied polymerization of the amino acid, 6-aminocaproic acid. They obtained a hard, gray waxy material—a low polymer—and a second product, a cyclic product called a lactam. Carothers wrote for the journals, before he learned the tricks of making his superpolymers, "...the lactam does not polymerize under the conditions of formation of the polyamide either in the presence or absence of catalysts."[35]

Carothers' patent lawyers certainly winced when they read his statement. Competing scientists' eyes could open with wonder. The implication that "it can't be done" damages, irreparably, a scientist's own institution. The reason is that if someone else does it, they will earn patent rights easily. They have an ironclad lock on an invention

---

[35]Carothers, W. H.; Berchet, G. J. *Journal of the American Chemical Society* **1930**, *52*, 5289.

because they have done what the literature claims is impossible. The lactam from 6-aminocaproic acid—called caprolactam (*see* reaction 9 in Appendix A)—actually polymerizes easily and makes a fine textile fiber, known around the world as nylon-6.[36] Researchers at Germany's I. G. Farben searched the literature in 1937 after learning through discussions of licensing with DuPont of Carothers' nylon. They quickly found Carothers' literature reference to caprolactam, and within a year the company was on its way to its own, exclusive nylon product.[37]   The very first polyester (*see* reaction 10 in Appendix A) Carothers described in the literature, in his first publication with Jim Arvin in 1929, combined phthalic acid as the diacid and ethylene glycol. They reported a glassy resin with a molecular weight of about 4000. Phthalic acid, through a compound called phthalic anhydride, was readily available to Carothers in 1929. Two acid groups are arranged side-by-side on a stable, planar, six-membered ring in phthalic acid. The acid groups are at 60° angles from each other. Each diacid introduces a sharp kink in the polymer chain.

In 1940, J. R. Whinfield and J. T. Dickson at the Accrington U.K. Laboratories of the Calico Printers Association, straightened the kink in the chain by using an isomer of phthalic acid, terephthalic acid, in which the two acid groups are 180° from each other. They became the inventors of the high-melting, strong, and durable fiber called Terylene, polyethylene terephthalate (*see* reaction 11 in Appendix A), known throughout the world as "polyester" fiber, a fiber that surpasses even nylon in worldwide use.[38] There is no way of knowing why Wallace Carothers never tried this combination.

---

[36]Chemists must be observant above all. The literature of chemistry is full of missed opportunities. A very young graduate student working for a former student of Speed Marvel, Prof. William Bailey, at the University of Maryland in 1956 isolated a ring compound called an oxazoline in low yield from a complex, tarry reaction mixture. He never attacked the thick polymer residue. Within 10 years an entire polymerization chemistry of similar monomers was unfolding, resulting in several useful products. I was the young chemist who first failed to identify this polymerized oxazoline.

[37]Hounshell, David; Smith, John Kenly. *Science and Corporate Strategy, Dupont R&D 1902–1980;* Cambridge University Press: Cambridge, England, 1988; pp 207, 302.

[38]Hayman, N. W.; Smith, F. S. "Major Advances in Polyester 2GT Technology, 1941–1990," In *Manmade Fibers: Their Origin and Development;* Seymour, R. B.; Porter, R. S., Eds.; Elsevier Science: London, 1993; p 369. Terylene was developed by ICI. Eventually DuPont purchased rights to polyethylene terephthalate and it became DuPont's "Dacron."

Chapter 9

A Synthetic Rubber and a Synthetic Silk: "Enough for One Lifetime"

Has this book turned from a biography of Wallace Carothers to a simple recitation of his accomplishments? For the moments just past, it has indeed. Little exists in the records that survive Wallace Carothers to direct us to the patterns of his days through 1930 and 1931, the two years of his astounding productivity. Little speaks of his life on Delaware Avenue, in the boarding quarters where he has lived since arriving from Cambridge. No women and no men appear to be particularly important to Wallace Carothers. Julian Hill provides a few notes. Carothers and Hill often played squash together. Hill called him a "wonderful companion, and when he was relaxed, after he had a drink or two, he was the funniest man alive. He had a very specialized form of understatement interspersed with classical quotations. He had read everything. I never knew that he specialized. He seemed to be an omnivorous reader." But at some time, perhaps as they dressed after a squash match, Carothers showed Hill the capsule attached to his watch chain. It contained a ration of cyanide. "I thought it was a matter of bravado," Hill said. "He had at his fingertips all the famous chemists who had committed suicide." Hill listed Boltzman and Emil Fischer and said, "the way he [showed it to me] was so gruesome."[1]

Twenty months elapsed from Arnold Collins' isolation of polychloroprene in mid-April 1930 until Carothers summarized DuPont's

---

[1]Julian Hill interview with Matthew E. Hermes, May 29, 1990.

synthetic rubber work at Yale in the week after Christmas, 1931. In that score of months, Carothers and his group invented synthetic rubber and defined the chemistry of dozens of its chemical derivatives. He wrote up and published a long first paper on the work and wrote more than a dozen related patent applications. He presented major reports to the Rubber Division of the American Chemical Society and to the American Chemical Society Organic Symposium.

Carothers discovered fiber-forming superpolyesters and fully characterized them. He introduced the fiber work to his peers at the Buffalo American Chemical Society meeting. He wrote 14 papers for the journals on polyesters. His *Chemical Reviews* discussion of polymer chemistry was a defining work in polymer science.

All of this writing demanded an intensity of mental focus. With his drive to compose and publish the record of his scientific accomplishments came the duty to assemble for each paper its list of relevant references, to read them, and to record their significance in the new work. We can see *Chemical Abstracts*, the encyclopedia of Carothers' science, and the bound journals, opened to the pages of the critical references, stacked one upon the other, week after week, in front of him.

But Carothers' work for DuPont was not the only chemistry he did. Arthur Lamb, editor of the *Journal of the American Chemical Society* named Carothers an associate editor. This nonpaying position was more than honorary. Under Lamb, the *Journal* developed its highly refined system of refereeing submissions. Lamb would send incoming contributions to Carothers and expected his new associate to examine these potential publications for errors of fact, procedure, good chemical sense, and clear usage. Each paper required the reviewer to be familiar with the specific field of the new paper or to do his own literature search to be prepared to review it with knowledge and objectivity.[2] Refereeing papers is an obligation of the chemist to the orderly structure of his science. But many scientists, fine chemists, look with dread at the package from the editor of the *Journal*. For most of us, scrupulous, objective refereeing is a painfully difficult task, draining a chemist's productive hours.

---

[2]Keyes, Frederick G. *Arthur Becket Lamb, 1880–1952;* National Academy of Sciences Biographical Memoirs; National Academy of Sciences: Washington, DC, 1953.

But Wallace Carothers evaluated 34 papers for Lamb from 1930 to 1931. He responded with long letters of careful criticism, adding his suggestions for improvement of the work and for additional experimentation.[3]

Wallace Carothers came to DuPont early in 1928 with misgivings over potential "neurotic spells." For nearly four years he labored toward scientific achievement. Carothers worked particularly hard these last two years. But no one, not Julian Hill or Sam Lenher or Elmer Bolton, saw in Carothers the diagnosis they would apply to him after he was gone. "He was manic depressive and he knew it," said Merlin Brubaker, years after Carothers' suicide.[4] Hill said in 1990, "A little lithium ion would have fixed him up." But Hill remembers that when Carothers wasn't relaxed, he was simply uncommunicative, calm, and withdrawn.[5] No one reports from Carothers' most energetic years that he was depressed or that his energy was touched with mania.

By the end of 1931, Carothers had achieved the recognition for his chemistry within the profession and, as a bonus, found DuPont had generated a widespread awareness of his work in the general public. The story of Carothers' rubber work found its way to the papers in Des Moines. And that was the occasion for Ira Carothers to say he had known nothing of his son's accomplishments beforehand and to comment that in high school his son Wallace was a "slow learner."[6]

■■

Carothers moved late in 1931. He was out of Delaware Avenue and living with three other DuPont scientists in a fine house on the Kennett Pike, the road Pierre duPont built to his estate in Kennett Square, just over the Pennsylvania state line. The little hamlet was called Fairville, and the house, Whiskey Acres. "It was a bachelor establishment that had kind of a circulating set of inhabitants," said Julian Hill. "I think there were always four men. Leigh Williams, Hans Svaneau, and Jack Reeves and I think there was an opening and Wallace stayed out there."

---

[3]Carothers' correspondence, 1930–1937. (Hagley Museum and Library Collection, Wilmington, DE, 1896.) Carothers' correspondence with Arthur Lamb consists of scores of letters.

[4]Merlin Brubaker interview with John K. Smith, 1982 and 1978. (Hagley Museum and Library Collection, Wilmington, DE, 1878.)

[5]Julian Hill interview with Matthew E. Hermes, May 29, 1990.

[6]*Des Moines Register*, November 5, 1931.

Fairville [was] just a wide place in the road beyond Center-
ville, Delaware. The history of it was there was a remittance
man, from a good family, a man of means, who inherited a
tract of land and maybe this house. In order to keep himself in liquor, he gradually sold off parcels. One of the parcels got to be known as Whiskey Acres.[7]

■ ■

Out in Maryville, Missouri, where Frances Spencer was raising her
son and teaching a large throng of elementary school children, she
read of Wallace Carothers' synthetic rubber. She and Wallace had not
seen each other, or corresponded, since Wallace wrote her in late
November 1929. Frances wrote Wallace in November 1931, congratu-
lating him on his rubber work. Frances' letter does not survive, but
her letter once again prompted Wallace to write. His reply on November 25, 1931, the evening before Thanksgiving, became the second of
Frances' carefully guarded cache of 14 letters from Carothers.[8]

These were not love letters. They ring of an accumulation of
painful truth. They tell of blind focus on Carothers' work, his desperate loneliness; they broadcast the hidden sights and sounds of discord
and turmoil in Carothers' life, magnifying them as he stands ever
more paralyzed to prevent oncoming fate. An objective, mature sum-
mation of Carothers' accomplishments does not track his mental and
emotional path. Carothers' emotions divert him even as he lifts his
pen to try to write a graceful, modest response to Frances' congratulations:

Fairville, Pa.
Nov 25, 1931

Dear Frances:
    This business of trying to write a personal letter has
become a little unusual and strange for me. There is so
much to be done that there never seems to be times for

---

[7]Julian Hill interview with Matthew E. Hermes, May 29, 1990.

[8]Mrs. Spencer held Carothers' letters for two years after I first contacted her.
During that time she wrote eloquently of her life at Tarkio and of her days
with Carothers. She preferred writing over talking; I never spoke with her on
the telephone. Her only son, David Spencer, professor of English at California State University, Bakersfield (CSUB), died in late 1991. At that point,
Frances directed that her daughter-in-law, Dr. Jeffry Spencer, also of CSUB,
send me the Carothers letters. Mrs. Spencer died June 6, 1992.

such matters as going to the dentist or getting a new suit, or having the car properly inspected at the right time, or for writing letters; or if there is time, the energies are completely exhausted and spent. The trouble seems to be that my eyes and imagination enormously exceed my capacities, with the result that I now spend my days at my desk almost buried in stacks of paper in a state of indecision concerning which of the seemingly endless list of enormously important things should first receive attention until the buzzing of the telephone or the arrival of callers gets the day really started in the more irresponsible business of doing nothing at all. You may have heard that chief duties of the modern big business are to contact, confer and react. To this may be added the matter of getting a picture and giving a picture. I am not of course a businessman or any kind of a business man, but it seems impossible to avoid some of the melancholy affections of that class when one is associated with a large corporation.

Carothers continued:

I had been intending to start out by saying that I was pleased to have your letter, and that I greatly appreciate the generous sentiments, but this less appropriate and rather egotistical introduction seems to have got started automatically.

He told Frances his version of the rubber story; Father Nieuwland received more of the publicity than warranted, he wrote. That was "political expediency." And, the man to whom most credit was due, "A. M. Collins, one of my assistants, didn't get much credit." Carothers noted it was too soon to tell whether something valuable would come of the work, but his rubber was new and seemed better than that described in "hundreds of prior patents in the field."

A suddenly fragile portrait of Carothers emerges in the letter. We see him at work:

In spite of the fact that I seem most of the time to have only sufficient energy to balance myself in a swivel chair by clinging feebly to the edge of a desk, we have been enormously lucky in our research so far. We have not only a synthetic rubber, but something theoretically more original—a synthetic silk. If these two things can be nailed down, that will be enough for one lifetime.

All of this work left little time for a life beyond the Experimental Station. At Whiskey Acres:

> There doesn't seem to be much to report concerning my experiences outside of chemistry. I'm living out in the country now with three other bachelors, and they being socially inclined have all gone out in tall hats and white ties, while I after my ancient custom sit sullenly at home.[9]

Frances wrote back, immediately this time, in what Carothers saw as a "skillfully fluid and...light vein" She described her son and her classes, and Carothers commented these must bring "significance to your days."

Wallace Carothers received her letter on December 8 and sat down that Tuesday evening at Whiskey Acres to respond. He wrote nearly 600 words that night in one long paragraph. He described for Frances his life as a scientist:

> My time is divided to molecules, which are abstract entities that one assesses but never sees. That is, theoretically my time is spent in this dry, frigid, and lofty manner; actually there is also the business of contacting, consulting and reading, which is equally unreal but has a slight flavor of hypocrisy. Incidentally the pictures that one gives in this invention have nothing to do with photography except in a figurative sense. The process consists of outlining (verbally) the states of a subject or field of knowledge; this outline becomes imprinted upon the abnormally receptive brain of the bearer, who then on his way home immerses his brain in some mental fixing bath; the impression thus becomes permanent & is carefully filed away in the capacious cranial recesses. Getting a picture of course is just the obverse of this process. The press reports alluded to have no pictures even in this highly figurative sense and are as a matter of fact composed almost entirely of hot air in the wrong key. Our rubber business is however plunging ahead in a most carefree fashion. The company has spent about $500,000 on it so far, and now they are out to cash in on it as rapidly as possible. I am a bit worried about the thing because I think that the people who have the responsibility for the technical development don't adequately appreciate the complications of the process they

---

[9]Wallace Carothers to Frances Gelvin Spencer, November 25, 1931.

are trying to commercialize. It looks as though they might throw away another $300,000 in a little while and then have to start over again more slowly. If this happens it will give the whole thing a pretty black eye. However there are some compensations in any event. The decision to push forward thus (in my estimation, privately) has made it possible to proceed with the publication of all our researches out of which the synthetic rubber, as a kind of a side issue, emerged. We shall probably have about 25 papers, and it looks as though I would spend most of the next year in the anguished but blessed (?) labors of authorship. The company has already got a lot of publicity out of the first two papers, which may compensate for part of the expense, and the visiting European celebrities in the field of chemistry seem to include my laboratory in the circuit of their American tours. As you may very well infer from this rigmarole my mind functions almost exclusively on topics of the shop.[10]

Again, Frances Spencer responded immediately to Carothers' letter, this time sending pictures of her son David. And it came to pass that on Christmas Eve, left alone in his room at Whiskey Acres, Wallace Carothers, while returning the pictures, began to unreel for his friend in Missouri, a state of mind of surpassing sadness, not only for the season he yearned to celebrate, but for the daunting task which lay just ahead. For he dreaded standing before his peers to give the paper at the American Chemical Society Organic Symposium in New Haven.

Whiskey Acres
Fairville, Pa
Christmas Eve

Dear Fran:
   I have to thank you for your very nice and prompt reply to my last letter. The pictures of David are returned herewith. He is a very bright and likely looking lad, and I don't blame you for being very proud indeed for him. I suppose you will be with him now and with your family enjoying the liberating, uplifting, gently sentimental and care-erasing influence of the one real holiday season. I send you my best and most sincere Christmas wishes.
   Wilmington appeared very gay tonight as I drove home through the dusk with the streets all strung with col-

---

[10]Wallace Carothers to Frances Gelvin Spencer, December 8, 1931.

ored lights, all its lights illuminated, and brightly lighted trees in front of the big mansions. The bright lights extended all the way out the pike (8 miles) to Whiskey Acres, but this somewhat squalid bachelor establishment has no Xmas tree, and the holly wreaths that some optimistic member purchased still lie disconsolantly on one of the front porches. Hans and Leigh have both set out and soon will be with their relatives in N. Y. & Va. respectively; Jack is wrapping packages in preparation for his return to the family mansion somewhere near Philadelphia, but our libations during and after dinner have somewhat distorted his schedule, and after a slight pause * * * * and a final glass of sherry I have just assisted him, his riding boots, his bags, and his packages into his very early 1928 Model A Ford Roadster; and the divine powers will doubtless amply protect him as they have now been accustomed to do at least twice a week for several years. Thus I am finally alone with the moronic comments emanating from the radio, and (sadly enough) not quite sober either. However, perhaps this slightly hazy state of mind is often all best in the circumstances. I am (damn, damn, damn) scheduled to make an address at New Haven on Monday the 28th—not only scheduled, but it appears that the press notices have already been written and are ready to be fed into the excessively prolific maws of the modern printing machinery whether the address is really delivered or not. You can imagine the state of my nerves and the necessity of some soothing influences or other. I spend the sacred festival tomorrow with local friends who have requested an early arrival to provide opportunity for adequate alcoholic preparations. Saturday (god willing) I go to Yonkers to see Eliz (who was married on the 19th) and thence to New Haven—and your prayers would be greatly appreciated. Best wishes.

Wallace.[11]

Carothers traveled to New York to visit his sister Elizabeth on Saturday, December 26. Her new husband, Robert Kyle, met the inventor for the first time that day. Carothers had not attended the wedding the week before. Robert Kyle said he thinks Wallace was "commissioned" by his mother to come visit the newlyweds. "He knew so much," Kyle said. "You know, art, literature, science.... And

---

[11]Wallace Carothers to Frances Gelvin Spencer, December 24, 1931.

he had a sense of humor." There was nothing Robert Kyle didn't like about his new brother-in-law that day.[12] Carothers stayed in the city, then went on to New Haven for his presentation to the American Chemical Society Organic Symposium on Monday, December 28. After the meeting, he did not return to Wilmington but lingered in New York, a city whose deluxe speakeasies were booming but whose half-empty hotels spoke of "Hard Times."[13] It was "quite gay," he later remembered. "I even dressed up for the second time in my life, and for the first time arrived at luncheon slightly tight from the libations of the night before.[14]

The two years of Carothers' great science paralleled two years of national decline. As Julian Hill and Arnold Collins made their discovery of fiber and rubber, the nation worried whether the market crash of the previous autumn actually signaled an economic depression. By the end of 1931, everyone knew. "Hard Times," they called it, capitalizing the phrase to give it life and even a dignity of its own. As Wallace Carothers was isolated at Whiskey Acres that Christmas Eve, President Hoover was wrapping for his grandchildren the modest presents he had bought a few days earlier, "a gasoline filling station and a war tank," gifts he purchased while mingling with the crowds at Washington's Woodward & Lothrop. Pierre Laval, *Time* magazine's man of the year and the premier of still-prosperous France warned Hoover, "A severe correction and disciplinary period is indicated." Perhaps the French premier knew the flow of Americans who sought France in the 1920s had drawn to a trickle. In 1929, hardly the peak year, 300,000 Americans took to the ocean and visited France. By 1931 the number had fallen to 100,000. As Carothers prepared for his talk in New Haven late that December, the city was stunned as the Broadway Bank & Trust shut its doors after its directors would not sign personal notes to keep the institution afloat. With the bank went the savings of many a Yale student and faculty member. All over New Haven banks required depositors to now wait 90 days for the withdrawal of more than $100.[15]

After his return to Wilmington, on the day after New Year's Day, Carothers wrote to Wilko Machetanz in California, describing the New Haven trip. He wrote on Sunday night, after a long week in New York, New Haven, then back to the big city. As usual, he was replying

---

[12]Robert Kyle interview with Matthew E. Hermes, May 2, 1990.

[13]*Fortune*, January 1932.

[14]Wallace Carothers to Frances Gelvin Spencer, January 11, 1932.

[15]*Time*, January 4, 1932.

to his friend's initiative. Mach wrote after reading of the rubber work; Carothers replied, thanking him for his "kind and eloquent epistle of recent date." He suggested Machetanz had a "somewhat exaggerated idea of the rubber business." "My own ideas about it get somewhat swollen at times," he wrote, slipping in a gentle pun—a critical property of his new rubber, in contrast to natural rubber, was that DuPrene did not swell up in fuel or oil—"and I am inclined to make extravagant remarks about it. We hope that it may turn out to be a success, but there is a fairly good chance that it may be a flop from the commercial point of view." Natural rubber could be purchased for a few pennies a pound at this point, making a synthetic extremely difficult to introduce competitively.

Wallace Carothers described his trip to New Haven, and it was not a pretty picture:

> I did go up to New Haven during the holidays and made a speech at the organic symposium. It was pretty well received but the prospect of having to make it ruined the preceding weeks and it was necessary to resort to considerable quantities of alcohol to quiet my nerves for the occasion. Thursday to Sat incl. I spent in New York going to shows and keeping fairly tight, and have just now returned home to suffer the consequences. My nervousness, moroseness and vacillation get worse as time goes on, and the frequent resort to drinking doesn't bring about any permanent improvement. 1932 looks pretty black to me just now.[16].

---

[16]Wallace Carothers to Wilko Machetanz, January 3, 1932. (Hagley Museum and Library Collection, Wilmington, DE.) I received this letter from Mrs. Barbara Osborn, Wilko Machetanz's daughter.

Chapter 10

"1932 Looks Pretty Black to Me Just Now"

Wallace Carothers began the New Year of 1932 with a singular task. Having presented the rubber work twice, at Akron in November and again in New Haven in the week after Christmas, he began to write a long series of articles completing for the journals the publication of this side of his science. By the end of February he completed 17 papers; all would be published by early 1933.

But he was rolling along on flattened wheels it seems, finding little of lasting fulfillment in his dogged determination to publish, "I have got far enough along in my work," he wrote Frances on January 11, "to realize that the enormous satisfactions hoped for can never be realized in that alone, and I don't seem to be prepared or even qualified for anything else."

Carothers' life was puzzling him. Once again, he looked to the past and found "baffled amazement." "One of Conrad's wise characters," he wrote, "says that what one gets out of life is a little knowledge of oneself that comes too late." A quote from T. S. Eliot, "fits the facts and appeals to me even more strongly...:

> After such knowledge, what
>   forgiveness? Think now
> History has many cunning passages,
>   contrived corridors
> And issues, deceives with whispering
>   ambitions,
> Guides us by vanities. Think now
> She gives when our attention
>   is distracted..."

The first line of the short poem, it is Eliot's *Gerontion*, reads:

Here I am, an old man in a dry month,
        being read to by a boy, waiting for rain.

Carothers did not quote this line, nor the last:

Thoughts of a dry brain in a dry season.[1]

A sense of fragile mortality pervades this new letter as Carothers recognizes himself in Eliot's frail man in a drafty, rented house. Carothers continues what he has begun in the Thanksgiving and Christmas letters to Frances. He resumes exposing to her more than his chaotic mental predicament, by opening two new themes. The first is his physical well-being. "Also there is something rather vaguely wrong with my health (the last diagnosis was in fact quite seriously vagotonia) and it doesn't seem to be possible for me to revise my way of living to conform to the dictates of common sense."[2] Carothers has not seemed to be in fragile health. "I play squash as much and as hard as possible," he wrote Frances.[3] He exercises on the courts with Julian Hill. Hill remembers seeing the almost transparent scar on Carothers' neck, visible only when his friend's face had flushed red after their games, where Carothers' goiter had been removed in an earlier operation. But that was all Hill found of concern.[4]

Carothers received a letter on January 11 from a man he didn't remember knowing, a man named K. Boettner. Boettner claimed to be a classmate of Carothers at Tarkio. In his letter, Boettner congratulated Carothers on his rubber inventions and sent along a note from his wife, Ila, asking about Carothers' marital state. "She is a complete blank to me," Carothers wrote Frances. "What does one do in such a case?", he wrote. "Or do such things never happen to sane people?"

Ila Boettner's question must have visited a moment in Carothers' mind as he continued his letter to Frances, for in the next paragraph, he opened his second new topic. "There isn't any news to report here," he wrote as he reported some rather interesting news. "I also drive back and forth to West Chester and Philadelphia quite a lot to see a girl. A perfectly hopeless case (my own) but we see all the shows in town...."

---

[1] Eliot, T. S. *Collected Poems, 1909–1962;* Harcourt, Brace & World: New York, 1970; pp 29–31.

[2] "Pathological overactivity of the vagus nerve." *American Heritage Dictionary.*

[3] Wallace Carothers to Frances Gelvin Spencer, January 11, 1932.

[4] Julian Hill interview with Matthew E. Hermes, May 29, 1990.

"Do you come East?" he closed. "I hope you will let me know if you do so I can ask you for a date."[5]

Frances Spencer fired back a reply, clearing up for the moment, Carothers' confusion about K. Boettner. Carothers did know him at Tarkio, if only in passing. There is no reference in Carothers' next letter indicating that Frances had commented on Wallace's new woman friend. But apparently Frances now offered to try to travel East during the summer, for Wallace Carothers wrote he hoped she would come to Philadelphia—or New York.

> I am beginning to acquire a beginners knowledge of Manhattan. I spent last weekend up there again (two already for one month) and was introduced to a real bohemian restaurant (a la New York) which is cheap and has marvelous cocktails and really drinkable wine (and delicious food.) Also a quite gaudy speakeasy which has fairly potable and very potent liquid refreshments and the most demoralizingly smooth and expert niger [*sic*] orchestra that I ever heard.

Wallace's specific description of the city was less than it seems. It was not, indeed, an invitation for Frances, but a lead in to the developing story. Carothers wrote on, without a pause:

> For the first time in my life I had a really tremendous yearning to dance, and of course the girl I was with refused to dance with me because my clumsiness would be a kind of desecration to such perfectly barbaric and pagan (or can't it be both?) music. I have seen six New York shows so far this month, and you can imagine or infer what causes produce such follies of extravagance. Also, sadly, things get worse rather than better—for both of us. She is 23, tall, slender, pretty, intelligent (though married) and she has the largest capacity for friendship of anyone I ever met. The trouble seems to be that I am slightly sub-human. At the start it was like an unspoken and indisputable act of God, or that isn't quite right either, perhaps like an undergraduates dream of perfect sympathy (spring time and sun rise) and now it seems to be angry sea salt and indignant clover. Or as must be evident words are quite failing me and I apologize for the defective sopho-

---

[5]Wallace Carothers to Frances Gelvin Spencer, January 11, 1932. This letter was not posted until January 19.

moric outburst. In fact it would be better to postpone further writing for another mood.[6]

He closed, pleading with Frances to write once again.

■ ■

For four years now, Wallace Carothers has bent to his research, producing two remarkable materials, newly won from the mystery of what can be. Only our gods know what may be possible for man to assemble from the stuff they have given us. But like Edison or Langmuir or General Motors' Thomas Midgley, who thrust at us tetraethyl lead and the new chlorofluorocarbon refrigerants that DuPont later owned and called Freons, Carothers shows us great new inventions, rubber and fiber, all at once drawn from the secrets of the laboratory.

But for four years Carothers is a clouded mystery, too. He is an enigma, a puzzling and complex man who has promised difficulties but who produces, in regular fashion, great science from a group of men who swear by him.

Now, in an instant of time by life's long standard, Wallace Carothers is writing his days for his friend in Missouri. This woman whom he has considered as a wife, but in whom he found qualities he detested, now must deal with a stream of dispatches, each one in turn more revealing and by sequence, more desperate.

Frances Spencer replied to Carothers' letter of January 25, asking about Wallace's work. He wrote back impatiently:

> I thought I told you once about my work. I manage to balance myself in a swivel chair by clinging desperately to the edge of an oaken table. This lasts for 7 1/4 hours a day. Physically my activities consist in shouting into a dictaphone, jumping up and down to go for a drink of water or milk, lighting cigarettes, answering the telephone and fitfully rushing into the laboratories. All very thrilling. At noon I usually drive home and sleep. After work I usually play squash or sit around and have too much to drink before dinner. There isn't anything in the way of recent results to report—unless 17 papers written during the last 2 months constitute a result. In the evenings I just sit and

---

[6]Wallace Carothers to Frances Gelvin Spencer, January 25, 1932. The postmark month, JAN, is clear on the envelope. The date is blurred but appears to be the 28th. Carothers' letter bears only Monday evening as a date. The Monday before the 28th was the 25th—about the earliest date possible for Carothers to reply to Frances' response to *his* letter mailed on the 19th.

brood, or else read Chemical Abstracts in a mood of excited absent-mindedness.

And without a paragraph or a pause, Wallace Carothers swung into the continuing story of the girl from West Chester; now neither a girl, nor unnamed, but a woman of vitality and life and interests, a woman named Sylvia Moore:

That is unless Sylvia is in West Chester. The week ends are usually a mess. Sylvia has been in Washington lately lobbying among the senators and people of social pull for sentiment and action in favor of the Geneva Conference (a work that she doesn't get paid for and, as far as the methods of her particular organizations go, doesn't believe in) and I had formed the firm and noble resolve *not* to go to Washington on Saturday. But Sylvia took a notion that she would come up. So she called on the telephone at 4PM and afterward, about 14 assorted callers arrived, and 30 bottles of beer were consumed beside a certain amount of whiskey, and I met the train in a state of morose indifference. A secretary of the league came along. A pretty but rather melancholy girl, and it seems her sorrow is a drunken husband from whom she is now separated. So we stopped at the Acres on the way to West Chester, and the sad lady drank rather too much, and I drank a great deal too much, and the boys arrived at 12:30 and the party lasted until 2AM and yesterday my condition was quite inhuman. We went out in the morning and chopped wood for the Camerons, and I couldn't possibly go back to the Winders for dinner, so Sylvia and Louise came back to the Cameron's. I couldn't contribute anything to the general gaiety but I lent them my nice shiny black coupe to use and they got banged into by some very dumb people, and the coupe is all smashed to the devil. So I have to get along with the second best car of the Winder family for a while, and it is a pretty bum car. And Sylvia has already dashed back to Washington with a bad arm and a banged head as though she were in military service. On inspection this seems to be a very idiotic letter, but if I were to go any further it would probably become (more) idiotic. So I will conclude. Perhaps you will give me some lessons on how to be philosophical about life while still (unfortunately) alive. Correspondence lessons would be greatly appreciated. I have to admit there is no justification for this request beyond the need.

Wallace.[7]

Frances wrote again. No answer from Carothers. She waited almost two months and wrote him, challenging him to announce himself should he still be among the living. Frances told him she would indeed visit the East. Carothers replied.

Tuesday Evening (May 3)

Dear Fran:

No I am not dead, but only moribund. Feeling rather feeble, smelly and cock-roach-like. Just why, I don't know. At any rate I go through at least a dozen violent storms of despair every day. This evening is comparatively serene. Spring is here too. The feelings and the stirrings of the soil are the same as those in Iowa, but the visible evidences are more extensive and spectacular here in Delaware. (This doesn't quite convey my thought, but at any rate we have hills and woods around this place, and the country side looks rather lordly). I have been out in the yard and have planted some radishes, lettuce and carrots,—first experience as a gardener. Dinner is over now with some good cold beer of considerable authority, and the boys are out on the lawn playing with the two puppies—very lively springer spaniels. One ought to feel like just vegetating. I wish I did but I really don't. Rather agitated in a febrile sort of way. I never seem to know 30 minutes ahead now a days whether I shall be able to sit up or not. I am glad you have decided to come East. Please let me know definitely when as soon as your plans are made so that I can keep the date open. I shall be glad to see you.

Wallace.[8]

With this letter, our accelerating vision of Carothers, his emergence from the shadows into a full examining light, reaches its apogee, where he will remain for just another month. Then the handle he gave us on his life will be slowly withdrawn, to reappear only in flashes of his own words and in the details his ever more knowledgeable friends will begin to collate.

But as for now, just five years away from his suicide, the famous inventor can find only self-degrading adjectives to differentiate his state of mind from death itself. He carries cyanide at all times. His continuing existence he finds unfortunate. Wallace Carothers' vital signs are not very good.

---

[7]Wallace Carothers to Frances Gelvin Spencer, February 22, 1932.

[8]Wallace Carothers to Frances Gelvin Spencer, May 3, 1932.

What could Frances Spencer, alone with her young son in northwest Missouri, do for Wallace Carothers? Had she heard, in quiet walks across the grounds at Tarkio, Carothers tell the beginnings of his vivid spiritual wounds? Are these letters from Wallace, now visible to us, but the retrievable remains of a longer correspondence dating from their parting? Mrs. Spencer wrote extensively of her time at Tarkio and of her relationship with Wallace Carothers. He read Swinburne to her, allowing the music of the poetry to cross between them. She wrote that Carothers preferred Nietzsche, and she saw Carothers' focus on the darkness of *Proserpine* as no more than a youthful dalliance, a flirtation with the seriousness of adulthood. She did not see him, then, drawn to the flame of his doom.[9]

Mrs. Spencer saved 14 letters from Wallace Carothers for more than 50 years. By their unity and context, they are the complete package. In the first, Carothers acknowledges they have not corresponded for a long time. Frances Spencer saved the letters. Would she have stopped after 14?

What could Frances Spencer do for Wallace Carothers? Did she read the letters in sequence as each new sheet arrived? Her friend by turn was clinging to his desk to stay upright. He was alone in the void of an empty Christmas Eve. The colors of the lights streaming by as he drove up Kennett Pike mocked his self-centered isolation. He progressed to drinking to moody insensibility, crushed by the meaningless content of his life. Sylvia offered him friendship and it drove him to despise himself even further. He chooses unfortunate life for now, "feeble, smelly and cock-roach-like" life. Does Frances Spencer know he carries cyanide?

But she can do nothing for him. She can continue to write. She can visit, perhaps as he has suggested. But she is the school teacher in a tiny hamlet far from these events. And "Hard Times" can strike Missouri, too.

Wallace Carothers indicated in a backhanded way to Wilko Machetanz in 1928 that he had once sought professional help with his neuroses. "The record of my tortuous wanderings during the past few years would make a rich case for the psychopathologists," he wrote, "but so far I have placed little of it in their hands."[10] If Carothers

---

[9]Frances Gelvin Spencer to Matthew E. Hermes, December 26, 1989; June 27, 1990; July 19, 1990; August 13, 1990; September 23, 1990; April 9, 1991; November 7, 1991.

[10]Wallace Carothers to Wilko Machetanz, January 15, 1928. (Hagley Museum and Library Collection, Wilmington, DE, 1850.)

sought help in his crisis of late 1931 and early 1932, he hid the effort in his revealing series of letters to Frances Spencer.

The public perception of the state of the psychiatric science—or art—was muddled in those years. Three giants of the near past, Freud, Jung, and Adler, once closely unified in the development of psychoanalysis as a manner of belief and treatment, now stood far apart in the swirling field of confusing concepts. With psychoanalysis, Sigmund Freud meant to draw from the unconscious the power to "cope with depressive states, even if they are of a severe type."[11] Freud acknowledged that C. G. Jung of Zürich and Alfred Adler of Vienna, "two former adherents," had left him to establish diverging schools of thought. Words seemed inadequate to characterize and classify normal mental states, let alone heal the anguish of the mentally tortured. Jung criticized Freud:

> Freud has unfortunately overlooked the fact that man has never yet been able single-handed to hold his own against the powers of darkness—that is of the unconscious. Man has always stood in need of the spiritual help which each individual's own religion holds out to him. The opening up of the unconscious always means the outbreak of intense spiritual suffering; it is as when a flourishing civilization is abandoned to invading hordes of barbarians, or when fertile fields are exposed by the bursting of a dam to raging torrents.[12]

In 1931, even the thoughtful person, stable, attentive to the tides of change and controversy swirling about psychoanalysis, could find no central direction for the "talking cure." Perhaps Frances Spencer, pondering how she should respond to the long, unconsciously revealing paragraphs she was receiving from Wallace Carothers, read *Scribner's Magazine* for December 1931. In "Psychoanalysis—The Inward Eye," George Draper wrote of the confusion with the modern state of psychoanalysis with the conflicting vision of Freud, Adler, and Jung, each given as absolutes. Dr. Draper pleaded with his readers that they might look past the eddy of confusion maintained by the resistance of both the Catholic and Protestant faiths to allow the study of the spirit. Let the people use the new techniques to deal with the problems of

---

[11]Freud, S. "Psychoanalysis, Freudian School," *Encyclopedia Britannica*, 14th ed.; Chicago, IL, 1939; Vol. 18, p 673.

[12]Jung, C. G. *Modern Man in Search of a Soul;* Bell, W. S.; Baynes, C. F., Trans.; Harcourt Brace & World: New York, 1933; p 240.

the mind. Let the psychologists join the physicians and ministers in a healing triumvirate.[13] How could Wallace Carothers, drinking away his idle time, his mind laced with anguish, despair, and depression, find his way through this confusion to the healing couch? Perhaps he, too, read *Scribner's* that crucial month. But he could draw no more direction, no more comfort, from the popular press than he could from his friend in Missouri. She knew everything, now, but she was powerless to help him.

For Wallace Carothers was in a physical and mental and spiritual nosedive. No physician, no preacher, no psychopathologist could help him. Carothers was stripped of any protection. *Time* magazine reported the wings ripped off a mail plane flying over the Midwest that week Carothers spoke in New Haven. The pilot rode the crippled craft, called a "Carrier Pigeon," perilously close to the ground, but at the last instant he freed himself from the tangled cockpit and parachuted safely to earth.[14] Wallace Carothers could not free his parachute. He was entwined in his mental morass, in free fall, unable to loosen the binding depression, vulnerable to the lure of the cyanide that fell with him. Hemingway's epigraph to Scott Fitzgerald could have been written for Fitzgerald's contemporary Wallace Carothers:

> His talent was as natural as the pattern that was made by the dust on a butterfly's wings. At one time he understood it no more than the butterfly did and he did not know when it was brushed or marred. Later he became conscious of his damaged wings and of their construction and he learned to think and he could not fly any more because the love of flight was gone and he could only remember when it had been effortless.[15]

What helped strip him of his reason, of his ability to line, on one side, the positives of his life—his secure job with DuPont in "Hard Times," the inventions that turned the heads of the world, the loyalty of his group, his substantial scientific reputation, his growing public stature—against his feelings of inadequacy and depression was the alcohol. Alcohol deprived him of reason. He was depressed and alco-

---

[13]Draper, George. "Psychoanalysis—The Inward Eye," *Scribner's Magazine,* December 1931, p 667.

[14]*Time,* January 4, 1932.

[15]Hemingway, Ernest. *A Moveable Feast;* MacMillan Publishing: New York, 1964; p 148.

hol was a depressant. Yet he found no power to rout the daily degradation, to return to the "dictates of common sense."[16]

There was little to help a man immersed in alcohol. An anonymous correspondent wrote for *Harper's Monthly Magazine* in late 1931 that she was a moderate drinker. But she went on, "The story of the moderate drinker is briefly this: there are almost no moderate drinkers." She found herself out of control. She could not stop even when just two drinks had the admirable effect of stilling "the jangle of New York's roar..." Drink impaired her work, her driving, her spirit, changed her personality and lied to her about the change, telling her nothing had changed. Prohibition had corrected nothing. The last 12 years of the great social experiment in the United States had altered not a whit her continual, unsuccessful trials with alcohol.

For now and for her, this woman reported an answer. It was abstinence from alcohol: sobriety. "What to do—that I don't know," she wrote. But for now, "...I am glad to be released—for some time at least—from the nagging problem of when to drink and how much. And how fraying this problem is! This drinking a little too much, hurting one's work, stopping drinking, being in good company and going on drinking."[17]

Harper's anonymous reporter is lost to us. Perhaps she may have been lucky enough to stay on the lonely track of sobriety in those years before any reasonable treatment for those obsessed with alcohol had developed. Frances Spencer and Wallace Carothers both read widely. We cannot know whether either came across the story of this drinker from New York. But Wallace Carothers continued his regular visits to his bootlegger, a farmer selling jugs from his mushroom house near Carothers' home at Whiskey Acres.[18]

Wallace Carothers mailed letters to Frances Spencer on May 17, 1932, and again on June 9. Then his letters ceased for almost a year. The theme of these last two messages went unchanged. He liked her letters, felt his contribution to their exchange was minimal, and continued in a "habit of hasty dabs and absent-minded glances" to do his work, sputtering along with a minimum of energy. He declined three "flattering" invitations to speak—one of them would have taken him to England. He admitted his rejections would put him in difficulty with his employer, but he envisioned a time, not too far distant, in which he would "retire to some fresh water college and justify the

---

[16]Wallace Carothers to Frances Gelvin Spencer, January 11, 1932.

[17]Anonymous. *Harper's Monthly Magazine* **1931**, *162*, 419.

[18]Julian Hill interview with Matthew E. Hermes, May 29, 1990.

original nickname of 'Prof'." His gloomy words, he admitted "just seem to sprout forth without any particular provocation." But he finally implied that with more than a thousand chemists out of work in the New York area, some earned more than he thought he ever would, a few of them "starving," he was not, after all, badly off.[19]

Frances soon canceled her trip East, citing a bank failure in her home town as the cause. Carothers responded that DuPont affairs looked "pretty black." DuPont had reduced salaries once again, and although his new rubber, DuPrene, was now on the market, the price of natural rubber had fallen to three cents a pound, providing for a very limited opportunity for the new synthetic, priced at 75 cents. "Summer is here," he wrote, "with its (for me) inevitably bad psychological effects..." Sylvia was headed soon for Chicago to work for the "League" on disarmament at the National Convention of the Republican Party, which would renominate hapless Herbert Hoover.[20] Carothers hoped he would reach the Midwest during the summer on the way to the American Chemical Society meeting in Denver in late August. He also hoped to go to Europe for vacation, but with the plunge of values on the stock market, he had virtually no savings any longer.[21] He never made either trip. He never saw Frances Spencer face to face ever again.

[19]Wallace Carothers to Frances Gelvin Spencer, postmarked May 17, 1932.

[20]I presume this is The League of Women Voters. The league does not have complete membership records for the period.

[21]Wallace Carothers to Frances Gelvin Spencer, postmarked June 9, 1932.

Chapter 11

A Struggle Over Research Objectives

> Our research program as a whole, particularly our fundamental and pioneering applied research, has been materially revamped with the object of effecting a closer relationship between the ultimate objectives of our work and the interests of the Company.[1]

With this text, buried in a long opening paragraph of his chemical department's 1932 report to DuPont's executive committee, Elmer Bolton withdrew the charter established by Charles Stine in 1927. Bolton did not abolish the fundamental program, not at all. But he aligned Wallace Carothers and the rest of the stable of researchers who were pursuing "basic science" with one of eight specific, business-oriented objectives, such as "Cellulose" and "Smokeless Powder." Under the section, "Resins and Coating Compositions," just ahead of a long report on "Wood Oil" and another on "Automobile Fender Enamel," Bolton wrote a short summary of the polyester fiber work, indicating disappointing commercial possibilities. He reported the chemical department was helping with the manufacturing problems on Carothers' DuPrene rubber. But the lofty goals—the attraction of staff, the publicity—were a thing of the past.

What replaced Stine's original program was a reduced effort and budget for fundamental work. Bolton described the progress of the

---

[1]Elmer K. Bolton to the executive committee. Annual Report—1932; January 6, 1933, p 1. (Hagley Museum and Library Collection, Wilmington, DE, Box 16.)

effort in reasoned terms and asked for $195,000 for 1933, down from $220,000 in 1932 and from $250,000 in 1931.[2] Carothers would receive $70,000 for the organic division, a reduction of more than one-third from 1931.

Elmer Bolton's transformation was driven by a series of factors. First, the original funding of the fundamental program by DuPont, a reserve of money put in place between 1927 and 1929, would run out early in 1933. To go on at all would require allocation of precious, new money by the company.

With the nation deeply depressed, and with DuPont's business threatened by the economic situation, budget cuts made sense. DuPont reduced salaries so that budget restrictions would not necessarily cause layoffs. Julian Hill said he felt particularly well treated by DuPont. Twice, his salary "increase" was an election by DuPont not to decrease his pay.[3]

Second, the devastating times made Charles Stine's basic science objectives obsolete. In Hard Times, good publicity was not needed to attract good people. There were no jobs. DuPont would not be hiring. Bolton wrote of a stable staff with negligible turnover and good morale. The depression gave DuPont what Stine had wanted to buy with his basic science. The men were glad to have work.

Third, from a research point of view, DuPrene was a peripheral success of the fundamental research program, even if the vanishingly small price of natural rubber made its commercial viability unlikely. But nothing else was on the horizon, and Bolton, who "did not consider that fundamental research could be administered logically as part of an industrial organization," could reasonably align the fundamental researchers within product interests.[4] The next step would be the elimination of fundamental research as a formal, separate program—completely.

In the summer of 1932, as Elmer Bolton began to signal the coming changes in Stine's mandate, Wallace Carothers attempted a thoughtful and objective summary of four and one-half years of his stewardship of fundamental organic chemistry. He advanced a plan for reduced fundamental science but strongly defended continuation of his work. The document came in the sticky buzz of the hot Wilmington summer, with Carothers threatened by his recurrent sum-

---

[2]Elmer K. Bolton to executive committee, September 28, 1932. (Hagley Museum and Library Collection, Wilmington, DE, 1784.)

[3]Julian Hill interview with Matthew E. Hermes, May 29, 1990.

[4]Charles M. A. Stine to Roger Adams, December 2, 1938. (Adams Papers, University of Illinois Archives, Urbana, IL.)

mer doldrums, aware he would not travel across the ocean to Europe, or west to the foothills of the Rockies to the Denver American Chemical Society meeting.

Most academic scientists can flip open an accurate, up-to-date list of their publications—a curriculum vitae—when asked. Carothers approximated. He claimed about 45 papers published or awaiting patent clearance before publication, along with book reviews, oral presentations, and "miscellaneous published items." And he had completed the science for a large number of additional publications. As a result of his work, Carothers added that Dupont had filed about 25 U.S. and 35 foreign patent applications.

Carothers would not attempt self-evaluation of his research progress in condensation polymers, the brief effort at making linear alkanes, and the new rubber discovery without first referring to the original objectives of the work. He reviewed the details of his hiring in 1927 and the "hypothetical examples" he had discussed with Stine, Tanberg, and Bradshaw. He restated Stine's original research goals, this time as his own:

> ...I came to the conclusion that "fundamental" research was what is more commonly called "pure" research. Its objects were to increase the body of scientific knowledge. It was not expected to yield any results of direct financial value. Its success would be gauged by the significance of the scientific contributions produced, and any financial profit that might accrue would be so much gravy.

Against these scientific aims, Carothers found his work on the alkyls "...better than average work in organic chemistry; but not of very great significance." The scientific results that had led to DuPrene were "...abundant in quantity but perhaps a little disappointing in quality." But he wrote of the condensation polymers, "...I should classify [them] as an unusually significant contribution to the science of organic chemistry."

This summer report by Wallace Carothers appeared to be an alltogether conciliatory text. He agreed that fundamental research in organic chemistry had not paid its way. He estimated $300,000 had been spent to date and chose but one-quarter of that amount as the dollar value of the publicity from the rubber and synthetic fiber work. Carothers took note of the depressed times, indicating that in a normal economic climate, "...fundamental research would have been successful from almost any point of view." He outlined the case for continuing his work. He did not want to abandon what might be financially successful in the future. He described his own limited abilities at commercial thinking:

> If I had been asked to do research on anything that I pleased with the mutual understanding that the object was to develop something that would bring in a direct profit, I should never have accepted the job. I never had any confidence in my ability to initiate and carry on research of this kind, and I still haven't any. There are certainly people that do have this ability, but I think that they are rather rare, and I doubt that there are any on the present fundamental research staff.

Wallace Carothers vented his current frustrations, writing that "...during the present transition period practically the only guide we have for formulating and criticizing our own research problems is the rather desperate feeling that they should show a profit at the end." However, in an uncharacteristic upbeat summary, he wrote, "On the whole, however, I should call the fundamental research in organic chemistry to date very successful work."

But Carothers suggested he had spent too much money over the years. He had allowed his group to grow to such a proportion that his own efforts were overly diluted by attention to too many projects. He chastised himself for hesitating to delegate "part of the mental responsibility of directing the work." He proposed changes to, not elimination of, fundamental work. He suggested two or three carefully selected projects, closely aligned with DuPont's interests and important enough to justify long-term study, be attacked with a reduced budget.[5]

Carothers had expanded his group from six chemists in 1929 to 13 by the middle of 1932. Martin Cupery and Harry Dykstra, both of Dutch descent, both graduates of Hope College in Holland, Michigan, and both with Ph.D.s, came to Wilmington and shared a laboratory working for Carothers. Cupery said the other chemists called him and Dykstra the "Dutch Gold Dust twins" in those days.

Martin Cupery recollected that Carothers talked to him on his second day at DuPont. "Now remember Cupery, you're not working for me, you're working with me," Carothers said. Martin Cupery said that Carothers was in the laboratory enough to make certain his scientists were not "going off on a tangent." His assignment was to

---

[5]Carothers, W. H. "Fundamental Research in Organic Chemistry at the Experimental Station," August 5, 1932. (Hagley Museum and Library Collection, Wilmington, DE, 1784, Box 21, pp 7, 10, 11.)

determine the structure of the polymers Carothers' group was making. Cupery had ideas of his own to chase from time to time. "We were very curious," he said. But he limited the diversions to a few hours at a time. Cupery reported to Wallace Carothers on the organization chart of the chemical department for four years. "And that was his attitude all along," Cupery said. "In all that time—never once did we have a cross word—never once did he find fault with me."[6] Harry Dykstra reported the same experience.[7]

Crawford Greenewalt called Carothers, "Kind, unprepossessing, gentle."[8] Greenewalt never worked for Carothers; he managed another group under Bolton. But he carried great influence. Crawford Greenewalt was ideally placed in Wilmington's socially sensitive hierarchy. In 1926, he married Margaretha duPont, daughter of DuPont's president Irénée duPont.[9]

Joe Labovsky was born in Russia in 1912. He survived pogroms, famine, and revolution and came to Wilmington with his father, a tailor, in 1923. The young man met Carothers briefly at a concert of Russian music at the Philadelphia Academy of Music in 1930. Carothers loved Russian music, Labovsky said. Labovsky worked at the Experimental Station one summer. He was sent to a small, underground rifle range built in the side of a hill above the Brandywine to study the surface of the liquid element mercury in a vibration-free environment. There he met Carothers again. He recognized the scientist with the deep, baritone voice. Crawford Greenewalt noted the work of the ambitious Russian immigrant, whose father made the clothing for many of the duPonts. Greenewalt arranged for Labovsky to go to the Pratt Institute under a scholarship funded by Lammot duPont. After his graduation in 1934, Labovsky tried to get a job with DuPont in New Jersey at the huge manufacturing site called the Chambers Works. He was digging ditches part-time at the Experimental Station when Carothers spotted him and remembered him. Wallace Carothers hired Joe Labovsky as his personal laboratory assistant.[10]

---

[6]Hal Cupery interview with Adeline C. Strange, August 2, 1978. (Hagley Museum and Library Collection, Wilmington, DE, 1985.)

[7]Harry Dykstra interview with Adeline C. Strange, August 2, 1978. (Hagley Museum and Library Collection, Wilmington, DE.)

[8]Crawford Greenewalt interview with Adeline C. Strange, 1978. (Hagley Museum and Library Collection, Wilmington, DE, 1985.)

[9]Hounshell, David; Smith, John Kenly. *Science & Corporate Strategy, DuPont R&D 1902–1980;* Cambridge University Press: Cambridge, England, 1988; p 272.

[10]Joseph Labovsky interview with Matthew E. Hermes, June 28, 1991.

■■

This Wallace Carothers, of August 1932, presenting an ordered and balanced account of his stewardship, claiming significant success, indicating needed improvements, and offering a workable compromise, seems at odds with the self-absorbed, obsessed, and threatened man with whom Frances Spencer had lately corresponded. This Carothers emerges as realistic and confident as he breathes life into his work. David Hounshell and John K. Smith write perceptively that:

> Carothers had established enough of a reputation for himself, both in and out of DuPont, that he could continue to work on his scientific studies. In spite of his concerns about what management expected of him, Carothers, following his theoretical interests in the mechanism of polymerization, moved his research away from linear, fiberforming superpolymers toward the study of cyclic compounds....[11]

The remarkable fact is Carothers seemed to grow in 1932, grow from the binding truth of the "industrial slavery" he long ago joked about with Jack Johnson.[12] He seemed to step lively across the chasms of depression. Carothers stopped writing Frances Spencer. Were these letters to Frances a symptom of his periodic illness? Did he not need to correspond with Frances when the neurotic spells lifted? Perhaps.

Whatever may be the truth, we can picture Wallace Carothers through much of 1932 and early 1933 dressed for work as always in his three-piece woolen suit. But he sits upright, not hunched and clinging. He is the DuPont Professor of Chemistry, carrying on a lively and active correspondence with Adams and Marvel of Illinois and Lamb of Harvard and Johnson of Cornell and Whitmore of Penn State and Gilman of Iowa State and James Conant, soon to be the president at Harvard. Perhaps Carothers is even smiling.

■■

Wallace Carothers' research moved away from the polyesters into an extension, a study of the cyclic compounds that could be obtained as alternatives to the long-chain polymers. From the beginning,

---

[11]Hounshell, David; Smith, John Kenly. *Science & Corporate Strategy, DuPont R&D 1902–1980;* Cambridge University Press: Cambridge, England, 1988; p 272.

[12]Wallace Carothers to John R. Johnson, February 14, 1928. (Hagley Museum and Library Collection, Wilmington, DE, 1842.)

Carothers postulated that long-chain polymers and rings would participate in an equilibrium. Conditions would be found to convert one to the other. Beginning in 1932, and extending well into 1933, Julian Hill studied this equilibrium reaction. The work was fun for both of them. The chemical field they would exploit had little history, some leading work from the Swiss chemist Leopold Ruzicka, and more than a little sex appeal.

Julian Hill converted a long series of the bifunctional acids into ring compounds by vigorously heating their polymers in the old molecular still. Hill would do the distillation and collect a small sample of liquid on the condenser. Most of his cyclic liquids were unstable and would convert quickly back to polymers, but if he worked quickly, Hill could catch a whiff of the odor of the cyclic compound. Hill was enticed by the odors, and as he worked through the series of dicarboxylic acids, making rings of 6, 9, 10, 11, 12, and 15 members, the odors changed. At five and six atoms in the ring, Carothers thought they resembled the odor of bitter almonds or menthol. (This was a curious comment, for few chemists know what the odor of bitter almonds is. Many have heard that the deadly gas, hydrogen cyanide, smells "like bitter almonds." Only a man familiar with the odor of HCN is likely to make this comparison.)

Rings from nine to 12 members smelled like camphor. But the larger rings, those of 14 or 15 atoms, gave the compelling and sensuous odor of musk. As the two scientists inched up the ring size, musk turned to civet at rings of 17 and 18 members. Rings of 19 atoms and above were odorless!

Musk is a greasy secretion from a glandular sac beneath the abdomen of the male musk deer. Its source was Tibet and the high plains of Kokonor, and its only supply depended on a mysterious series of transfers and adulterations before the musk reached Shanghai. Civet is secreted by both the male and female of the cat of the same name. From Abyssinia, civet came packaged tightly in animal horns to the perfumeries of Europe. Both materials have powerful odors, but when diluted, these rare, difficult to purify materials become the basic materials providing "'life' and diffusiveness" in fine perfume.[13] Ruzicka determined that the natural odorants were cyclic compounds called ketones. Natural musk has a 15-member ring; civet has a similar structure in a 17-member ring. Other macrocyclic compounds also have attractive odors.

DuPont's industrial scientists became, for a time, perfumery experts. Carothers and Hill playfully wrote their science as if they

---

[13]*Encyclopedia Britannica*, 14th ed.; Encyclopedia Britannica: Chicago, IL, 1939; Vol. 17, p 507.

were connoisseurs of fine wine. They disclosed for the *Journal of the American Chemical Society* the "vague and indefinite" odors in the small rings and the "camphoraceous or minty note" in the medium-sized compounds trending to a "woody or cedar-like quality" and finally to a "nuance of musk" as the ring size increased.[14]

Each of the scientists brought home from work that nuance of musk. They brought home, on their hands and clothing from their days in the chemical laboratory, the sensual fragrance of the night-time perfumes to their startled wives and friends.[15] Polly Hill asked her husband, Julian, "Where the hell have you been?"[16] We do not know what Sylvia Moore asked Wallace Carothers.

The work had practical meaning. Carothers was able to extract from the structure of the dozens of compounds Hill made the principles governing the odors. And he and Hill prepared one 17-membered ring compound, tetradecamethylene carbonate, that was stable. DuPont made up some of the fragrant cyclic carbonate and sold it for a time as a perfume ingredient called Astrotone (*see* 8 in Appendix A). Merlin Brubaker remembers the artificial musks. He claimed he put a dab of Carothers' compound on a handkerchief and washed it 15 times but could still smell the odor. He remembered that DuPont made about 300 lb of the Astrotone but couldn't sell it in the United States. A Frenchman solved the problem. He bought the artificial musk at a reduced price, diluted it with alcohol, and added a trace of diethylamine to the product. Diethylamine smells like fish oil. That did it, Brubaker's story goes. For years a well-known French perfume was made from this concoction.[17]

Julian Hill remembers Elmer Bolton thought the perfume work was a "hoax." "The musks fatigued the nose very rapidly," Hill said. "If you smell them in the bottle, you can't smell a thing. The way you did it was to dilute it with alcohol. [A perfumer] has books of thin strips of blotting paper that he dips and that's the only way he can smell it. Otherwise your nose fatigues and has to rest for quite a while

---

[14]Hill, J. W.; Carothers, W. H. *Journal of the American Chemical Society* **1933**, *55*, 5039.

[15]Hill, J. W. "Wallace Hume Carothers," In *Proceedings of the Robert A. Welch Foundation Conferences on Chemical Research;* Milligan, W. O., Ed.; Houston, TX, 1977; p 245.

[16]Julian Hill interview with Matthew E. Hermes, May 29, 1990.

[17]Merlin Brubaker interview with John K. Smith, September 27, 1982. (Hagley Museum and Library Collection, Wilmington, DE, 1878.)

before you can smell them again. Bolton never did smell the stuff because his nose fatigued so fast."[18]

Carothers derived from the relationships between ring closure and polymerization an important understanding of the mechanism of his favorite polymerization systems. In September 1933, he sent to the *Journal of the American Chemical Society* papers XIX through XXII of his studies on polymerization and ring formation.[19]

■ ■

In pursuit of one of the starting materials for the large ring work, Wallace Carothers turned to his friend Jack Johnson at Cornell for help. Johnson had prepared a dicarboxylic acid called brassylic acid by treating an unsaturated fatty acid with ozone and then with hydrogen peroxide. Johnson was considering his work for inclusion in Roger Adams' *Organic Syntheses;* he gave Wallace Carothers the directions for its preparation so that Carothers could check his work. Carothers wanted this 13-carbon dicarboxylic acid to continue the series of ring compounds.[20]

Within months, Dr. Ed Spanagel converted Johnson's brassylic acid to an ester called ethylene brassylate (*see* reaction 12 in Appendix A). This cyclic compound contained 17 atoms in the ring. It replaced Astrotone as DuPont's commercial musk.

Carothers cooperative success with Johnson came amid a carnival of university contacts. Johnson had visited Carothers in Wilmington in March 1932; Speed Marvel had been in town in May for one of his periodic consulting visits.[21, 22] Carothers was working with Adams and Gilman on editorial policies for *Organic Syntheses.* He sent catalyst samples to Conant, volunteered to review some of Johnson's manuscripts on carbon atom rearrangements, and wrote his friends asking

---

[18]Julian Hill interview with David A. Hounshell and John K. Smith, December 1, 1982, pp 35 and 36. (Hagley Museum and Library Collection, Wilmington, DE, 1878.)

[19]Mark, H.; Whitby, G. S. *Collected Papers of W. H. Carothers on High Polymeric Substances;* Interscience: New York, 1940; pp 202–235.

[20]John R. Johnson to Wallace Carothers, May 31, 1932. (Hagley Museum and Library Collection, Wilmington, DE, 1842) and John R. Johnson to Wallace Carothers, June 2, 1932. (Hagley Museum and Library Collection, Wilmington, DE, 1896.)

[21]John R. Johnson to Wallace Carothers, March 20, 1932. (Hagley Museum and Library Collection, Wilmington, DE, 1842.)

[22]John R. Johnson to Wallace Carothers, May 31, 1932. (Hagley Museum and Library Collection, Wilmington, DE, 1842.)

for support for Charles Stine as American Chemical Society president.[23–25] Johnson asked for advice once again, seeking a way to get a short communication into the *Journal* ahead of anticipated competition from Henry Gilman at Iowa State.

Johnson footnoted this last letter, written in November 1932, chiding Carothers for his use of the word "elegancy" in describing an experiment. Johnson had seen this usage in a paper Carothers had just published with Don Coffman, Carothers' first DuPont assistant back in 1928 when Coffman was a graduate student at Illinois, taking a summer off at DuPont. The graduated Coffman worked for Carothers now.[26] Carothers defended his usage in jocular fashion, noting that he had not believed he had used the word until he went back and examined the paper. Carothers attributed his usage to his "perverse stubbornness," suggesting he had a difficult time convincing the local DuPont management that any experiment was elegant. "I was slightly consoled," Carothers continued, "by discovering that one of our Illinois masters of prose uses the word 'innocency' in one of his very literary papers in the *Atlantic Monthly*."[27]

Wallace Carothers wrote to Professor Frank Whitmore at Penn State, comparing notes on the origins of Father Nieuwland's work on divinylacetylene. Whitmore had sent Carothers a letter from the European chemist, Lespieau. The Experimental Station mail room, after opening the letter from Whitmore, had mutilated both Whitmore's and Lespieau's letter with a rubber stamp. Carothers apologized to Whitmore noting, "I have a special arrangement with the mailing room here not to open my mail but to send it directly to me. This rule is very scrupulously observed except in the case of personal things where it really makes a difference." Carothers discussed rearrangements of the carbon atom skeleton with Whitmore and invoked the new theories of electronic control of these maneuvers being developed by Christopher K. Ingold at University College, London.[28] But he found Ingold's explanations unsatisfactory. He outlined for Whitmore the details of the reac-

---

[23]James B. Conant to Wallace Carothers, March 29, 1932. (Hagley Museum and Library Collection, Wilmington, DE, 1896.)

[24]Wallace Carothers to John R. Johnson, June 20, 1932. (Hagley Museum and Library Collection, Wilmington, DE, 1842.)

[25]John R. Johnson to Wallace Carothers, November 7, 1932. (Hagley Museum and Library Collection, Wilmington, DE, 1842.)

[26]Ibid.

[27]Wallace Carothers to John R. Johnson, November 18, 1932. (Hagley Museum and Library Collection, Wilmington, DE, 1896.)

[28]Brock, William H. *The Norton History of Chemistry;* W. W. Norton: New York, 1992; p 507.

tion of hydrochloric acid with vinylacetylene and hoped Whitmore could explain the results on the basis of his own electronic theories.[29]

Adams asked Carothers for his opinion on a paper about hindrance of "free rotation" in certain bulky organic compounds. Sheer size of groups in a molecule can prevent them from turning like a wheel on an axle. Defining these restrictions interested Adams. But he wanted Carothers critical appraisal of his manuscript before publishing.[30]

In February 1933, Carothers sent Jack Johnson a manuscript of his own for Johnson's review. "Your paper on rearrangements and the accompanying note have upset me a bit," Johnson wrote on February 20, 1933.[31] With Carothers and Johnson working along parallel lines and communicating regularly, conflict was inevitable. Johnson believed Carothers had plagiarized some of Johnson's ideas in his manuscript. Johnson praised Carothers' work lavishly and attributed his own concern to a priority interest in the rearrangement mechanisms Carothers had discussed. Johnson seemed almost subservient in his letter to Carothers and offered a series of compromise solutions to the conflict. Carothers wrote back the next day. He began, "Apparently I have made a complete mess of things." The manuscript Johnson disliked was already in the hands of Arthur Lamb at the *Journal*. Carothers would withdraw it, of course, and proposed to Johnson the changes that he hoped would appease his friend. He admitted his plagiarism, but compared it to that of the unconscious copying attributed to Mark Twain while suffering from a fever in Hawaii. Twain had published an original poem, only later to find it had been written originally by Oliver Wendell Holmes.[32]

Twenty-four hours later, Johnson's response to Carothers' apology was in the mail. "I have done grave injustice and must make amends," Johnson wrote. "After reading your paper slowly and with digestion I am convinced that my original emotion of panic was merely disappointment at a comparison of your general presentation,

[29]Wallace Carothers to F. C. Whitmore, January 4, 1933. (Hagley Museum and Library Collection, Wilmington, DE, 1896.)

[30]Roger Adams to Wallace Carothers, January 11, 1933, and January 24, 1933; Wallace Carothers to Roger Adams, February 3, 1933, and February 6, 1933. (Adams Papers, University of Illinois Archives, Urbana, IL.)

[31]John R. Johnson to Wallace Carothers, February 20, 1933. (Hagley Museum and Library Collection, Wilmington, DE, 1784, Box 18.)

[32]Wallace Carothers to John R. Johnson, February 21, 1933. (Hagley Museum and Library Collection, Wilmington, DE, 1842.)

telescopically and in elegant form with a thing of which I have been doing microscopically to a much more limited extent." Johnson called his first letter a "petty outburst," brought on by his obsession with understanding the rearrangement chemistry he studied.[33] The two men remained fast friends.

■ ■

This incomplete narrative describing Wallace Carothers' ongoing correspondence with the leaders of academic organic chemistry in this country places him among them as their peer—respected, sought after, a leader of thought. But his correspondence is not a seamless account of his life. As before, in the period of his great inventive burst within DuPont, any continuous chronicle of his personal activities disappears. When Wallace Carothers was well, functioning with crisp assurance and direction, we see it in the spirit of his work.

But Carothers was under pressure away from the Experimental Station, and as the spring of 1933 began, his upbeat mood soured. Sam Lenher, now married well, said, "My wife and I enjoyed Wallace's company in a personal way." But they hedged their approval. "Wallace had mistresses. We knew his mistresses. Two of them were friends of ours." One of the ladies, Lenher said, "was interested in affairs," and told the Lenhers of her relationship with Carothers. "They were very attractive women," Lenher said, but Carothers wasn't seriously interested. "I thought Wallace was a philanderer. And I didn't respect him for it. This is opinion on my part," Lenher said in this 1990 interview. "Fold back the time."[34]

Crawford Greenewalt zeroed in on the "dame," Sylvia Moore. Sylvia was "terrific, bright, sexy, good looking." Greenewalt continued, "Everyone knew her husband, Thomas Moore—nice guy." The affair was a "potential scandal," Greenewalt said.[35]

Greenewalt worked with Carothers, was his friend, but as he was married into the DuPont family that ran the family business, he served as conduit for information both ways. The "potential scandal" would surely be in some manner of Greenewalt's making. Should he report to Lammot duPont of the seemingly public indiscretions of

[33]John R. Johnson to Wallace Carothers, February 22, 1933. (Hagley Museum and Library Collection, Wilmington, DE, 1784, Box 18.)

[34]Samuel Lenher interview with Matthew E. Hermes, August 24, 1990.

[35]Crawford Greenewalt interview with Adeline C. Strange, 1978. (Hagley Museum and Library Collection, Wilmington, DE.)

DuPont's leading scientist, he would certainly earn the task of pressuring to get the situation corrected. By now, Wallace Carothers was "desperately in love" with Sylvia Moore. Julian Hill suggests the arrangement was frustrating for Carothers because she "played him on the end of a string, and enjoyed pulling it."[36] Julian never mentioned, or never knew, that in the spring of 1933, Sylvia and Thomas Moore entered the courts in West Chester, seeking a divorce.

■ ■

On Wednesday night, May 10, 1933, Wallace Carothers sat at his typewriter in his room at Whiskey Acres and wrote two letters. Wilko Machetanz had written him once again and Carothers, who never initiated correspondence with Mach or Fran and always was left to respond to their initiative, thought about their long separation:

> Letters such as I am capable of composing seem very inadequate and futile for the need of communicating with you. We have been separated now for so long a time that it is difficult even to know where to start. The only reasonable way of getting caught up with the inexorable progress of time is to have an actual meeting face to face. But for me, being an industrial slave, it is difficult to make any steps leading to such a consummation. As far as the last year is concerned there is nothing of material importance to communicate. I still struggle along as a Group Leader, which is to say, a kind of a Clerk. I never had any talent for clerical matters, and haven't developed along those lines to speak of. Research lately has been rather foul on account of the depression. My budget hasn't been cut. We are still spending money like nothing at all; but an atmosphere of anxiety has arisen which causes us to scrutinize topics from a misty standpoint of what may be ultimately practical.

Carothers noted he had harbored the thought of eventually leaving DuPont for a "really good University appointment.... I still have a yen to get back into some University, preferably Harvard." He mentioned to Machetanz he had a "lively correspondence with Fran (Gelvin) Spencer" in the summer of 1932. "Correspondence ceased and hasn't been resumed," he wrote in the passive voice, distancing himself from responsibility for the end of their communication. "Do you suppose she would mind if I were to write again?" he asked Fran's

---

[36]Julian Hill interview with Matthew E. Hermes, May 29, 1990.

half-brother-in-law who lived 2000 miles west of Fran, in Visalia, California. "I suppose there wouldn't be much point in it," Carothers concluded, "but the letters were very nice while they lasted."

And Carothers made an offhanded reference to an arrangement that would seem insignificant to Machetanz but that would bring profound trauma into Carothers' own life. "At present," Carothers wrote, "I am planning to bring my father and mother to Wilmington and find a country place to settle down. This is pretty definite...''[37]

Carothers provided greater detail for Adams in his second letter that night:

> Business back home in Des Moines has collapsed completely in the past couple of years; my father and mother are still there; and it is only in the last couple of months that I have discovered how badly off they have been. On top of the financial worries, my father has been working very long hours at tasks that are sheer drudgery. He is completely worn out and both he and my mother need a change of scenery. I have made plans to bring them to Wilmington with the idea they could get out in the country here. It is practically impossible to find country places for rent. One has to buy. I have just about set on a place, not ideal of course, but one that seems possible in view of the limited supply of cash.

The problem lay in that most of the attractive land and homes were owned by the duPonts, Lairds, and Bancrofts; the first families of the Brandywine Valley. Prices were not seriously reduced from boom times. Carothers noted, naively, "In the 20 to 40 000 dollar class there are two or three real bargains—places that can be had for 14 to 18 000. However they are much bigger than we need, and buying one of them would mean going very deeply in debt."[38]

But Carothers expressed a serious dilemma to Adams, in far more detail than he had with Mach. "News of Conant's election to the presidency at Harvard arrived Monday when I was at Niagra [*sic*] Falls," he wrote. Conant's election triggered thoughts that Carothers would be a worthy successor to Conant at Harvard. It initiated his latent sense of the correctness of the academic life at Harvard, particularly

---

[37]Wallace Carothers to Wilko Machetanz, May 10, 1933. (Fred Machetanz.)

[38]Wallace Carothers to Roger Adams, May 21, 1933. (Adams Papers, University of Illinois Archives, Urbana, IL.)

as his reputation grew and he acted at DuPont more in the style of an academic professor than an industrial scientist.

Conant's move from chemistry had a psychological effect on Carothers because any thoughts of Harvard conflicted with the plans to help his beleaguered parents. But he was uncertain he could provide practical help to DuPont in the future. "It is no doubt a manifestation of excessive conceit to jump, in writing, from Conant's election to my own chances of getting back to Harvard. I don't try to excuse it on any basis of fact; it is merely wishing as far as I am concerned. However I suppose that they will want some one..."[39]

Adams responded to Carothers immediately. He had known of the upcoming change and suggested Harvard was looking for a "man with some biochemical training." But he let Carothers know he would make Harvard aware Carothers was available. Adams commiserated with Carothers on his house-hunting efforts, "I can see that the situation you are in is rather a difficult one and that you are anxious not to make a wrong decision, even though property may not be as high in value as it was a few years ago."[40]

Carothers' anxiety over Harvard passed quickly. He wrote Adams on May 21 from Whiskey Acres, "Sorry that I put you to the trouble of replying to my silly letter. It was the result of a brain storm which passed over rather promptly." And, he continued, "...for the present it might be better to say nothing to the people at Harvard."[41]

But Adams had already "dropped a line to some of the people in Cambridge merely mentioning in a general way what the situation was. It will do no harm I am sure one way or the other," Adams assured Carothers.[42]

■ ■

Wallace Carothers' May 10 letter to Machetanz soon triggered a note from Frances Spencer to Carothers. The timing suggests Machetanz or his wife, Fran's half-sister Tort, wrote Frances in Missouri, letting her know Wallace was agonizing over the end of his correspondence with his old friend. But one fact remained unchanged.

---

[39]Ibid., May 10.

[40]Roger Adams to Wallace Carothers, May 16, 1933. (Adams Papers, University of Illinois Archives, Urbana, IL.)

[41]Wallace Carothers to Roger Adams, May 21, 1933. (Adams Papers, University of Illinois Archives, Urbana, IL.)

[42]Ibid., May 26.

Carothers would not write first. On May 24, 1933, Carothers wrote Fran Spencer for the first time in almost a year:

> I *didn't* go to Colorado last year; the date of the meeting interfered with the date of my measly fortnight of yearly vacation which was spent in the Pennsylvania mountains just 200 miles from here. I made several starts at explaining this to you in writing, but they all looked rather funny and were dumped in the waste basket. Then finally with one thing and another I got out of the habit of writing letters entirely.

Carothers responded to Fran, who boasted in her letter of having sold a poem. He wrote:

> I have lately had one published myself. Forget the name of the magazine, but it was something like Technical Library Notes.... Unfortunately I was not apprised of the fact that this poem was to be published and the thing was incomplete insofar as I had at the time not provided it with a title. Also since the thing was deliberately pirated, the publishers did not trouble even to append my name as the author in spite of the fact it was well known to them that I had written it. I have decided to call it VI lines. Here it is:
>
> ### VI Lines
> I have not stolen from its place your book
> Nor swiped, nor in any other manner took;
> Nay I have never saw the blasted thing,
> Know not its color neither blue or ping.
> In fact your book I didn't even covet;
> I have a private persnl copy ovet.
>
> This was originally dedicated to our librarian at the experimental station; but in view of the scarcely honest way in which she pushed it into print, this dedication has been withdrawn. The nature of the subject matter is such that I can hardly transfer the dedication to you. There however will be differences of opinion concerning its high quality and that it is obviously the work of a sensitive and gifted nature. The real reason for presenting it here is to convince you that I deserve consideration as a fellow poet; that I can justly lay claim to a degree of license greater than that granted to ordinary mortals, and that I can expect from you a more exalted sympathy than you would extend to a mere average human being. Also it justifies me in requesting a reprint of your poems. Incidentally I recently

read a rather bright story bearing on the subject of poetry. It seems that a scarcely literate college graduate in the bond business bet some of his high brow friends that he could get an original poem of not less than eight lines published in the *New York Times*. He won his bet without any difficulty at all by the following method. He wrote to the questions and answers department enquiring whether anyone could give the second verse of a poem which started with the four lines so and so. In the next weeks number he replied under a different name giving the next four lines.[43]

Of the 14 letters Wallace Carothers wrote to Frances Gelvin Spencer between 1929 and 1933, this is the only one not hand written—it was typed—and the only one not filled with distress. Carothers seems lively and well-adjusted. But the letter is incomplete. As I received it from Frances Spencer's daughter-in-law, Dr. Jeffry Spencer of the California State University, Bakersfield, it consisted of two yellowed pages with a single word ripped from the second page. But the letter was not closed on the second page; it went on, but the subsequent page or pages were missing.

Five weeks later, on June 28, Carothers wrote Frances once again and began to tell a troubled story:

> Wednesday a.m.
>
> Dear Fran:
>
> I can't remember ever writing to anyone before at this hour. It is 6 a. m. The birds are crying out with songs of cheeriness, my stomach is crying out with pangs of collapse, and my mind is in a bad state of unrest and dissatisfaction. Also this is one of my last mornings at Whiskey Acres. My father and mother arrived in Wilmington Friday; the goods came yesterday afternoon. Within a day or two I shall be living at Arden, Del. the notorious single-tax colony founded by Upton Sinclair and now populated by imitation bohemians, Jews of a certain type, etc. I shall be living in a tiny house badly arranged with the doors radiators and windows all in the wrong places. My selection of the place was a peculiar example of feeble and self-stultifying perversity. I knew for several days before signing the lease that it would be awful even with the furniture installed. But I went ahead and took it anyway. As matter

---

[43]Wallace Carothers to Frances Gelvin Spencer, May 25, 1933.

of fact the whole scheme was pretty silly. I not only don't
have any affection for my father but find it exasperating
and sometimes sickening merely to be in his presence,
and my mother and I seldom understand each other. They
don't know a soul in Wilmington and are not likely to find
any chronies among my few friends. And the moving etc.
is costing a lot of money — just about all I have.

On these considerations Whiskey Acres doesn't look so
bad. The place is untidy and not very well arranged; but
the house is decently put together; the floors are flat; the
rooms are of decent size; one can have good company if
one wants it, or be alone if one prefers; our combined but-
ler and housekeeper has his points of weakness, but he
can put up a good meal and serve it graceful. Oh well
damn it there is nothing to be done about the matter now.
As you will see it is really too early in the a.m. for the composition of a letter. Good luck and power to your studies.
If you should write to me again my address will be RFD#2
Wilmington.

Wallace[44]

Arden truly was a bizarre choice. The small cottage had a high,
peaked cedar shake roof and siding in the "English Tudor" style. But
the house was modest, as Carothers suggested.[45] And Arden, well, was
different. It was less than 10 miles from the DuPont building, but
Arden was like a distant, forgotten moon to the Wilmingtonians. No
one went there, except to its Guild Hall, in which local craftsmen
worked by day and an occasional dramatic performance was staged by
night. The paved roads bypassed Arden; a total lack of zoning
reflected the staunch independence of its founders. The permissive
miniculture of Arden was the antithesis of the duPont's ancestral Delaware in which Wallace Carothers worked and in which he might
expect his Presbyterian parents would be comfortable.

The Arden house soon became tense. Wallace Carothers wrote
Frances Spencer a short, tightly worded letter on August 7, just a month
after the new arrangement began. This once, he seems to have written
spontaneously and he was not answering one of Frances' letters:

Dear Fran:
It is long past my bedtime but I feel in a viciously com-
municative mood. Perhaps if the power to manipulate a

[44]Ibid., June 28.

[45]Suzie Kyle Terrill showed me a picture of the Arden house that Ira
Carothers had taken in 1933.

pencil can be sustained I can work some of the choler off on you. I just returned from a week in N. Y. Visiting mostly — at Ithaca and Clinton. In a way it was lucky since I spent three days of worst heat of the season immersed in beer and Lake Cuyahoga. Also I had 24 solitary hours in a nice Inn, in Montrose, discovered accidentally. I devoted these hours exclusively to the quaffing of beer, sleeping and reading Ann Vickers. Best 24 hours I have had in a long time. Immediately upon arriving in Arden this evening I discovered a domestic detail undone that provided the opportunity for falling into a furious silent rage. Matters are now somewhat improved since I have had some beer and two hearings of the Bach concerto for $\underline{4}$ pianos. Also the Beethoven piano concerto in G major. Did I tell you about my Gramophone Robot. It is a miracle. It will play two dozen records in the selected order (turning them over as the occasion requires) completely automatically. You just have to sit back and listen. And it is really worth listening to, because what it brings out is really music and if you want you can have it loud enough to rattle the door-knobs at certain frequencies.

I got hooked for the synthetic rubber speech at Chicago after all, & so should probably be there about Sept 14 to 21. Are you definitely set against going to the Exposition? I hope not. Please give my very best compliments and wishes to Tort.

Wallace[46]

Wallace Carothers went to Chicago to attend the American Chemical Society meeting in September, which ran during Chicago's Century of Progress exposition. He roomed with Jack Johnson. Frances Spencer did not visit him there. For Carothers, the highlight was a reunion with his sister Isobel. He arranged for Isobel, along with two of her former classmates at Northwestern, Louise Starkey and Helen King, to put on a comic performance for he and his friends at the end of the meeting.

Isobel Carothers, Louise, and Helen were now well known around Chicago and had a nationwide audience as radio performers. They did a daily "sketch" called "Clara, Lu, 'n' Em," which had started locally in Chicago but was now broadcast across the country on radio's Blue Network. Isobel played the widow Lu in a high-pitched voice, sounding at times like a hen. The domestic trials faced

---

[46]Wallace Carothers to Frances Gelvin Spencer, August 7, 1933.

by the trio spun out into a continuing story. Cyrus Fisher, who reviewed radio for the literary magazine, *The Forum,* commented on "Clara, Lu, 'n' Em" in June 1933, placing them at the top of a new, but vapid genre. For "Clara, Lu, 'n' Em" had been radio's first-ever soap opera.[47]

"One of the most firmly rooted radio traditions is that morning week-day programs edify and delight an audience composed exclusively of women," Fisher wrote. "Apparently these neglected ladies will accept anything. Programs seem to progress in quality from morning to afternoon to evening in direct ratio to the number of males listening in."

Nevertheless, he found Isobel's show "brilliantly inspired," at least in sarcastic comparison to the rest of the offerings in the "sweeping and dusting hours." Miss Starkey emotes enthusiastically as "Clara," Fisher wrote, "Miss Carothers pleats some humor into "Lu," and Miss King sends her "Em" skittering into frequent confusion. All three seem unable to decide whether they have a burlesque, a satire, or a realistic version of the lives of their wretchedly tiresome small-time gossips. The medley of interpretations does give variety but not much else, despite its ratings as one of the most popular morning effusions."[48]

We don't know how the three women may have modified their "sweeping and dusting" entertainment for presentation to the chemists, virtually all men, bustling about and looking busy, attending the Chicago convention.

Early in October, Carothers responded to Frances' inquiries in the last of the 14 letters he wrote to her. He wrote that his "domestic situation has improved. At least it doesn't look as dark as it did. Chiefly a question of mental adjustments." He continued:

> There isn't anything in the way of news here. Happenings in the laboratory are rather mild and there is nothing at all outside. Sylvia has a job in N.Y. and has been gone for 2 weeks now. I come home, sometimes play a couple of games of deck tennis with a neighbor then after dinner put 15 or 20 records on the phonograph and drink a few

---

[47]Edmondson, Madeline; Rounds, David. *The Soaps, Daytime Serials of Radio and TV;* Stein and Day: New York, 1973; pp 38, 39. *Tune in Yesterday: The Ultimate Encyclopedia of Old-Time Radio, 1925–1976;* Prentice-Hall: New York, 1976; p 136. The Museum of Broadcast Communications, 800 South Wells, Chicago, holds tapes of four "Clara, Lu, 'n' Em" episodes from 1934 and 1935.

[48]Fisher, Cyrus. "Radio Reviews," *The Forum* **1933,** *89,* 382.

dozen bottles of beer while pretending to work until about 10 o'clock when the well known phenomenon of gassy stupefication accompanying large volumes of beer sets in. Then I go to bed. The family does all the errand running, pays all the bills, shines all shoes, buys the beer, writes letters, and in short everything except getting a haircut, which has lately ceased to be necessary, is done, so that I never have to do anything except go to the lab and come back home. I look forward to a quiet winter during which there will be at least a grand pretense of working. Oh—I nearly forgot to mention that there has been one red letter day recently. Richard Willstätter, the great organic chemist of the world was in my office for an hour and a half last week. Bolton the chemical director of the Company, one of his former students, persuaded him to stop off on his way from Chicago (where he had gone to receive a prize) back to Germany. Afterwards there was a tea party arranged and conducted by your correspondent who had never theretofore attended a tea party. Fifty-six guests were present (no women). It went off rather well too. Anyway Willstätter in his field corresponds to God in some others and I was enormously set up when our meeting turned out to be so agreeable—naturally as far as one could infer. A different kind of experience with eminence occurred about 3 weeks ago when Professor Katz from Amsterdam wrote from N. Y., when he docked there, requesting permission to call. He has a moderate renown as a specialist in x-rays & physics applied to rubber & similar materials. I had vaguely expected a small swarthy jew and was surprised to meet an enormous blond Dutchman reminiscent of a character out of Rabelais, and a middle-aged wife, an absolutely typical Bostonian. The Professor had once been a practicing psychoanalyst in Russia. It was in this connection that he had met his wife, an ardent pupil and disciple of Adler. There was at first some fright and embarrassment on my part through the doctor's insistence on giving us a set of lectures he had prepared to present this fall at Leningrad by invitation, but the lectures, somewhat abbreviated, turned out to be excellent, and after that we had a grand time explaining and raving about the marvelous things in my lab.

He asked Frances about Machetanz, because he feared from the tone of a recent letter that Mach was "horribly pessimistic and melancholy." And he asked, as an afterthought, for news about her. Then, for the last time, he signed off on a letter to Frances Spencer, writing simply, "Wallace."[49]

On the surface, nothing changed radically in the year of 1933 for Carothers. He started off well. But he lived a difficult work life, acting the professor in an industrial setting. Wallace Carothers agonized in May 1933 over his decision to bring Ira and Mary Carothers to Wilmington and to live with them. He fell near to despair in late June as the date for his parents' arrival neared. He lived a difficult home life, maintaining an intemperate agenda and continuing the affair with Sylvia Moore under the tight-lipped gaze of his parents.

One matter that Wallace Carothers let slide in his correspondence with Frances Spencer had to do with Sylvia. Wallace neglected to mention to Frances that Sylvia and Thomas Moore had filed for divorce and the decree was finalized on July 7.[50] In April and May, the very months Carothers was choosing to reestablish himself in a home surrounded by his parents, his lover was proceeding to free herself. She would be a single woman within days after Ira and Mary Carothers moved in with their 37-year-old son.

What a strange decision that move was! Wallace Carothers wrote enough about his father to assure us he wanted no place with him. He wrote of his father to his best Tarkio friend, Wilko Machetanz, in 1928:

> When I was a kid, I looked upon my father and upon such persons as preachers as incredibly remote, wise and exalted beings. Such beings were not necessarily to be worshipped. They were just on an inaccessible plane. But they were to be feared. That was about the only emotional reaction. When I grew up I realized that my father was lacking in many desirable qualities. But much of the old patterns remain. I can sit down with a preacher and realize that he is vapid, vain, bigoted, vulgar and lacking in intelligence; but this realization never comes without a feeling of anger

---

[49]Wallace Carothers to Frances Gelvin Spencer, October 4, 1933.

[50]Chester County, PA, Archives. The divorce record is dated 1933, Number 60 of the April term of the court. That record gives no details save that the Moores paid Master and Examiner Joseph N. Ewing $50 to arbitrate the settlement. The divorce was granted on July 7, 1933, by Chester County Judge W. Butler Windle, Jr. The file indicating which party filed and the alleged grounds was destroyed years ago. There is no indication in the record that Sylvia Moore restored her maiden name.

as though he had been deliberately perpetrating a hideous hoax just now revealed.[51]

Yet of all the possible solutions to the family's financial plight, Carothers chooses the most complex and difficult. He could have simply sent them money or paid their rent or offered to move them to live with Robert and Elizabeth Kyle in Ohio. But he convinced his parents to uproot and move east, and moreover to share a house with him. And Carothers did so at the time when he finally had the opportunity to move to a legitimate relationship with Sylvia Moore. It was as if living with and supporting his parents would establish a buffer, a shield, from the difficult possibilities now looming for him. Some time in the winter of 1933 or the spring of 1934, Wallace Carothers actively began to consider taking himself, once again, to the healing couch.

---

[51]Wallace Carothers to Wilko Machetanz, January 15, 1928. (Hagley Museum and Library Collection, Wilmington, DE, 1850.)

Chapter 12

The Invention of Nylon

In the spring of 1934 Wallace Carothers invented a new series of polymers, one of which came to be known as nylon. The new materials came quietly, came from a subtle extension by Carothers of his previous success, and came as a natural consequence of the peculiar relationship formed at DuPont between Dr. Elmer Bolton and Dr. Wallace Carothers.

Elmer Bolton defined the process of research and development for DuPont as no man had before. He was focused and forthright. As Hard Times deepened, he was willing to scuttle the fundamental program and to realign the research process completely. He responded to his management's clamor for instant results as the nation's business collapsed. Bolton's changes to a visible, practical set of research goals actually protected his staff from the long night of unemployment by making the value of their work instantly apparent to the duPont's who ran the company. The staff would take his direction without grumbling, happy to be among those still employed. Bolton knew his men, carefully chosen by Tanberg from the best chemistry departments in the land, were a prime asset, more valuable to DuPont than its buildings or equipment.[1] And his most precious resource was Wallace Carothers.

Bolton named Adams and Conant and Willstätter, the great scientists on both sides of the Atlantic, as his friends. But Bolton said later, looking back over almost half a century, "Carothers read from

---

[1] Elmer K. Bolton interview with Alfred D. Chandler, Richmond D. Williams, and Norman Wilkinson, 1961. (Hagley Museum and Library Collection, Wilmington, DE, 1689, pp 28, 29.)

the depths of organic chemistry such as I have never seen. I think that he is the smartest organic chemist that the DuPont Company ever had."[2]

At the end of 1933, Elmer Bolton made his annual survey of the research work of his charges at DuPont's chemical department and reported to his superiors on the executive committee. He highlighted first, "the successful use on a large plant scale of our organic deluster-ant in medium luster yarn." He was writing of a modification of the cellulose-based yarn, rayon. DuPont's Rayon Company depended on this yarn and its close chemical relative, rayon acetate, which DuPont called Acele. Even during the Depression, fibers for fashionable clothing provided an upscale, profitable opportunity.[3]

The *New Yorker*, in its pre-Christmas issues for 1933, published three, full-page fashion advertisements, purchased in part by DuPont. A sculptured crepe rayon cocktail costume from Lord & Taylor, Fifth Avenue, was available in the evening shop at $59.50.[4] Something simpler, a rayon Shalisheer print, was at Franklin Simon & Co. for $16.95.[5] For $39.75 at Franklin Simon in Acele, "Youth in the grand manner is epitomized in every line of the evening gown of luscious Crepe Suzzara.... The soft, heavy-as-cream texture is beautifully dull...."[6]

Silk was shiny, lustrous. The continuous filaments woven into soft, drapable fabrics reflected light. Cotton, linen, wool—all short, staple fibers, carded and twisted into yarns in modern equivalents of ancient practice, then woven to fine fabrics, absorbed light. They appeared "dull." No practical reasons favored a dull over a lustrous surface, but fashion, faced with the absence of a condition for any reason, desires to have it. Since 1926, "dull" had been "in."[7]

Rayon, the fiber made by dissolving the cellulose from wood chips in a bath called viscose and then spinning out a continuous filament by feeding the viscose into an acid solution in which the cellulose polymer is insoluble, was inherently shiny and lustrous like silk.

[2]Ibid., p 20.

[3]Hounshell, David; Smith, John Kenly. *Science & Corporate Strategy, DuPont R&D 1902–1980;* Cambridge University Press: Cambridge, England, 1988; p 167.

[4]*New Yorker*, December 2, 1933, p 3.

[5]*New Yorker*, December 23, 1933, p 3.

[6]*New Yorker*, December 2, 1933, p 8.

[7]Hounshell, David; Smith, John Kenly. *Science & Corporate Strategy, DuPont R&D 1902–1980;* Cambridge University Press: Cambridge, England, 1988; p 167.

In turn, the cellulose rayon producers learned to dull rayon by adding fine, particulate titanium dioxide pigment. But DuPont failed to be first in developing this technique and could not dull its rayon with the titanium pigment unless they licensed competitors' patents.

So DuPont responded by dulling its Acele rayon acetate; that they could legally do. And in 1931, the Rayon Company authorized funds for Elmer Bolton's chemical department to find a delusterant for rayon, one which would make a competitive DuPont dull rayon, not requiring the expensive licenses.

In the long history of DuPont research and development, this work is a simple grace note on the company's considerable success. But Bolton's emphasis expressed the mood of the company at the end of 1933. As in 1932, Bolton's annual report ties in closely with the practical problems facing DuPont.

Stine's fundamental research program is still alive but not as an independent entity. It survives as a parasite, as a series of asterisks noting those projects still funded under special money but included in a long list of current departmental interests.

Elmer Bolton was not fixed solely on the immediate and practical. He still looked ahead to the potential for new synthetic fibers as he had once searched for a synthetic rubber. He inserted in the middle of the long 1933 report, "The tremendous advance in the artificial silk industry in recent years has emphasized the importance of developing an entirely new textile fiber, in our opinion one of the most important speculative problems facing the chemist today."[8] Later he said:

> Well, I'll tell you, we had done a lot of work on these polymers. And the trouble was that the properties were not outstanding as a result of the particular things that were combined. There was solubility in alcohol, low melting point and a lot of things that would mitigate against any commercial use of such polymers. Carothers was very much interested... in scientific publication. I used to say to him, "Wallace, if you could just get something with better properties, higher melting point, insolubility and tensile strength, you would have a new type of fiber. I am sure, just looking at the success of rayon and acetate rayon that there would be a lot of room for a fiber of entirely different properties...look it over and see if you can't find something. After all, you are dealing

---

[8]Elmer K. Bolton "Annual Report;" January 2, 1934. (Hagley Museum and Library Collection, Wilmington, DE, 1784, Box 16.)

with polyamides and wool is a polyamide...." Well, it got him back in the field. He came one day and said, "I think I've got some new ideas."[9]

Carothers had backed away from fibers throughout much of 1932 and 1933. Carothers believed he had thoroughly researched condensation polymers. His theoretical concepts acted as a guide for future research. He had established a rational vision of the structure of natural polymers, contributed the general theory of polymer synthesis by condensation methods, developed a system of nomenclature, shown how properties of the polymers were related to their structure, and made a new series of synthetic fibers.[10] He was in no hurry to move further into the field. He had traversed it thoroughly in nearly six years of intense attention.

Shortly after New Year's Day of 1934, Carothers wrote enthusiastically to Jack Johnson. He had just returned from Ithaca, where the American Chemical Society Organic Symposium was held at Cornell. This was the meeting Carothers had dreaded so fearfully two years earlier at Yale, when he had presented his rubber work. But this time, at the end of 1933, the meeting seemed to serve as a tonic. "At the time your letter arrived my head was practically on fire with theories. I had to write 6 new projects and I stuffed them full of theory. The prize one is this." And he went on to write of new descriptions of the detailed mechanism of the polymerization of acetylenes. His tone was joyous. He wrote in an uncharacteristically tongue-in-cheek manner, "Your congratulations on this theory will be eagerly welcomed...."[11] Carothers seemed to continue on a high plane, as interested in the structural and mechanistic chemistry Jack Johnson was developing as he was in his own work. The Organic Symposium had given Carothers time to sit and think, to bask a bit in the respect of his peers, and perhaps to tear away from the persistent drum roll of his mind—the critical, challenging thought that once upon a time he had come upon one good idea, but since that time, his intellectual contribution had been worth little.

Carothers did not speak at the Organic Symposium, and that took away any pressure regarding his attendance. Scientific meetings

---

[9]Elmer K. Bolton interview with Alfred D. Chandler, Richmond D. Williams, and Norman Wilkinson, 1961. (Hagley Museum and Library Collection, Wilmington, DE, 1689, p 21.)

[10]Wallace Carothers to Arthur P. Tanberg, August 4, 1934. (Hagley Museum and Library Collection, Wilmington, DE, 1784.)

[11]Wallace Carothers to John R. Johnson, January 9, 1934. (Hagley Museum and Library Collection, Wilmington, DE, 1842.)

can be a superb tonic for the mind. At the important meetings, the great minds gather, highly charged minds whose energy flashes and crackles. Even the finest scientist is wed in daily bondage to the localized, restricted problems of his own work. Carothers called himself a clerk when that happened. But when a fine chemist who loves chemistry hears another scientist describe well constructed, clever, new science, brilliant science, he can feel a transformation. Time moves slowly as the hearer begins to trace, literally, the exhilarating new chemical models in his own mind. The glow of dazzling science lasts. Consciously or unconsciously, our presence in the hall, our hour of trancelike attention, infects us with a reserve of mental strength. Our own work reflects the new vision, the uncompromising truth that there is great, but unknown, work yet to be done.[12]

At this now uncommon peak of creativity, Carothers turned to the old problem of a practical synthetic silk. Carothers came to Bolton with new ideas on polyamides after his attendance at the Cornell meeting. He picked up the fiber story himself:

> But in early 1934 I decided it was worth one more effort on the following lines. Previous amides had apparently been too high melting to deal with, but if we were to go to longer unit lengths we could probably reduce the melting point to any desired degree. By using carefully purified ester of an amino acid, chemical balance would be assured and the position of the equilibrium would probably be sufficiently favorable to make superpolymer preparation possible without the molecular still.[13]

From Carothers' very first weeks at DuPont, he had understood that the chemical reaction of difunctional monomers, his xAx and yBy and xAy, was not limited to forming polyesters. The scheme could produce long chains of other atoms.[14] Polyesters were fashioned by the reaction of diacids and diols, xAx and yBy with the loss of water,

---

[12]In 1960, shortly after he received the Nobel Prize for Medicine, I watched Prof. Arthur Kornberg of Stanford describe his first enzymatic synthesis of DNA. The magical figures and diagrams of that darkened room stayed with me, almost verbatim, for decades.

[13]Wallace Carothers to Arthur P. Tanberg ''Early History of Polyamide Fibers,'' February 19, 1936. (Hagley Museum and Library Collection, Wilmington, DE, 1784, Box 18.)

[14]Wallace Carothers to Charles M. A. Stine, March 1, 1928. (Hagley Museum and Library Collection, Wilmington, DE, 1784.)

or by the heating of a hydroxy acid, xAy, one containing both an acid group and a hydroxy group. Water would be lost, too, in this process. Carothers knew another type of extended chain, a polyamide, could be made by the reaction of a diacid and a diamine as the xAx and yBy. Or a polyamide could be formed by the loss of water from an amino acid. He and Berchet and Hill had already made polymers from an amino acid, from aminocaproic acid, years ago. But Hill was unable to make fibers. As predicted by Carothers in 1928, the melting point of the polyamide was very high, so high that as Hill had heated the polyamide to soften it so he could draw fibers from the sticky mass, the polymer had decomposed to a charred and clearly useless, brittle accumulation.

Carothers now concluded he would attack polyamides again, this time hoping to reduce the melting point to some workable level— higher than the polyesters, high enough so the material might function as a fiber, but low enough so the polymer could be processed and handled without disintegrating. He knew he could reduce the melting point by lengthening the carbon chain between the amide groups. After all, he and Hill had determined that a long, straight chain of carbon atoms seemed to melt at about 100 °C. If they inserted more carbons into the chain of the amino acid, the melting point should drop back, reliably, toward that low but marginally useful value Carothers had found for the long-chain hydrocarbons. So Carothers directed Dr. Don Coffman to make aminononanoic acid, an acid with eight carbons between its terminating functional groups. This acid would be three carbon atoms longer than the aminocaproic Berchet and Hill had studied earlier; as a polymer, it should melt far lower and might be stable enough to permit the scientists to make a fiber.

To make the polymer, Carothers suggested a new trick. On March 23, 1934, Carothers directed Don Coffman to make the ethyl ester of aminononanoic acid and attempt to polymerize the amino ester rather than the amino acid. He expected the following chemistry:

$$n \, H_2N(CH_2)_8CO_2C_2H_5 \rightarrow$$
$$-<NH(CH_2)_8CO_2>_n^- + n \, C_2H_5OH$$

The amino end of the amino ester molecule would react with the ester end of a neighboring molecule and split out, not water, but ethyl alcohol. In their attempts over the years, Berchet and then Hill had experienced difficulty removing the water produced in their polymerization attempts on aminocaproic acid. The switch to the ester was based on Carothers' old principle, that an xAy molecule could be heated to produce a polymer with the splitting off of a portion of each end of the original molecule. Carothers knew aminononanoic ester would not cyclize. If cyclization occurred, it would result in a 10-membered ring, and 10-membered rings didn't form easily. In the dynamic mass of writhing molecules, amino groups would glide into, impact, and split off the ethanol molecule of the ester group. As carbon chains were lengthened, one end of any particular molecule becomes, on average, more distant from its other end in the three-dimensional mass of molten material. There was little chance an amino group would attack its own ester, literally bite its own tail, and cyclize. Carothers would get a polymer. The only question was, would the polymer reach a molecular weight of 10,000 or above, the magic number Carothers had determined was required to give the fiber properties of a superpolymer.

Don Coffman worked in the laboratory from April 4 until May 21, just to make the aminononanoic ester. He finally obtained a small sample of sufficient chemical purity to go forward to make a polymer. On May 23, 1934, he made the polymer without incident. Coffman was convinced from the first he had made a superpolymer. He observed a characteristic property of the high molecular weight materials. The thin-walled flask, in which he prepared the polymer, shattered as the highly adhesive, molten mass cooled and shrank. The next day, Coffman

> heated [the polymer] in a bath at 200 °C just above its melting point. By immersing a cold stirring rod in the molten mass upon withdrawal a fine fiber filament could be obtained. It seemed to be fairly tough, not at all brittle and could be drawn to give a lustrous filament.[15]

Don Coffman spent the rest of May and June, "spinning and testing." He moved deliberately to characterize his new polymer. He made additional polymer samples, converted them to fibers, and determined their melting point, heat resistance, solubility, and tensile strength.

The results were stunning. The polymer was indeed strong, much stronger than silk. It melted at nearly 200 °C, far above that required

---

[15]Hounshell, David; Smith, John Kenly. *Science & Corporate Strategy, DuPont R&D 1902–1980;* Cambridge University Press: Cambridge, England, 1988; pp 244, 245. Wallace Carothers to Arthur P. Tanberg "Early History of Polyamide Fibers," February 19, 1936. (Hagley Museum and Library Collection, Wilmington, DE, 1784, Box 18.)

for a fabric. Coffman put his new polyamide into a test tube of dry-cleaning solvent. The old polyesters would soften and dissolve quickly. Coffman pulled the fibers out of the solvent. They were robust, flexible, unchanged. The polymer was soluble, but only in exotic solvents like hot phenol.[16] This first polyamide fiber was a fantasy made real.

Bolton and Carothers and the chemists struggling with their science in the quiet laboratories along the corridors of the old Purity Hall could only dream, as the weavers and merchants over the millennia had dreamed, for a material like silk, for a fiber of royal class and fashion, for a yarn not held in bondage to the plump Eastern silkworm with its insatiable demand for properly desiccated mulberry leaves.

Silk, spun secretly in the East by *Bombyx mori,* reeled from its enveloping cocoon and woven to fine textiles, had gained the attention of Aristotle. Justinian the Emperor set the ladies of the empire to weaving the filaments from the Orient at looms installed in the palace at Constantinople. Silk was worth its weight in gold. A hollow bamboo tube, carried across Asia in the year 550 brought the silkworm to Constantinople and the practice of sericulture to the West.[17] Carothers wrote Frances Spencer late in 1931 that he had made a synthetic silk.[18] He was premature then. But now he had produced in the flasks of his Wilmington laboratory a synthetic fiber with attributes clearly surpassing that of ancient practice.

Polyamides were a class of materials, and Coffman's polyamide was likely not the best possible candidate. Other polyamides might have even better properties. Carothers and Coffman had obtained the starting material only through chemical heroism; aminononanoic ester would always be an expensive product requiring sophisticated handling to achieve the extreme chemical purity needed to make the high molecular weight polymer. Carothers knew there might be a better polyamide as yet uncreated.

Carothers used the shorthand he had developed for polyesters for identifying the polyamide structure. He named the polymer by identifying the number of carbon atoms in the units of the polymer chain. Thus, Coffman's polymer from aminononanoic ester was

---

[16]Wallace Carothers to Arthur P. Tanberg "Early History of Polyamide Fibers," February 19, 1936. (Hagley Museum and Library Collection, Wilmington, DE, 1784, Box 18, p 7.)

[17]*Encyclopedia Britannica,* 14th ed.; Chicago, IL, 1939; Vol. 20, p 664.

[18]Wallace Carothers to Frances Gelvin Spencer, November 25, 1931.

polyamide-9. If he used the alternative synthesis scheme and started from a diamine and a diacid, he would name the polymer by the number of carbon atoms in the diamine portion first, then the diacid.

It was easy for Carothers to lay out plans for additional work. There were dozens of combinations of diamines and diacids to be tried. In June 1934, he assigned W. R. Peterson to join Coffman in the work. By late July 1934, Peterson had made another truly astounding polymer, a polyamide from the five-carbon diamine, pentamethylene diamine, and the diester of the 10-carbon diacid, sebacic acid. The DuPont chemists called it, according to Carothers' scheme of nomenclature, polymer 5-10 (*see* reaction 13 in Appendix A).

■ ■

But by the time Dr. Peterson made polymer 5-10, Wallace Carothers was missing from the Experimental Station. It took a day or two for his absence to become a concern for Elmer Bolton and Arthur Tanberg. They looked for him at home, they checked with the men still living at Whiskey Acres, but no one had seen him. There was little to sense of danger in the life of Carothers in the early summer of 1934 as his group made the discoveries that would become nylon. But in July 1934, as this seed of Carothers' greatest conception began to germinate and, as the silkworm is drawn to the mulberry leaf, the seed began to draw the resources of his department to its growth, Wallace Carothers, DuPont's inventor of nylon, disappeared.

Chapter 13

Carothers Disappears

Writing the biography of a man who was prone to vanish carries some challenges. Perhaps Carothers' disappearance in the summer of 1934 was his first unexplained flight. But others were to come. And as the years passed and his mental state worsened, he left less and less of a trail. Little is known and less is revealed of his life as it went forward. He was anonymous to the world he passed through. He was famous, but only in the tight circle of organic chemists. He was quiet, uncombative, compliant.

No group of 12 will come to unanimous agreement on Wallace Carothers based on the evidence I present of his life during 1934. I expect a hung jury. This circumstantial assembly comes from what I believe I know about Carothers and what I know about the others who did or may have influenced him. Take what you believe and leave the rest behind.

■ ■

Carothers lived in 1934 through a changing culture, brightening a bit from the worst of Hard Times. I posit the culture affected him, he reacted to it. Carothers had a bootlegger. It seems consistent that the end of Prohibition would be important to him. Carothers admitted depression and admitted overdependence on alcohol. We know that. I must assume that his depression and his drinking were part of his hidden times, and that he would continue to talk about depression and alcohol with those with whom he was comfortable. His difficult relationship with the now divorced Sylvia Moore was alive at the end of 1933; it was over by the end of 1934, and Sylvia Moore has passed from our view. I cannot find a trace of her. I judge Sylvia was a critical issue in the changes that occurred during the year.

Carothers once heard the violinist Spalding, at Illinois. He still went to concerts, played the classics on his automatic phonograph. Carothers read everything, we know that. He knew the contents of the *Atlantic Monthly*. I will presume he read the other magazines, *Scribner's Monthly*, *Vanity Fair*, *Harper's*, etc. He could reference Alfred Adler to Frances Spencer and had, in his past, visited the "psychopathologists." If he knew Adler, he knew Freud and Jung, and where to go to get help in America or Europe.

■ ■

Soon after New Year's of 1934, after Wallace Carothers returned from the American Chemical Society Organic Symposium at Cornell, where he had stayed with Jack and Hope Johnson, he wrote Jack. He sat in his upstairs room in the Arden home, his automatic phonograph churning through a loud program, four minutes to a side:

> This is a tardy but double barreled communication. First about the bread and butter, the wine and lobster and the delicious hospitality generally, will you please read or show this expression of my profoundest gratitude and thanks to Hope.

Carothers went on to outline a struggle he had endured on the Delaware, Lackawanna and Western Railroad in getting home from Cornell to Wilmington. In his letters, it was Carothers' style seldom to break to a new paragraph. Here he simply marked his new thought with the editor's sign, mindful of his new reflection, but he kept on writing without a separation, on the same line.

> The journey from Ithaca was rather amusing. The poor engineer had so many cars to pull that he couldn't get up over the tops of the mountains in Pa. He tried, awfully hard, but when he would get close to the top the wheels would slip. He coasted back and made several more tries but with no success. The railroad was very generous about the matter though. We didn't have to get out and push. They ordered another engine. I consoled myself with a little scotch (at 40¢ a sip) & had a half a bottle of wine with dinner. We made Philly 4 hours late and I dashed on to Wilmington, arriving about 1 a.m. and stopped off at the Hotel. Speed was there & the next night I met Mrs. Speed. I hadn't the vaguest notion of what she would be like but immediately decided she is just what I expected. Rather nice looking. I think she has been in library work of some

kind. She seems to have read everything current. We did
not have Champagne with dinner. I don't think she is
quite that type, or Speed either yet.[1]

Wallace Carothers and Jack Johnson both were the type to relish
champagne with dinner. Carothers suggests Speed was not ready
"yet." Ready for what? Ready perhaps to join the formal acknowl-
edgement that the nation had just taken a stutter-step in some new
and uncontrolled direction. Wallace Carothers, Jack Johnson or
Speed or anyone else, could now freely ask for champagne with his
dinner, for celebration or any other purpose. They no longer had to
depend on their bootleggers.

For late in the Western night of December 5, 1933, the state legis-
lature at Salt Lake City voted for the 21st amendment to the U.S. Con-
stitution. Utah's vote, the 36th state ratification of repeal, sealed the
end of Prohibition, the total ban on sale of alcoholic beverages, which
had been the law of the land since the approval of the 18th amendment in 1919. Many of the states had legislation already in place to
allow the sale and consumption of alcohol. Carothers could now
drink any drink he wished to order in the hotels and restaurants in
the state of Delaware. And, by law, if he was not already drunk, he
could purchase a single bottle in a liquor store or grocery store, from
nine in the morning until midnight.[2]

For many, repeal restored a sense of correctness. As Frank Crowinshield's December 1933 *Vanity Fair* ("The Kaleidoscopic View of
Modern Life") put it,

> Vanity Fair recognizes that there is a place for alcoholic
> beverages in the field of gracious living. It recognizes, too,
> that fourteen years of prohibition have done much to
> impair the proper appreciation of quality and to lose to us
> the ritual and the art of correct service and use.

Repeal brought a sense of prosperity, too. Crowinshield's edito-
rial was folded between 10 consecutive, full-page advertisements for
spirits: for Hennessy Cognac, for Martell's Brandy, for Johnny
Walker, for Teacher's Highland Cream, for Moët and Vat 69 and
Dixie Bell Gin and Piels and Pabst and Schaefer's ("Our Hand Has
Never Lost Its Skill").

---

[1] Wallace Carothers to John R. Johnson, January 9, 1934. (Hagley Museum
and Library Collection, Wilmington, DE, 1842.)

[2] *Time*, December 4, 1933, p 13.

That moment of repeal was almost symbolic. After all, Carothers and anyone else who wished to drink during Prohibition found the means. Martini and Rossi's vermouth advertisement made a clear comparison:

> If you think you've had real vermouth in the last 13 years you've been chasing rainbows. Only two kinds were on sale in this country during prohibition—bootleg and nonalcoholic. The bootleg was terrible—made in some cellar on the other side of the tracks usually...

But on the night before Utah voted repeal, the American author, William Seabrook, from rural Maryland but at heart and in practice a roamer of the world, put himself into the hands of his editor at Harcourt, Brace & Company, Alfred Harcourt. He was obsessed with his need to be locked up in a place where he could no longer drink, no longer bargain for a drink, no longer get to a drink. From Paris, he had begged Harcourt to help him, and on his editor's instructions he isolated himself in a first class cabin on the *Europa* for the four-day crossing from Cherbourg. Now he sat in the editor's flat looking out over Gramercy Park in New York, gulped down a final Scotch whiskey, then said, "Okay. Send for the wagon and net," and allowed the man to cart him off to Westchester County. Harcourt saw that Seabrook was secured behind the impassable walls of the huge mental hospital in Westchester called Bloomingdale's. Seabrook had his health, he wrote, and he had money, but he was sick from drinking and he was sick of drinking and he had tried to stop drinking—and he couldn't. He was dying from his drinking and he knew it, and he didn't want to die, right now. The big mental hospitals generally shunned alcoholics. The psychiatrists and psychologists, trained in the way of Freud and Jung and Adler, didn't know what to do with alcoholics. But Harcourt was persuasive at Bloomingdale's, and Seabrook was admitted and there he stayed, for seven months. Diagnosis: "chronic, acute alcoholism." With marked "neurasthenic symptoms."[3]

The text and content of public thought at the end of 1933 stretch to more than just the end of Prohibition. We see symbol and prophecy and transition from the dreariness of Hard Times to a glamorous and exciting moment. This was a span of time of seeming artistic and cultural and scientific permanence and a time of political revolution.

---

[3]Seabrook, William. *Asylum;* Harcourt, Brace & Company: New York, 1935; pp vii–xiii. Seabrook, William. *No Hiding Place;* J. B. Lippincott: Philadelphia, PA, 1942; p 344.

The musicians and artists and film stars Carothers now heard and watched were the group of men and women who transcended the coming mid-century madness; men and women whose careers arose in the 1930s and who matured and ripened to old age as the century passed. We heard them and revered them in our own time. Steichen's photographs from the pages of *Vanity Fair* picture young Jascha Heifetz and Leopold Stokowski and José Iturbi. Nathan Milstein and Gregor Piatigorsky and Vladimir Horowitz, in full dress, of course, talk and smoke in a sophisticated portrait. Albert Spalding, the violinist Carothers once heard at Illinois, stands at the center of the group in a pose for Steichen. Toscanini is there, as is Yehudi Menuhin.[4]

Katharine Hepburn glowers in a portrait from the front cover of Crowinshield's April 1934 issue. Inside are Crawford, Colbert, Dietrich, Mae West, and Gladys Swarthout of the Met. And James Jacobsen, the 1933 national ping-pong champion.[5]

Ernest Hemingway published *Death in the Afternoon*. F. Scott Fitzgerald, after eight years of short stories and essays, completed *Tender Is the Night*, and *Scribner's Magazine* began a three-part serialization of the novel in January 1934.

The New Yorker profiled Albert Einstein ("...a strange compound of cosmic wisdom and worldly inexperience") and young Dr. Armand Hammer ("...an outstanding example of the blessings of thwarted ambition").[6, 7] New York taxicabs had radios now, and a reporter visited New York's airport at Newark (for Fiorello was mayor, but LaGuardia was not yet an airport). United Airlines promised 22 hours from New York to Los Angeles or San Francisco for $160. But when clouds rolled across the runway, "a radioman goes on the roof, plugs in a microphone, ...and waits for the drone of the plane groping above the ceiling. He talks with the pilot, tells him when his engine noise is overhead, and helps him down step by step."[8]

Adolf Hitler, in caricature, peers quizzically from dead-center of the cover of that December 1933 *Vanity Fair*. He is wrapped loosely in a ball of yarn. Mussolini, Uncle Sam, and symbolic figures of France and Britain tug at the strings. "The tangle in Europe," the sketch is captioned. The excitement of revolution pervades the accompanying manuscript. "Hitler is of course the hero of heroes." Bemedalled

---

[4] *Vanity Fair*, **1933**, *41(4)*.

[5] *Vanity Fair* **1934**, *42(2)*.

[6] *The New Yorker*, December 2, 1933, p 23.

[7] *The New Yorker*, December 23, 1933, p 18.

[8] *The New Yorker*, December 2, 1933, p 85.

Herman Göring is "the most dramatic and dazzling figure in Germany today..." Joseph Goebbels "is the super-salesman of Naziism..." But not without warning, the magazine's Rosie Gräfenberg advises:

> While the tide of revolution rises higher and higher, the condemned classes hold on to everydayness as to a raft.... While Hitler was already at work to drive them out, the Jewish and left-wing intellectuals gathered, according to old habit in the Eden Bar in Berlin.
>   One can recognize the condemned classes by their failure to sense danger.[9]

■ ■

Throughout 1934, Carothers continued a lively dialogue with Johnson on the structure of a simple organic chemical, ketene dimer, which Johnson was studying. Johnson was a consultant for DuPont, and Carothers recognized DuPont might have commercial interest in the new compound. In the days before instrumentation allowed careful, clinical observation of a molecule's interior, chemists had to guess from the chemical reactions of a specific species what its structure was. Clinical medicine was like that, too. Before medical testing gave direct diagnosis of disease, physicians were limited to diagnosis by symptoms. Johnson had proposed two structures: a linear structure, acetylketene (*see* 9 in Appendix A), and a cyclic arrangement, methylenepropiolactone (*see* 10 in Appendix A). Carothers first favored the cyclic arrangement, then was convinced by Johnson's experimental work to change his mind. Their discussions, filled with careful literature citations provided by Carothers, carried on throughout 1934. The cyclic structure is correct.[10]

On May 16, 1934, Wallace Carothers wrote Jack Johnson from Arden. The hated Arden house was empty. After a year in Wilmington, Ira and Mary Carothers had returned to Des Moines that morning, ending the strange but generous experiment that Wallace Carothers had initiated to deal with his parent's extreme financial necessity. Carothers wrote:

> I am still in Arden. The folks left this a. m. I went out before dinner to do a little boulder hopping and fell

---

[9] *Vanity Fair* **1933**, *41(4)*, 50.

[10] Wallace Carothers to John R. Johnson, February 5, 1934; May 16, 1934; and May 23, 1934. (Hagley Museum and Library Collection, Wilmington, DE, 1842.) Wallace Carothers to John R. Johnson, August 2, 1934. (Hagley Museum and Library Collection, Wilmington, DE, 1784.)

smack in the creek. So I feel that the new regime has been properly anointed. I move in to a new apartment in town July 1st. It will be big enough to offer you a bed and corner for study when you visit this fall.[11]

■ ■

When we see something in another person we do not understand, it is likely we do not have all the facts. Wallace Carothers disappeared in mid-July. "Have lately been in a very large (metaphorical) explosion...," he wrote Johnson later.[12] That is almost all we know directly from Carothers. Essentially all the rest is hearsay, speculation.

■ ■

"One day when I went into the lab, Tanberg called me in," Julian Hill said. "[Tanberg] said, 'Carothers has disappeared. But we think he's in Baltimore. Would you try to see if you could find him'." Julian Hill remembered the details of the trip.

"So I guess we suspected where he was, because he had been going down to see this guy. I think his name was Hohman. Terrible fool I thought." Hill told how Dr. Hohman's office had referred him immediately, without question, to a small hospital that was run by Dr. Hohman. Many psychiatrists had their own, small institutions in those days. Carothers was there, at a little place called the Pinel Clinic, located in an old mansion on St. Paul Street, not far from the Phipps Clinic of Johns Hopkins University. Phipps was the center of American psychiatry.

So I went there, and there was Carothers in one of these rooms, sitting there in his dressing gown, looking like death warmed over. I said, "All right to take him out for lunch?" Oh, yeh, sure, so we went out to lunch. It wasn't a very gay occasion but it was very friendly. Nothing overt that you could see was wrong with him. He was in a terrible funk—he had gotten into one of these depressed periods, he had gotten into his car and driven to Baltimore and put himself in the hands of the psychiatrist; that's where we found him. The guy didn't know what to do with him.[13]

[11]Wallace Carothers to John R. Johnson, May 16, 1934. (Hagley Museum and Library Collection, Wilmington, DE, 1842.)

[12]Wallace Carothers to John R. Johnson, August 2, 1934. (Hagley Museum and Library Collection, Wilmington, DE, 1784, Box 18.)

[13]Julian Hill interview with Matthew E. Hermes, May 29, 1990.

Carothers reached Baltimore on Wednesday, July 18. He carried with him a letter from Roger Adams that asked Carothers for help on a manuscript and asked him for a short historical summary of his past research for "presentation to others."[14] Adams was preparing the case for Carothers' election to the National Academy of Sciences. Carothers replied from Pinel.

> 2938 St. Paul St.
> Baltimore Md.
> Thursday night

Dear Roger:

Sorry to be so slow in replying to your kind letter. Unfortunately a complete reply is impossible at the moment. I came down here yesterday afternoon to consult a doctor about certain manifestations and after a half hour of consultation he brought me here to a nursing home—which appears destined to be my residence for several days. Meanwhile the possibility of making any plans has evaporated. In regard to the paper, I am rather anxious to see it—mostly for my own instruction. Stereo—beyond the most elementary aspects—has always been somewhat of a mystery to me, and I am looking forward to your paper as a light on dark places. Would it be too much to ask you to write again when it is ready about the possibility of forwarding it. Perhaps even if I am down here I shall be able to sit up and dig into it. The outline unfortunately cannot be produced at the moment—and I certainly am in no prizeworthy condition if that should be the point.

My doctor here is Dr. Hohman—a rather young fellow and I never saw any one capable of arousing so much respect and confidence in so short a time. He opened a conversation this afternoon with a few very pertinent questions concerning the structure of the benzene ring. That seems fairly good for a psychiatrist—although it is scarcely an adequate indication of his qualities. At any rate if anyone can do anything for this present nervous collapse he can do it. Already he has restored some of my deflated interest in chemistry. Incidentally, there is just one thing I have wanted to say, and that is I realize more and more

---

[14]Roger Adams to Wallace Carothers, July 11, 1934. (Adams Papers, University of Illinois Archives, Urbana, IL.)

keanly [*sic*] all the time how completely I am indebted to you for everything I have professionally!

Wallace Carothers[15, 16]

Roger Adams replied to his student, his defeated student, who pictured himself prone, and having to draw himself up even to read. His finest student, now somehow unworthy of reward, now withdrawn under the nursing hem of a caring clinic.

Adams did his best, in reply, but Roger lived on practical, New England values. What was apparently true, must be true to all. Why would Carothers feel this way? What could Adams really offer more than the self-evident truth of his student's pre-eminent standing in his chosen field?

> Dear Wallace:
>
> I was certainly surprised to get a letter from Baltimore and to hear that you are spending a few days cooling off. I hope that you will benefit by it.
>
> The paper on stereochemical possibilities will not be ready for a week or two, and besides I won't burden you with it until you are back on the job, since it is not necessary to get it to the Journal until some time toward October.
>
> I do hope you can let me have the resume of your work some time before the first of September. I expect to be at Box 62, Greensboro, Vermont, and will be there until the first of September.
>
> In regard to your complex that has led you to get a deflated interest in chemistry, I can only say that I know of no one who could justify such a feeling as this as little as yourself. If you don't realize it you should that your reputation in chemistry is very widespread, and that your ability to handle and make useful a wide fund of information in different branches of chemistry is the envy of all your friends, including myself. I can add also that the du Pont Company, as you well know, admire you and your work as they do no other chemist in the Company. If I have been able to help you in any way especially, I feel that I have not been able to do half as much as you rightfully deserve.

---

[15]Stereo or stereochemistry is the spatial arrangement of atoms.

[16]Wallace Carothers to Roger Adams, July 19, 1934 (date inferred by reference to Adams' letters of July 11 and 24). (Adams Papers, University of Illinois Archives, Urbana, IL.)

So cheer up, you have nothing to be discouraged about, and I hope you will be soon away from your present abode, and will either be on vacation or back at work.
With best wishes

Sincerely yours,[17]

■ ■

Carothers' psychiatrist knew that Carothers had things to be discouraged about. Everyone did. Everyone had a history, a complex biography, a combination of physical health and intellect and education and artistic drive and spiritual life. Things could go wrong with any of them.

■ ■

Dr. Leslie B. Hohman was four years older than Carothers. He received his M.D. from Johns Hopkins in 1917, then studied psychiatry in Vienna. But he wouldn't visit Freud. He disagreed with Freud.[18] He returned to Hopkins and Phipps as an assistant professor of psychiatry. He continued to train and learn under Dr. Adolf Meyer.

Dr. Meyer ran the Phipps Clinic at Hopkins for 32 years. His clinic became world-renowned for its education of psychiatrists. Meyer was called the "Dean" of American psychiatry.[19] It was Meyer who coined the word "depression" as a clinical term for what was once called "melancholia."[20] Meyer believed that mental illness is a result of learned, inappropriate behavior patterns. Psychiatry was a process of reeducation in which healthy behavior patterns replaced the old, ineffective behavior.

Meyer was getting to be an old man in 1934. He was 68, but he was treating a young patient in Baltimore, a patient exactly Carothers' age, 38. The patient was uncooperative because he believed he was healthy and he believed it was his wife, alone, who needed the psychiatric care. The patient was F. Scott Fitzgerald.

---

[17]Roger Adams to Wallace Carothers, July 11, 1934. (Adams Papers, University of Illinois Archives, Urbana, IL.)

[18]Dr. Hohman died January 28, 1972. This information appears in his obituary in the *Baltimore Sun,* January 30, 1972.

[19]*Encyclopedia of Psychology;* Corsini, Raymond J., Ed.; John Wiley and Sons: New York, 1984; Vol. 2, p 370. *See* "Meyer, Adolf" and "Dean of American Psychiatry."

[20]Styron, William. *Darkness Visible, A Memoir of Madness;* Vintage Books: New York, 1992; p 37.

Fitzgerald moved with his troubled wife, Zelda, to a run-down, musty house called La Paix, on the Turnbull estate north of Baltimore in the spring of 1932. The house was near the Phipps Clinic, and Zelda was admitted there periodically. The Fitzgerald's lived at La Paix much of the next two years until Zelda burned out part of the house trying to destroy old clothes in an upstairs fireplace.[21]

Adolf Meyer was Zelda's psychiatrist, but Scott didn't fool him. He was arrogant, demanded to be part of his wife's therapy, saw her role as inferior to his great artistry, and diagnosed her problem as a jealous demand to share, with her modest talents, his literary acclaim.

> Meyer considered it a dual case. He wanted Fitzgerald to face his drinking, to be treated for it if necessary, but Fitzgerald balked at psychotherapy...partly from the artist's instinctive distrust of having his inner working tampered with.[22]

Meyer could see Scott Fitzgerald's role in Zelda's disintegrating life.

> [Meyer] felt that what was involved was not simply a question of Zelda's *case*; it was Scott's life as well. Zelda, of course was his patient, but Meyer saw Scott as someone who, though unwilling, also needed help. He didn't want Scott to function as a sort of boss to Zelda, nor as a psychiatrist-nurse. He wanted a closer understanding of both of the Fitzgeralds, but he was certain that could be achieved only if Scott gave up alcohol.[23]

■ ■

Dr. Meyer's protege, Dr. Leslie Hohman, lived in the fashionable inner city of Baltimore, on North Calvert Street. He was an impeccable dresser. He was drawn to artists and the theater. Alfred Lunt and Lynn Fontaine were among his friends. The lights burned late from his home on Sunday nights. Hohman invited the interesting and prominent to his home for discussions and debates and fine food. He was a gourmet chef. "Whenever you went to Dr. Hohman's house, he

---

[21]Turnbull, Andrew. *Scott Fitzgerald;* Charles Scribner's Sons: New York, 1962. Turnbull describes the Fitzgeralds at his family's estate. *See* p 237 for the incident of the fire.

[22]Ibid., p 233.

[23]Milford, Nancy. *Zelda, A Biography;* Harper & Row: New York, 1970; p 272.

offered you a drink," his daughter-in-law, Mrs. Anne McKenzie, said. "Unless you were an alcoholic, in which case you were lucky to get water. He didn't want to start anything." Hohman was a practical man. "I really miss not having a psychiatrist in the family," Mrs. MacKenzie said, 20 years after his death. "He was extremely practical. A very kind man. He was outspoken and told you like it was."[24]

Dr. Hohman remained friendly with many of his patients long after he had seen them. That gave him a unique opportunity. He distrusted what he learned from these ailing people when they came to see him; at that moment, both the patients and the families were full of distress. They were incapable of an objective look at their problems. That's why they came. "While the patient is actually sick," Hohman wrote, "he is self-accusatory and self-debasing and the family and friends are for the time being selfless in attitude. The truth of the oft-expressed opinion that you are apt to see people at their best when they are sick, is one of the factors which often obscures correct history-taking." But as Hohman became friends with his patients he began to notice more and more about their families, "the life situation which precedes the actual attack."

Hohman wrote in 1938 about one successful man he had known since 1928. The man came to Hohman because he wasn't sleeping or eating well, he couldn't concentrate on his work as a commodities broker, and he had made two serious attempts at suicide. Hohman learned his patient had been "tremendously active erotically in the typical man-about-town manner." But he had married well. He drank too much. His mother and his mother's sister had killed themselves.

During the man's course of treatment with Dr. Hohman, his depression disappeared, without Hohman ever finding a cause or a behavior that had set the man off. They became friends, and in time Hohman learned that at the time of the first depression, the man was in a flirtatious relationship "with a very pretty woman of his social set," which he had neglected to discuss with Dr. Hohman. The man's wife had brushed it off as unimportant. Hohman learned their marriage had been arranged to please the man's father; it was a sign of respectability and social placing.

Several years later, the man came to Hohman again, once again depressed and suicidal. Hohman quickly found the man was drinking heavily now and had resumed the relationship with the same woman. She had invited him to travel with her to another city. The man was indecisive. Hohman wrote for the *Journal of Nervous and Mental Diseases*:

---

[24]Anne MacKenzie interview with Matthew E. Hermes, November 27, 1992.

By this time I knew the patient very well and realized how much his position in the community and the conventional morality of the group meant to him. I was sure that his attitude was heavily weighted by the fact he felt he must live up to the ideal as set by his father. This attitude did not admit of any "backdoor" behavior. When he was confronted by erotic desire on the one hand and the father ideal of conduct on the other, depression entered.

I advised the stopping of alcohol because intoxication took down his barriers of control and made for susceptibility and then urged that he make the decision to give up the flirtation and the proposed clandestine meeting saying that it violated too much of his life pattern. He followed both injunctions and the depressive attack stopped.

Two years later the same story was repeated, depression again recurred for four weeks, and the same therapy stopped it.[25]

Dr. Hohman cited two other cases, instances of depression which he ascribed after a long look as "backdoor" behavior. In two of the three of the cases, Hohman forced the patients back on to their own "organized pathways of life;" in a third, he condoned an extramarital "adolescent romance." But Dr. Hohman made clear, "I did not feel that an analysis of the deeper *genesis* of their personality problems would serve them any real purpose."

■■

So it is hard to imagine the urbane Dr. Hohman approaching his new patient, Wallace Carothers, with anything more than the practical ambition of finding what his behavior was, identifying what parts of his actions were inconsistent with what Wallace Carothers believed, and suggesting he change his behavior. He would not sit with Wallace in long psychiatric huddles and probe his unconscious. He would not act as empty husk or hollow shell into which Wallace Carothers would empty his complex relationship with his father and his mother.

Rather, Leslie Hohman would take Carothers' history. He would ask about his work at DuPont, learn of his fame. Carothers might reveal he had just experienced a new breakthrough toward a synthetic silk. He might do so in an off-hand way, because he was depressed and would belittle his own accomplishments. Hohman

---

[25]Hohman, Leslie B. "The Abortion of Recurrent Depressive Psychoses," *The Journal of Nervous and Mental Diseases* **1938**, *88(3)*, 273.

would hear Carothers tell of his depression, of the experiment, now ended, of bringing his parents to Delaware, of the failure of the home at Arden, and of his new apartment in the stylish Wawaset Park section of Wilmington. He would hear of Sylvia Moore and of the antagonism of the management at DuPont over this enduring relationship they still felt was scandalous. Hohman would ask Carothers whether he drank, whether he used drugs. Those questions were part of standard history taking in the 1930s. Carothers might distort the extent of his drinking, many people for whom alcohol is important do that.[26] Because for many men, the drink is sacrosanct, beyond the reach of anyone. They do not reveal their drinking because alcohol is their friend, their stability. Their biggest fear is alcohol may be taken away from them. "Just a couple," he might say. We do not know if Wallace Carothers carried the small capsule of cyanide with him into the Baltimore clinic, whether he told Dr. Hohman he carried the poison, whether he surrendered the capsule to the trusted physician.

Hohman would hear Carothers' history and like the man. Like him as a patient. This was the kind of patient he and Dr. Meyer were used to seeing. Well spoken, extremely intelligent, famous in his own way, placed socially, fond of Russian music, well read, yet depressed. He would light the streets of Carothers' soul, or as much as Carothers would share of his soul, and try to identify what it was that made Carothers climb into his red Oldsmobile convertible and flee DuPont and Wilmington, hurry into Maryland, cross the Susquehanna, drive through the rolling countryside on the 70-mile ride down the Baltimore Pike into Baltimore and into Dr. Hohman's small clinic.

Carothers' flight to Dr. Hohman at least carried with it a sense of optimism. And he was asking for help from a man whom he immediately liked. What triggered Carothers' flight?

It is a long and grasping stretch to assume Wallace Carothers read any particular book or magazine, in the absence of some reliable citation. But he did read everything. And it stretches credulity even further to assume reading leads to doing.

Carothers lived in a vibrant, intellectual world, surrounded by the best of culture and learning. He may have read *Scribner's Magazine* in May 1934, and he may have remembered some of his own words, many years before, when he saw them repeated almost verbatim in the magazine.

Scribner's Magazine followed its serialization of *Tender Is the Night* with an anonymous tale in its May issue. "I Drank My Way to Psychia-

---

[26]Appel, K. E.; Strecker, E. A. *Practical Examination of Personality and Behavior Disorders;* Macmillan: New York, 1936; p 9.

try" went the title.[27] "Liquor had me practically beaten," the author wrote. In his own mind, he attributed his sprees to a negative and pessimistic and selfish outlook, resting on the beliefs of Nietzsche and Schopenauer. But psychiatry led the man back, back in time toward a series of problems stemming from the "same neurotic root." He wrote of uncovering that "in my youth, some part of my emotional development was arrested." The death of his father when he was 13 pushed him into a vague, hostile world, a world in which the young man was uncertain, alienated, "detached from the general current of life." "I would feel like a strange onlooker," he wrote.

The crux of the story was, "that alcoholism was not a congenital weakness, not organic as to cause, but just one of hundreds of other symptoms of the same basic neurosis." Liquor was the wellspring of life for the man, the only medium through which life went on, the source he must use to overcome timidity, fear, anxiety, and lack of self-worth. He called this "insane." But he found, with his psychiatrist, the basis for his emotional insecurity lay well back in his childhood, with a powerful aunt who taught him to fear his mother's death, too. He finally learned the impact of his aunt's foolish harangue was to make him anxious and withdrawn. He needed alcohol to counteract that withdrawal. With this understanding, he was now free. The man still drank, he wrote, but now he controlled his liquor; it did not control him.

Carothers wrote Machetanz long ago of his "completely arrested development of personality on a certain side."[28] He recounted how he stood as an adult observer, angry at the "hoax" of his friends' passage to adulthood. He described how his father was inaccessible, how he feared the man.

Did Carothers identify the text of the *Scribner's* story as his own tale? "My development was arrested," wrote the now optimistic patient. "Completely arrested development," Carothers once wrote. Did the man's ambition to identify alcoholic drinking as a symptom of a neurosis, now controlled, give Wallace Carothers a gleam of hope? Did he see some manner of relief, release, some profit to be gleaned from going into care? It is not likely, in the modern perception of alcoholic drinking, that the man's analysis of his own affairs was totally correct. But what does it matter? Sufficient parallels exist to suggest the story may have sent Wallace Carothers off to see Dr. Hohman.

---

[27]Anonymous. "I Drank My Way to Psychiatry," *Scribner's*, May 1934, p 361.

[28]Wallace Carothers to Wilko Machetanz, January 15, 1928.

■ ■

By August, Carothers was gone from Baltimore and Dr. Hohman's clinic. He was in the small town of West Tisbury on the island of Martha's Vineyard. He was staying for a time with Julian Hill and his wife, Polly Butcher Hill, at the home of Polly Butcher's family. You cannot tell from the Butcher place that you are on an island. It is plain country, a clapboard house behind a picket fence set in southern New England's fields. In the waning summer days of late August, Carothers went down to the water, he hit tennis balls with Julian Hill. Most of the time, he sat about dressed in his shirt and tie and dark three-piece woolen suit. He read and he wrote.

Carothers sent a letter from the island to Roger Adams on August 21, 1934:

> Dear Roger:
> My forced holiday has been considerably prolonged and has finally ended up here on the island where I am just concluding a long week end with the Butchers (Mrs. Julian Hill's family). The weather has been perfect; the accommodations are rather luxurious; there has been lots of pleasant company, tennis and swimming. Chemistry and many other things seem to be important again.
> I am enclosing an outline of researches. I am not sure it is the kind of thing you want—it seems a little vague and extravagant to me. I should be glad to attempt a revision if this edition does not seem satisfactory.
> I hope that you and Mrs. Adams are having a pleasant holiday. Shall expect to see you at Cleveland or even sooner.
>
> Sincerely,
> Wallace Carothers[29]

■ ■

Julian Hill remembers that Carothers told him a bit about his treatment with Dr. Hohman. Dr. Hohman told Carothers he should marry.[30]

Was marriage Dr. Hohman's prescription intended to direct Wallace Carothers back to "an organized pathway of life?" What did Dr.

---

[29]Wallace Carothers to Roger Adams, August 21, 1934. (Adams Papers, University of Illinois Archives, Urbana, IL.)

[30]Julian Hill interview with Matthew E. Hermes, May 29, 1990.

Wallace Carothers with the Hill family. Wallace Carothers with Polly Hill, Margaret Butcher, and Julian Hill. Carothers holds Louisa Hill. (Helen Carothers.)

Hohman discover in Wallace; what did he point to as behavior antagonistic to his values—the backdoor behavior causing his depression? Sylvia Moore was divorced but apparently was out of Carothers' life now. Had Carothers, as Sam Lenher told it, continued these "affairs."

After Carothers returned from Baltimore, he "had tried dating different gals around town," Polly Butcher Hill said. "I guess when he thought it would be a good idea to get married. And he even dated my sister. It was a bust. It was a mess. It didn't work. If you can imagine one unbalanced person to another. It's not good. He dated Margaret."[31]

---

[31]Polly Hill interview with Matthew E. Hermes, May 29, 1990.

Wallace Carothers on Martha's Vineyard, August 1934. (Helen Carothers.)

Julian Hill in 1990. (Matthew E. Hermes.)

One of the Butchers took photographs that August weekend on the Vineyard. In one of them, Julian and Polly Hill and Wallace Carothers and Margaret Butcher stand on the front lawn of the Butcher place, in front of the picket fence with its sturdy, granite fence posts. Julian and Polly look confident and relaxed, Margaret Butcher, Tessie they called her, is a block of a woman, plain and sturdy, dressed in shorts and saddle shoes.

Wallace Carothers' suit is rumpled and he is holding a little girl in the picture, a blond, barefoot, tiny toddler, Julian and Polly's 14-month-old daughter, Louisa. And Carothers is smiling. An uncomfortable smile, to be sure, as if he wishes to hand off Louisa. But he is smiling.[32]

■ ■

Polly Hill said, "Wallace was a guy you don't mess around with. You don't treat him lightly. He was an important person. He was

---

[32]I photographed Julian Hill—tall, gray, aging, but as comfortable as he was in 1934—standing in front of the same fence in 1990.

deserving of everything you could do to help." She became frustrated with her description. "Alright, let's try again," she said.

> He was a good friend of Julian's. He was a wonderful man. He was not the usual run of human being and it took all your imagination (I was very young and inexperienced), it took whatever I had to try to reach him. And it was worth trying, because whenever anything came out it was worth hearing. There was a lot of silence and quiet, which is all right with me. But he was an important person in your life. You don't play games with that. You don't pretend, you don't ignore. You do what you can. You reach out to the extent you are able or think he wants. He inspired your devotion.[33]

■ ■

There is a wearing sense of tiredness in the maintenance of Wallace Carothers. His periodic desperation exhausts and controls all his friends. They begin to struggle and talk among themselves, assuming foolishly there is something *they* can do to reverse or prevent Carothers' decline. In late August 1934, Carothers returned to DuPont and he watched, for a time, the rest of the invention of nylon.

---

[33]Polly Hill interview with Matthew E. Hermes, May 29, 1990.

Chapter 14

A Year at Wawaset Park

Wallace Carothers came home alone from Martha's Vineyard to his new apartment at 7th and Greenhill, in the Wawaset Park section of north Wilmington. The apartment complex was on the south side of Greenhill Avenue. Greenhill could be called the boundary between Wilmington proper to the south—the manufacturing and banking city dominated by DuPont—and the estate country to the north, the large homes and palatial duPont manors scattered among the low rolling hills. The estate country fanned north out of Wilmington in an arc. The arc was scored into segments from west to east by the roads leading from Wilmington into Pennsylvania—the Lancaster Pike, the Kennett Pike, the Concord Pike—and by the Brandywine River with its meander toward its source in the north.

At Arden and Whiskey Acres Carothers did not have a long drive to work at the Experimental Station, but now, in Wawaset Park, he was less than two miles away. He would drive out to Pennsylvania Avenue, make a left and then a right at Rising Sun, then down to the Brandywine and across to the station, built there in the angle of the river, just below the old DuPont powder mills.

Carothers met a new neighbor. William Latta Mapel moved to Wilmington and to Wawaset Park as executive editor of the *Wilmington Morning News* and *Evening Journal.* He was just 32, six years younger than Carothers, and he brought "an immense stock of boyish enthusiasm and idealism plus a certain amount of naivete about the real world" to his new job. Mapel had a "joy in living," a "hearty laugh."[1]

---

[1] Reese, Charles L., Jr. "Christiana Farewell: A Newspaperman's Memoirs of the News Journal Company, 1922–1977," *Delaware History* **1988**, *23*, 72. Mapel arrived in Wilmington in August 1934. He was gone from the *News-Journal* papers by mid-1936.

Mapel later called Carothers the closest friend he ever had. He remembered Carothers coming for dinner often, dropping in unannounced. Evelyn Mapel was a Tarkio graduate—an immediate bond for Carothers. Mapel said he and Evelyn would join Carothers for Friday concerts in Carothers' second-floor, front apartment.[2] Perhaps it was wise each record took but three or four minutes per side before changing, because Mapel was an impatient man. "Bill Mapel took charge of things," Polly Hill said. "Bill Mapel took charge of things," Julian Hill echoed, "including all of his friends."

"He took charge of us for the period he was there—we became absorbed in his circle—it was exciting, you couldn't get out," Polly Hill said. "I'm having something happen tonight. You've got to come. You've got to do this, you've got to do that!"

> He was a charismatic individual, and he grabbed ahold of Julian, he thought he was sort of a pet, somebody he thought he needed to know more about. He was a strange bird to the hilt. Bill Mapel was big. Stuck out all over. Gave ultimatums to his wife all the time. One was there always had to be ham in the ice box. His son was always to wear pants with suspenders, not one of those things that looked droopy. He'd have these ultimatums. And we'd look at Evelyn. Poor Evelyn. She coped.[3]

Wallace Carothers seemed to cope, too. Carothers fell into a steady, almost homey routine in the fall of 1934. Julian and Polly Hill lived nearby on Coverdale. They invited Carothers to dinner regularly. Once a week, Polly suggests. The Hill's service—candles and fine linens—were not part of Carothers' normal experience. There was nothing special about it for the Hills. Carothers was their friend. Julian Hill and Polly went to Carothers' apartment on Friday nights, too. Polly remembers them as their "Friday night drinking parties." But they were basically music nights. The Hills came, along with the Mapels and John and Lib Miles. Miles was a physicist from Princeton. He worked at the station doing fundamental testing on Carothers' new fibers.[4] The Miles loved music. They were Southerners, a rarity at the Experimental Station. John Miles and Carothers, both deep-

---

[2]William C. Mapel interview with Adeline C. Strange, July 1978. (Hagley Museum and Library Collection, Wilmington, DE.)

[3]Polly Hill interview with Matthew E. Hermes, May 29, 1990.

[4]Hounshell, David; Smith, John Kenly. *Science & Corporate Strategy, DuPont R&D 1902–1980;* Cambridge University Press: Cambridge, England, 1988; p 259.

voiced, sang together in a soulful bass. Leigh Williams from the old bunch at Whiskey Acres was there too. They sat and sipped while Wallace took them through his record collection.

Polly Hill complained the sophisticated record changer of which Carothers was so proud, but which often needed repair, had a particular affinity for ruining her records. But Carothers was adept at repairing the balky changer. Polly talked of Carothers' extensive record collection. "He'd sort of go wild," Polly said. "I remember for a while all he bought was Bach, then he got hip on Gilbert and Sullivan and he bought all of those."[5]

■ ■

Carothers' reputation brought him a job offer from outside DuPont. In late 1934, Robert Maynard Hutchins, the youthful and controversial president of the University of Chicago, contacted Carothers with a specific proposal. Hutchins, a rumpled, abrupt young man of 35, who could be seen at Chicago's football games in a "wrecked and beer stained fedora hat" like the one Carothers was fond of wearing, was already in his fifth year as head of Chicago.[6] He offered Carothers the position as chairman of the chemistry department there.[7]

Just six and one-half years before, Carothers had left Harvard as an instructor, at the very bottom of the academic ladder. The success of his researches now brought him in direct confrontation with a real academic opportunity. Hutchins had reorganized Chicago in 1931 into a college division for general education and required all students to pass a comprehensive exam. They were then admitted to one of four degree-granting upper divisions. These upper divisions were the university. Hutchins abolished course and attendance requirements. Hutchins' revolutionary changes seemed to satisfy both the major benefactor of the university, John D. Rockefeller, and its student clientele. By 1934, Rockefeller had donated more than $35,000,000 to Chicago. The campus was in the midst of dramatic good times. Over the past five years, over the years of gloomy depression, the physical size of the campus had been tripled and the student population now exceeded 13,000.[8] Chicago built a new chemistry building. It could

---

[5]Polly Hill interview with Matthew E. Hermes, May 29, 1990.

[6]Mayer, Millton S. "Rapidly Aging Young Man," *Forum* **1933**, *90*, 308.

[7]Hounshell, David; Smith, John Kenly. *Science & Corporate Strategy, DuPont R&D 1902–1980;* Cambridge University Press: Cambridge, England, 1988; p 245.

[8]*Encyclopedia Britannica*, 14th ed.; Chicago, IL, 1939; Vol. 5, p 457.

be for Carothers. At Chicago he would join great scientists at the Quadrangle Club. Hutchins championed scholarship; he saw Chicago as a place for unhindered scientific inquiry.

But Carothers chose not to go. Why would Wallace Carothers refuse to return to the Midwest, to the city where his sister Isobel flourished, to a place where he certainly could start afresh, where he could flee the last traces of the affair with Sylvia Moore, where he might be closer to his roots, to Frances Spencer? At the University of Chicago he would have the total scientific freedom he said he wanted. Carothers had studied at Chicago in the summer of 1922. He knew it as a diverse place. It held a position of leadership in law and philosophy and political reform. It was a lively place now, running with a current of liberal reorganization led by the brilliant young Hutchins. Just a year before, Carothers had cast a line toward the Harvard position opened up by Conant's election as president in Cambridge. He had withdrawn that idea, almost as quickly as it came to him, but here was real opportunity, solidly presented and apparently meeting his needs as he had expressed them.

There is no record that Carothers talked to Bolton, or that DuPont offered Carothers any incentive to stay in Wilmington. But stay he did. He rejected Hutchins' approach in November 1934, citing the "exciting stage" of his scientific work.[9]

■ ■

Carothers appeared at work with regularity in the fall of 1934. His group labored on the artificial silk problem. Polyamide 5-10 was the target. Don Coffman built new machinery for melt spinning the polymer. W. R. Peterson found a new way to make the 5-10 superpolymer. In October 1934, he switched away from the diester pathway Carothers had suggested, back to the combination of the diacid and the diamine. By mixing the two components in exact proportions, without heating, the two components reacted to form a solid salt. Everyone knew that, and knew the salt had to be heated to make the polymer. But Peterson isolated the salt and purified it. Peterson's method let nature do what the chemists found very hard to do—combine precisely exact equivalents of diacid and diamine to make polymers. The salt he isolated left behind the small excess of diacid or diamine that would ruin his attempt to get high molecular weight. But by the end of 1934, Peterson's method was the preferred route to the polyamides.

---

[9]Wallace Carothers to R. M. Hutchins, November 4, 1934. (Hutchins Papers, University of Chicago.)

Peterson's method led to a new problem none of them had ever seen before. The new synthesis gave long chains, so long that at times Peterson was unable to make fibers. They had reached an upper limit to the chain length in the quest for a useful fiber.[10] Peterson struggled with a solution to this problem. He tried a slight excess of diamine, of diacid. Neither technique was totally satisfactory. In early 1935, Peterson proposed to add a small amount of acetic acid as an excess acid as a chain terminator, capping the chains at a predictable length. Carothers opposed Peterson's scheme to adulterate the polymerization. But, once again, Peterson's solution came to be the preferred method. Dr. William Hanford, who came to DuPont just after Peterson's methods were adopted, called Peterson a "hero" for challenging Carothers and insisting Carothers' exactly matched amounts of diacid and diamine were inappropriate.[11]

Paul J. Flory, a newcomer from the chemistry department at Ohio State University, joined DuPont in the summer of 1934 and began working for Carothers. Flory was a physical chemist, adept at calculations. He stayed near his desk and began figuring whether there was a way to learn more about the chain lengths of the polymers the other chemists were making out there in the laboratory. Each molecule in the sticky mass of polymer must be different, he reasoned, and he set about by hand to evaluate the distribution of individual chain properties. (Flory stayed with this work long after Carothers was dead and he himself was gone from DuPont. He went to Stockholm in 1974 to receive the Nobel Prize in Chemistry for his work.)

Wallace Carothers hired Joe Labovsky, the young Russian immigrant whom he had met at the Philadelphia Academy of Music in 1930, to be his personal laboratory assistant. The duPont's had financed Labovsky's education at the Pratt Institute; now he had returned to Wilmington. Carothers saw him working on a part-time job at the station, out on the grounds. In November 1934, Labovsky came inside, into the laboratories. Labovsky remembered Carothers as "elegant, scholarly, dignified." He called his boss "Doc." Among all the doctors of philosophy, Carothers was the one they all called "Doc." Labovsky and Carothers talked about Turgenev, about Dostoyevsky. They never talked American literature. Labovsky and Carothers went to concerts at the academy to hear the Philadelphia

---

[10]Wallace Carothers to Arthur P. Tanberg, "Early History of Polyamide Fibers," February 19, 1936. (Hagley Museum and Library Collection, Wilmington, DE, 1784, p 8.)

[11]William E. Hanford interview with Matthew E. Hermes, July 9, 1990.

Orchestra under Leopold Stokowski. They drove to Philadelphia in Carothers' convertible. Carothers was a "lousy" driver, Labovsky said.[12]

■ ■

Carothers was satisfied with polyamide 5-10, Elmer Bolton was not. At times, the first striking sample made in the course of a great invention turns out to be the best the scientists ever get. So it was with Carothers' polychloroprene. Carothers and his group tried for two years to find a better synthetic rubber than Arnold Collins' first accidental preparation. But it was not to be. The chemists could not identify a better composition than the long-chain polymer isolated from the unexpected, volatile oil.

Roy Plunkett's accidental discovery of poly(tetrafluoroethylene) (Teflon) was never improved on. DuPont's Plunkett noticed the pressure in a gas-filled cylinder of the compound tetrafluoroethylene had dropped to zero. He cut the tube open and found the glistening, white polymer, converted from the gas.

But invention could go slowly. Langmuir's Mazda lamp was the product of nearly 50 years of test and try after Edison's first incandescent bulb. Midgley's tetraethyl lead, which led to high-quality "Ethyl" gasoline, came from a long series of laboratory trials. And Bolton had lived through painful, incremental advances in the development of DuPont's dyestuffs. So Elmer Bolton insisted that Carothers do a systematic study of new, potential polyamides. Any chemist could concoct a series of potential candidates by simply combining a whole series of diamines, one at a time, with a similar series of diacids. A limitless number of samples could be visualized. Somewhat arbitrarily— but with clear anticipation that as the carbon chains in diacid and diamine were lengthened, the important melting point property of the polymer would fall below useful levels—Carothers began to make as many of the polyamides by combining diacids and diamines containing two to 10 carbon chains as he could. There were 81 potential candidates in this series.

Carothers' old friend, the luckless Gerard Berchet, got the job to run through the list of candidates. In 1930, Berchet had allowed a series of critical samples to sit on a shelf, unattended, while Collins invented polychloroprene. He had failed to get a polyamide from aminocaproic acid. Now in 1935, Carothers put Berchet to work on

---

[12]Joseph Labovsky interview with Matthew E. Hermes, June 28, 1991.

the polyamide project. Berchet made many of the diamines himself and purified them so they could be polymerized. He cranked out polymer after polymer. He went beyond the simple combinations of linear diamines and diacids and made examples of six other types of potential polyamide polymers (*see* reaction 14 in Appendix A).[13]

On February 28, 1935, Gerard Berchet recorded a simple experiment in his laboratory notebook. He initialled the experiment, and Harry Dykstra witnessed the record:

> *Adipate of hexamethylene diamine*
> 7 g. diamine, 8.8 g. acid and 20 cc. m-cresol heated 215°
> for 3 hours. Water came off during the first half hour. The
> temp. was then raised to 255-60° and the cresol distilled
> off in vacuum. The residue solidified at one time but then
> melted again at 265°. It was heated under 1 mm at 265° for
> 3 hours. On cooling (overnight) the polymer broke the
> flask by contracting and showed a tenacious adherence to
> glass.
> It was a very hard, horny solid melting at 252-254°. It was
> very readily spinnable. Sample turned over to D. D. Coff-
> man.
> There was obtained 12.5 g. of polymer, yield 90%.
> 3/1/35 GJB 3/21/35 H. B. Dykstra[14]

This half-ounce of polymer was polyamide 6-6, from hexamethylenediamine and adipic acid. The pearly, lustrous mass became DuPont's nylon.[15]

Polyamide 6-6 was an attractive candidate from the first. Perhaps the most striking property of polyamide 6-6 was its dramatically high melting point. Those of us who have handled and studied nylon over the last half-century will automatically quote a melting point for the polymer. It has become a standard, a reference, a comparison for other polymers. A number that a chemist memorizes. Each of us may remember a slightly different value, because our desire for exactness in the matter exceeds the accuracy of the actual phenomenon. I

---

[13]Wallace Carothers to Arthur P. Tanberg, "Early History of Polyamide Fibers," February 19, 1936. (Hagley Museum and Library Collection, Wilmington, DE, 1784, p 9.)

[14]Gerard J. Berchet, copy of a DuPont Notebook page. (Anne Knepley.)

[15]Hounshell, David; Smith, John Kenly. *Science & Corporate Strategy, DuPont R&D 1902–1980;* Cambridge University Press: Cambridge, England, 1988; p 259.

quote 256 °C. My *Polymer Handbook* lists 265 °C.[16] But the high melting point of the 6-6 polymer scared off Carothers and the other scientists in the laboratory. No one, at first, was willing to believe the polymer, which had to be processed to a fiber by heating for a long time, above its melting point, would withstand this kind of heat without charring to a useless mass.

Work on polymer 5-10 went on. By the spring of 1935, Don Coffman took polyamide 5-10 filaments he had prepared by forcing molten polymer through a tiny hole in a steel plate—a spinnerette. He made a thread by twisting 24 of the filaments into one strand. The thread was heavy and stiff, at least by the finer standards of silk, but the DuPont plant at Waynesboro, Virginia, used it to make some of the first knitted fabric samples.[17, 18]

This was the beginning of the true development of Carothers' polyamide fibers to realistic commercial products, the first of many small steps. The chemists made a knitted fabric first because a knitted fabric requires only one yarn. A woven fabric is much more difficult to prepare. Looms weave a large number of warp yarns, or long-direction yarns, by firing a fill yarn between alternately separating banks of warp yarns. The preparation of the warp—beaming—is a tedious process requiring large amounts of yarn to be wound separately on the beam. No such amount of the precious research yarn, twisted from the Experimental Station laboratory, existed.

In the summer of 1935, Elmer Bolton made a critical, far-sighted decision. He told Carothers that he was to work on polyamide 6-6 as the commercial candidate. None other. Polyamide 5-10

---

[16]Brandrup, J.; Immergut, E. H. *Polymer Handbook;* Interscience: New York, 1967; p III-39.

[17]An old, French method is used to measure the physical size of fiber strands. It is nearly impossible to get meaningful measurements of the thickness or diameter of fibers. This is particularly true of most natural fibers—wool, cotton, and linen—in which short sections called staple are twisted into useful, but irregular, yarn. The French term *denier* is used to characterize yarn. The weight of 9000 meters of yarn is its *denier*. The first polyamide 5-10 yarn was 123 *denier* and consisted of 24 twisted yarns; thus, its *denier* was approximately 5 *denier* per filament (dpf). Continuous natural silk is about 2 dpf.

[18]Wallace Carothers to Arthur P. Tanberg, "Early History of Polyamide Fibers," February 19, 1936. (Hagley Museum and Library Collection, Wilmington, DE, 1784, p 10.)

had fine properties, to be sure, but polyamide 6-6 was better. And more importantly, the two six-carbon monomers required could be made from a single feedstock, benzene, present in and obtainable from crude oil. Bolton insisted. He accepted no alternative from his technical managers who were enamored of the ease of handling polyamide 5-10. Bolton anticipated the monomers for polyamide 5-10 would always be expensive; the 10-carbon fragment came from castor oil. There would never be enough castor oil for the market Bolton anticipated.

Elmer Bolton suggested that over the long pull, benzene would be available and cheap. His decision took the scientists down a hard road. First, polyamide 6-6 was very difficult to process to a fiber. As the chemists heated the polymer to a molten mass and forced it through a spinnerette, the tiny, extruded filaments would break. The entire objective was to make long continuous filaments—miles long—that could be twisted to yarn. Additionally, although Bolton had the vision of the two monomers, adipic acid and hexamethylenediamine, coming smoothly from benzene, the diamine process did not exist. Not at all. But Bolton gave Carothers clear and unequivocal direction, and Carothers complied.

■ ■

Helen Everett Sweetman was a tall young woman with a shy smile. She initially worked at the Experimental Station the first summer Carothers was there; she was just 16. She carried mail to all the scientists, dropping it off in the laboratories. Helen liked the job. She got to meet everyone. Her father, Willard Sweetman, was the chief clerk at the station. This was the kind of service position that was underrated and underappreciated among the scientists. Willard Sweetman sent his daughter to the University of Delaware, where she majored in chemistry. She graduated in 1933 and returned to DuPont, to the station, where she worked in the patent section preparing patent applications.

Carothers' work dominated the patent section. DuPont filed more than 50 patent applications naming Wallace Carothers as inventor. The patent document differs substantially from the text, in the chemists hand, written in the laboratory notebook describing his work. The notebook is an unfolding of events, with no premonition or hint of what the final result may be. Sometimes in Carothers' group, the chemist would record what Carothers had suggested he do. Donald Coffman entered in his laboratory record Carothers' precise instructions on polymerization of the nonanoic ester—the first viable polyamide.[19]

The patent is a totalization and unravelling and transcribing of the invention record, in prescribed, formal notation. It begins with a specification of all that has gone before and shows how the present work advances knowledge. The invention must be new, it must be useful, and it must not be obvious to others "skilled in the art." The patent document lists a long series of experiments to show the most applicable embodiment of the invention. And it concludes with a list of carefully drawn statements called "claims." The claims are the heart of the patent. They define the precise matter of the invention, the specifics of the invention, the sum of knowledge that the inventor, or in Carothers' case, his assignee, DuPont, will now own for 17 years. During that period, no one else will be legally able to make for commercial purposes, use, or sell the described invention without DuPont's permission.

When Carothers was hired by DuPont, he was asked to sign a standard form, assigning all rights to anything he patented to the company. That practice had become conventional with the end of the days of the individual inventive genius in his own laboratory and the origination of the first, large industrial research laboratories at General Electric and Bell Telephone. The deal was that the large company would invest in its group of scientists, providing them the means for their work and steady employment. This group consideration would be payment for the unpredictable, inventive bursts the companies might expect from their individual scientists.

If one particular scientist was extraordinarily inventive, he would be financing the collective work of the group. To this inventor, the rest owed a great debt.

DuPont did not expect Carothers to prepare the patent document, with its peculiar style and format. The patent section—attorneys and chemists who specialized in deriving a patent application from chemists' notebooks—wrote the documents. But the papers went back and forth from patent section to chemist with revisions and additions. Few chemists enjoy wading through patent applications. They come to the scientists' desk in large manila folders. They sink to the bottom of the pile as journals and notes and books

---

[19]Hounshell, David; Smith, John Kenly. *Science & Corporate Strategy, DuPont R&D 1902–1980;* Cambridge University Press: Cambridge, England, 1988; p 662.

and ashtrays and staplers and coffee cups accumulate. But you can always see them. The edges of the legal-size folders protrude from the stack, always a reminder that the editing drudgery remains undone, a nagging talisman of guilt.

Young Helen Sweetman had the job of tracking the progress of Carothers' applications through the patent group. "He had many patent applications," Helen Sweetman Carothers said, "and couldn't keep track of them. Whether he had them or I had them. He was a very serious man."[20] The tall young woman was often at Carothers' side, pointing at the stack of folders, hoping he would move the ashtray off the top of the stack, perhaps asking him shyly to look at an application her bosses told her was particularly important.

■ ■

Who defines a man's mate, a woman's choice? Books and music and fine dining drove the demanding intellectual climate of the Lenhers and Hills and Berchets and Miles and Mapels—the young, socially wistful scientists and professionals who stood just outside the rigidly controlled access to Wilmington's elite. Polly Hill and Lib Miles and Ruth Berchet and Evelyn Mapel came to Wilmington with their husbands. They adopted the tone and style of the place, a good place, full of culture and entertainment and service. They joined the Junior League, the acting groups, the musical ensembles, the good life.

For the single men, the right girl and the right marriage were vital. Sam Lenher married well, married the daughter of a General Electric executive. Crawford Greenewalt married Margarethe duPont. He stepped across the invisible barrier and was folded into the duPont family, with all that transition implied. Greenewalt would become the president of DuPont.

Carothers, the famous scientist, this fine man whom they all knew and whom most loved, remained single. Sylvia Moore? "She *was* a girl," Gerard Berchet said. "Bright, intelligent, wow. My wife Ruth always said that if Wallace had married Sylvia he never would have killed himself."[21] "She was terrific, bright, sexy, good-looking." Greenewalt said. "The affair broke up because it threatened [her]

---

[20]Helen S. Carothers interview with Matthew E. Hermes, March 5, 1990.

[21]Gerard Berchet interview with Adeline C. Strange, 1978. (Hagley Museum and Library Collection, Wilmington, DE, 1985.)

marriage.... It was a potential scandal," Greenewalt said in 1978, forgetting perhaps that Sylvia's marriage was over by the middle of 1933.[22]

In 1934, Wallace Carothers' friends caught glimpses of the inventor and his youthful patent assistant, Helen Sweetman, riding along, top down, in his convertible, her auburn hair blowing behind as they rode. She was pretty, she was gentle, she was a fine tennis player.[23]

But Carothers' friends were not pleased. "She was numb," Polly Hill said. "Well that was the impression I got. Was that fair? She never had anything to say except pleasantries, she never had any objections to anything." "She wasn't a very spontaneous girl," Julian Hill said. "She was much younger, she was not intellectual in any way up to ours, you couldn't talk to her about anything," said Lib Miles. Helen was just chasing after the great scientist, Lib Miles believed.[24] Crawford Greenewalt said Helen was pleasant, with a simple background. She had a college education but was intellectually inadequate and had little social rapport.[25]

Helen Sweetman Carothers remembers herself modestly, as an average young woman with no outstanding abilities, as an avid but undistinguished tennis player and golfer.[26]

Carothers may have withdrawn from Sylvia Moore willingly because their liaison spoiled of its own burdens. Perhaps Carothers fled Sylvia under pressure from his Wilmington management. Perhaps she ended the relationship, pointing to Carothers' choice of his parents over her at the very moment she ended her marriage. Dr. Hohman apparently urged Carothers to marry. Sylvia Moore was once a potential wife for Wallace Carothers, but they did not marry.

But Sylvia left her mark on Carothers' friends. Each of his close friends had a stake in the inventor's health, in his mental condition. They truly depended on his leadership. From their chemistry hours had grown an appreciation of Wallace Carothers beyond simple respect. Berchet said he was a wonderful human being. These great friends labored with Carothers' depression, watching, predicting,

---

[22]Crawford Greenewalt interview with Adeline C. Strange, 1978. (Hagley Museum and Library Collection, Wilmington, DE, 1985.)

[23]Poly Hill interview with Matthew E. Hermes, May 29, 1990.

[24]John and Lib Miles interview with Adeline C. Strange, July 1978. (Hagley Museum and Library Collection, Wilmington, DE.)

[25]Crawford Greenewalt interview with Adeline C. Strange, 1978. (Hagley Museum and Library Collection, Wilmington, DE, 1985.)

[26]Helen Sweetman Carothers to Matthew E. Hermes, July 19, 1994.

measuring, judging the impact of each of the events of his life on his mental condition. Although they stood outside the relationship between Wallace and Sylvia, they established her as a standard for Carothers. If Wallace could not have Sylvia, he must have a woman like her.

■■

We live physical lives, consuming space on the planet: eating, sleeping, acting, working, making love, playing squash, bowing the violin. Some of this is very public. Some of this is obsessively private. We talk and think and read and argue and define and command the language and its clever nuances. We debate politics and sex and war and literature and taxes and government. We do most of our intellectual work in order to share it. Sometimes to impress or convince or just shed light. We are perhaps best known for our public, intellectual product.

But beyond our physical lives lies a deep spirit, an ephemeral basin in which we hold our fears and anxieties, our longings and trepidations, our faith and ultimately our true power. We are not merely intellectuals, grown in stature as our ideas mature, defined only by what we think and say. We are also men and women of action, physical action. Doers of good and evil. And we are deep spiritual beings, relying sometimes on a deep faith, despairing sometimes under oppressive fear.

Two things went wrong when Carothers' friends made judgments about Wallace and Sylvia and Wallace and Helen. First, they could not know truly of Wallace and Sylvia. The Hills and Berchets and Greenewalts were looking at the outside of the affair. Carothers did not detail his doings with Sylvia with Julian Hill or any of his friends as he had with Frances Spencer. Hill and Berchet and Greenewalt saw the exterior, the painted face, the facade, the structure as Carothers and Sylvia would manage it. Carothers' friends could not live the inner life, the hidden physical and spiritual days of the man or the woman, any more than Carothers could peer into the hearts of Polly Hill or Lib Miles or Crawford Greenewalt. All Carothers' friends knew he despaired regularly. What told them his desperation was relieved with Sylvia? He had been with her for more than two years. And he did not get better.

Carothers' friends could not know truly of Wallace and Helen. If fragile life is made of physical and mental and spiritual parts, must all of these three come into line like the earth and sun and eclipsing moon for there to be an attraction? Perhaps Helen Sweetman and Wallace Carothers drew together simply, under the gravity of tennis or physical love or spirit or song, apart from the rigorous scrutiny of the intellect, the intellect so satisfying to these DuPont scientists. Perhaps Carothers told his young companion very little about himself as they moved along a pathway of friendship and attraction, both of them wondering and quiet and puzzled and even fearful. Perhaps he told her everything and encumbered her with a long history of subjective trouble kneaded through the reality of his brilliant success.

And maybe it was Helen who needed Wallace and let him know it and gave Wallace a fleeting opportunity to respond with kindness. After all, it was thoughtfulness and a concern for others, despite his own problems, that Carothers' friends loved in the man. Had they forgotten, in their single-minded focus on their friend, that his tie with Helen would go two ways? Could they doubt he would give that same attention to this woman's needs as he would their own? And that she, too, might love him for it?

Carothers was rebuked in his relationship with Helen Sweetman, the woman he would marry, by the high expectation of his friends. These friends loved Carothers but thought they knew best for him. They put him in an impossible position. They scowled at his affair with Sylvia, but defined her as the right woman for him. They frowned at his licit relationship with Helen, certain she was an unsuitable choice.

■ ■

"Be glad you're neurotic," wrote New York psychiatrist Louis Bisch for the *Reader's Digest* in January 1935. "Jung says that all neurotics possess the elements of genius," he wrote.

> I do find that neurotics who are encouraged to understand and appreciate themselves have vested sources of happiness, peculiar and distinguished abilities and a driving force that carries them high above the average. The genius doesn't worry because he is different from his fellows: he capitalizes on his restless neurotic urge by writing a book, painting a picture or building a fortune.

The downside of this *Reader's Digest* condensed version of life was that "the worst neurotics (plagued by suicidal impulses, natural terror of bodily ailments, of closed places and the like)" had far less hope. "The unsuccessful neurotic...does nothing to compensate for his brooding discontent. He thinks himself a worm because he is not like everyone else."[27]

Our condensed version of Wallace Carothers' life shows he had great success in the fall of 1934 and the first half of 1935, developing the exciting polyamides that all now knew would be a truly revolutionary synthetic silk. His friends' fondest memories are of the nights sitting along the walls of his Wawaset apartment, with music and conversation and drink. Carothers began to see in Helen Sweetman the wife that Dr. Hohman had suggested for him.

Wallace Carothers disclosed little, but he had spells, highs and lows. He wrote Jack Johnson at the end of 1934:

> I have been more or less up and down since leaving you; apparently the general decomposition that entered a galloping stage this summer and was afterwards partly halted got a start again. However there has since been some improvement doubtless; I managed to survive a week end houseparty mostly among strangers at Ardmore from Friday through most of Sunday.[28]

Perhaps it was really all as simple as the self-admitted neurotic Dr. Bisch described:

> To get well, the mild neurotic must first be taught to respect himself. He must banish shame, guilt and self-accusation. He must be shown there are thousands of others like him, that he is not hopelessly out of kilter, that his fellow-neurotics have to their credit more of life's honors than those regarded as normal. He is in superb company with Alexander, Caesar, Napoleon, Michelangelo, Poe, Walt Whitman....

---

[27]Bisch, Louis E. "Be Glad You're Neurotic," *Reader's Digest* **1935**, *26(153)*, 13.

[28]Wallace Carothers to Jack Johnson, December 1934. (Hagley Museum and Library Collection, Wilmington, DE, 1842.)

Chapter 15

A Gathering of Seven Friends

If we would place on a small table Arnold Collins' spherical mass of polychloroprene, still swollen with water, still scented with the sharp odor of unreacted chloroprene, and Gerard Berchet's irregular knob of polyamide, with shards of glass still clinging to its surface, we would see in these simple objects the enormous, but complete, sum of Wallace Carothers' life work in science: the first synthetic rubber and the first synthetic fiber. For with the synthesis of polyamide 6-6 in early 1935, Wallace Carothers productive science ended.

Try to move forward, 30, 40, 50 years, and somehow after all these years, enshrine Collins' spongy ball of neoprene and Berchet's nugget of nylon in an elegant display in a case at the DuPont Experimental Station and spirit into the room a convention of Carothers' scientific friends. Ask Sam Lenher and Julian Hill and Gerard Berchet and Speed Marvel and Roger Adams and Jack Johnson and Paul Flory to come back and talk about the inventor and about themselves and about DuPont and what had happened over the decades since Carothers' chemistry ended. All but Adams were alive 45 years after Carothers' death. Marvel, Hill, Lenher, and Berchet lived for more than half a century after Carothers ended his life. They would have come, all of them.

Julian Hill and Sam Lenher and Gerard Berchet never worked for anyone else but DuPont. Lenher was a proud and precise and constant servant of the company for 48 years. He was fixed on success and rose quickly through DuPont's management after he was suddenly handed wartime responsibility for the huge manufacturing site at Carney's Point, New Jersey, which DuPont called the Chambers Works. Lenher served on DuPont's executive committee. In 1990, the silver haired, bent old man, with the clearest of blue eyes, limited by a stroke to but an hour of visiting, took painstaking care in composing his answers to my questions. He was cooperative with the facts he chose to give, with little regard for motive or analysis. He was not completely forthcoming. There were things I would not learn. His son, George Lenher, told me Dr. Lenher would never reveal why DuPont sent him to Hitler's Germany before the war to talk to the German chemical giants. George Lenher asked me to ask his father to describe his mission. Sam Lenher was so well prepared for a task in Germany— loyal, Berlin educated, intelligent. But Sam Lenher would say nothing to me of his assignment of more than 50 years past. Dr. Lenher sipped but half of his single, scheduled, noon-time Scotch as he told me haltingly, swallowing the feeling, how he had learned just days before Carothers killed himself, how desperately incapable Carothers' new psychiatrist felt he was of preventing Carothers' suicide. Dr. Sam Lenher died in 1992 at the age of 88.[1]

Julian Hill left Carothers' group before the polyamide work began in earnest. He had his own group of chemists supporting the rayon business. Hill, of all Carothers' friends, was a quiet and philosophical man. He asked good questions, took facts as he saw them develop in the laboratory and related them to nature. He was an expert at how a spider spins its web, he took great care in understanding the mechanism of how individual filaments of rayon were made. He became an assistant director of the chemical department and sat on its quasidemocratic governing board, the steering committee, through the end of World War II.

Hill became DuPont's secretary of the committee on educational aid. "It was a great job, because nobody really cared where I was or what I was doing," he said.[2] Hill traveled the country, visiting the academic scientists who longed for DuPont research support. DuPont's riches and Hill's solid connections and his calm and easy manner gave him access to whomever he wished. He remembered lunch with Conant, the president of Harvard, as developing "more ideas, good, bad and indifferent, in the course of a half hour than all the other people I spoke to in the whole time I spoke to them." As we will learn, Julian Hill stayed in contact with Carothers' parents for more than a score of years after Wallace's death.

Julian Hill stood tall when I visited him in 1990 at the same home on Martha's Vineyard that Wallace Carothers visited in 1934. He

---

[1]Samuel Lenher interview with Matthew E. Hermes, August 24, 1990.

[2]Julian Hill interview with David A. Hounshell and John K. Smith, December 1, 1982. (Hagley Museum and Library Collection, Wilmington, DE, 1878, p 61.)

leaned a bit against an elongated cane pole that came just short of the top of his head. He contracted polio in 1948 and has limped noticeably since then. I took his picture in front of the same maple tree, the same granite post where more than a half-century before, Carothers had stood for the camera. Julian said he was a bit older than Sam Lenher, a little closer to 90. He practiced violin daily, bowing his seventeeth century instrument. He had played in the Wilmington Symphony years ago. Now he played in a chamber group every week, joining his neighbors in the classic repertoire. Of his time with Carothers he wrote, "In any event it was a privileged period of my life that I treasure. I'm glad that Polly got into the picture soon enough to know and love him."[3] Julian Hill died January 28, 1996 at the age of 91.

Gerard Berchet was led out of France in 1926 by Carothers and Jack Johnson. He accepted a promise of entrance into the Ph.D. program at Colorado University. Berchet spoke no English; he met his wife, Ruth, who taught French. They came to Wilmington in 1929 where Berchet would work for Carothers. They were poor; they scraped by. Ruth lived in New York for a time after they married. She worked at Macy's. Berchet remembered his first two changes in pay were reductions of 10%.

Gerard Berchet was the sort of man who placed his work in a properly balanced perspective. He was proud of his involvement with nylon and neoprene and would send a photocopy of the critical page describing his synthesis of polyamide 6-6 whenever asked. But he was more of a civil servant type than an obsessive inventor. He served DuPont for years in the patent section, in the midst of a cascade of invention that flowed from the chemical department after the war.

Gerard Berchet preferred to be called Jerry. He was a voracious reader. His daughter remembers growing up in a house of music. Berchet, a handsome man with rich, curly hair, joined a small acting troupe in Wilmington—the Brandywiners—in 1932. The group staged musicals at P. S. duPont's outdoor theater amidst the sculpted gardens at Longwood. Berchet played in 47 of the Brandywiner productions over 50 years. He acted at the Robin Hood Theater in Arden; Ruth Berchet did makeup.[4] Berchet had a passionate love for his adopted homeland, "a country where the majority of men and women were obliging, kind-hearted, open-handed, generous to a fault, and didn't even seem to know it."[5] Although, in his late career,

---

[3]Julian Hill to Matthew E. Hermes, June 17, 1990.
[4]Anne Knepley interview with Matthew E. Hermes, July 11, 1994.
[5]Berchet, Gerard, "Pre-Kendal Memories," June 26, 1982. (Dennis Berchet.)

he could be an outspoken francophile, finding virtually everything in his native France better than things in the United States.[6] Once, in 1978, he remembered the day he learned of Carothers' death. He and Paul Flory were working together in the laboratory. "When the news came, we just sat there, we couldn't..." Berchet could not finish the sentence.[7] Dr. Gerard Berchet died in 1990 at the age of 87.

Paul Flory knew Carothers for only three years. He came to DuPont from Ohio State in 1934 and fled DuPont quickly after Carothers died. He went to the University of Cincinnati, then to three industrial companies endeavoring to develop new synthetic rubbers for the war effort. He was at Cornell, the Mellon Institute and, for the last quarter century of his life, at Stanford. Flory's own list of honors and tributes fills a page. In 1974, he was awarded, for his polymer work, the Priestley Medal of the American Chemical Society, the National Medal of Science, and the Nobel Prize in Chemistry. He wrote of his days at DuPont, "...it was my good fortune to be assigned to the small group headed by Dr. Wallace H. Carothers, inventor of nylon and neoprene, and a scientist of extraordinary breadth and originality."[8] Flory sat for the video camera in 1982. "I didn't know what a polymer was," he said in measured tone. Carothers convinced Flory that polymers could be studied scientifically. "Carothers died in 1937," Flory said, "and that was one of the most profoundly shocking events of my life. His sudden death pulled the rug out from under my hopes and aspirations and plans. That changed the situation completely. That was the cause for my leaving DuPont. He encouraged me. My inclinations were somewhat different from his. I was his physical chemist. It was an extraordinary opportunity."[9] Dr. Paul Flory died in 1985 at the age of 75.

Jack Johnson stayed at Cornell for the rest of his academic career, until 1965. DuPont contracted him as a consultant until he retired. He was elected to the National Academy of Sciences. Carothers and Marvel and Johnson were great friends. Johnson wrote Marvel on the fourth of July in 1974. He wrote of the weather and his vegetable garden in Vermont. And he added, "P.S. 50 years ago I was just return-

---

[6]Burton C. Anderson interview with Matthew E. Hermes, August 20, 1994.

[7]Gerard Berchet interview with Adeline C. Strange, 1978. (Hagley Museum and Library Collection, Wilmington, DE, 1985.)

[8]Flory, Paul J. *Statistical Mechanics of Chain Molecules;* Hanser: New York, 1989; p v.

[9]Flory, Paul J. "Reflections by an Eminent Chemist," The Eminent Chemist Video Tape Series; American Chemical Society: Washington, DC, 1982. This tape is available from American Chemical Society, Education Distribution Center, P.O. Box 2537, Kearneysville, WV 25430.

ing from France and getting set to start teaching at U. of I. You and Wallace and I were together there the following two years."[10] In 1978 he talked at length with Mrs. Adeline Strange of Wilmington, who went to Brattleboro, Vermont, to talk to him about Carothers. He remembered the suicide book Wallace had bought in France in 1926. Sometime later, after they had returned to the United States, Carothers gave him the book, Johnson remembered. Professor Johnson dug through the stacks of books and papers now in his home in Vermont and found the book. He and Mrs. Strange looked through it together. "Isn't this a charming book," Mrs. Strange said sardonically. Then there was a pause. "I wish I could have gotten down there to see him before it all happened," Jack Johnson said, his strong voice rising and breaking.[11]

Johnson withdrew to a farm in Townsend, Vermont, where he died after a lingering illness in 1983. Dr. Jack Johnson was 83 years old when he died.

Speed Marvel was truly a marvel. He taught the same collegiality as Adams and graduated scores of Ph.D.s with polymer focus. One was William Bailey. Bill went to Maryland and taught me polymer science. I am Marvel's scientific grandson. Marvel said Carothers taught him everything he knew about polymer chemistry and so I am Wallace Carothers' chemical great-grandson.

Marvel gave the United States a practical synthetic rubber, styrene–butadiene, for tires during World War II through his participation in the synthetic rubber program. Overseas natural rubber was unavailable; neoprene does not make good tires. Marvel retired from Illinois in 1961 as he was approaching its mandatory retirement age of 70. He moved to Arizona and its university in Tucson. And there he worked at chemistry nearly every day for a quarter of a century more. He consulted for DuPont for 60 years.

I remember him as a kind, round man. He visited me in my laboratory at DuPont once when I was a young man. I had several chemical questions for him. Consultants need time and involvement in order to help. Marvel's plate was full; my questions were superficial. We never met again.

---

[10]John R. Johnson to Carl S. Marvel, July 4, 1974. (Hagley Museum and Library Collection, Wilmington, DE.)

[11]John R. Johnson interview with Adeline C. Strange, July 1978. (Hagley Museum and Library Collection, Wilmington, DE, 1985.) A tape of this interview shows Mrs. Strange was a remarkably agile and talented interviewer. She acquired from Prof. Johnson a wonderfully full picture of Carothers and brought to public view, into the Hagley Collection, Wilmington, DE, a large body of documents Prof. Johnson had collected.

Marvel was an enthusiastic fisherman with a passion for muskie. He convinced Johnson and Carothers to wet a line with him from a canoe on Wisconsin's lakes. He could not swim, but his friends assumed, "in case of an upset, Speed would have floated because of his density and phlegmatic character."[12] In his later days, he became a world-class bird watcher. He had a life list of more than 650 species and could recognize most by their call. As he became very old he could not venture into the field, but stayed perched in a vehicle, glasses in hand.

Marvel said if he had been there the day Carothers killed himself, he could have saved him.[13] Dr. Carl S. Marvel died in 1988 at the age of 94.

Roger Adams gave more of himself to help Carothers in the last two years of his student's life than any of these men. Adams remained head of the Illinois chemistry department until 1954. He was active as research professor at Illinois for a time and had a couple of postdoctoral assistants until 1967. In 1963, his wife Lucile died and Roger Adams slumped a bit. He found life alone, "dreary." He died of cancer in 1971 after attending a meeting of directors of the Battelle Institute in Columbus, Ohio. Dr. Roger Adams was 82.[14]

■ ■

These were remarkable men, seven men born to science, schooled to the highest level, Midwestern in values if not birth. Seven old men living their lives, their friendships seeming to have a permanency, an unbreakable continuation. They were so alike, and so different from their friend Wallace Carothers. Each of the seven married in his twenties. For most, stable marriages and families went on to the end. Scarcely a man avoids depression and anxiety when pressed by problems—illness, death, financial insecurity. And most men find resources, spiritual strengths, to fend off the nighttime demons. They do not yield to the dragons of the mind more than a foothold. They do not prepare a place for the demon to stay. These

---

[12]Robert M. Joyce to Barbara Pralle, June 14, 1995.

[13]Burton C. Anderson interview with Matthew E. Hermes. Carl S. Marvel interview with David A. Hounshell and John K. Smith, May 2, 1983. (Hagley Museum and Library Collection, Wilmington, DE, 1878.) Carl S. Marvel interview with J. E. Mulvaney. (Chemical Heritage Foundation, Philadelphia, PA.)

[14]Tarbell, D. S.; Tarbell, A. T. *Roger Adams, Scientist and Statesman;* American Chemical Society: Washington, DC, 1981; p 176.

men shared a "healthy-mindedness" of which William James had written in *The Varieties of Religious Experience:*

> Systematic healthy-mindedness, conceiving good as the essential and universal aspect of being, deliberately excludes evil from its field of vision; and although when thus nakedly stated, this might seem a difficult feat to perform for one who is intellectually sincere with himself and honest about the facts, a little reflection shows that the situation is too complex to lie open to so simple a criticism.

James added in explanation, "To the man actively happy, from whatever cause, evil simply cannot then and there be believed in."

And when evil truly comes, the healthy man reconfigures evil with an attitude that forces him to fight what he can and accept what he cannot fight.[15]

These seven friends were different from Carothers. It came as no accident that they lived to an average of 85 years, whereas Carothers died at 41. These seven friends could fight the sinister in their lives. They battled it, turned it, and sometimes made peace with it. But part of the battle plan for the healthy minded is denial. They would gain comfort at times by ignoring the dark side. Some of these seven friends might be men whom William James called "once born." These are men for whom the days come in a natural flow, men whose values are clear and consistent. Their lives consist of plus and minus—but "Happiness and religious peace consist of living on the plus side of the account."[16]

These friends could not understand Wallace Carothers. At his best, Wallace was the quickest of them all, mastering their science in a way so complete and with such clarity of expression that the best of them—Flory and Marvel—thought of Carothers as their true mentor. Hill and Berchet knew they would never work for someone so generous in support and help, one so modest in claiming recognition. If it was music or literature or politics or even the racquet sports that Carothers chose to master, they knew he would master it.

But the moods, the periods of despair, the withdrawal they could see, which he never would discuss with them, were maddening; they made no sense. For Carothers' seven friends, talent and success simply outweighed whatever trials could arise. There was no reason for

---

[15]James, William. *The Varieties of Religious Experience;* Longmans Green and Company: New York, 1902; pp 88, 89.

[16]Ibid., p 167.

Carothers to be "down." The friends attempted the simple remedies that would have been effective for them. But nothing worked, and Wallace Carothers, still single, still heading his research group at the chemical department, moved day-by-day closer to his own chosen death. In the summer of 1935, neither Carothers nor his friends knew he would take the cyanide in less than two years.

Chapter 16

A Second, Superb Idea

In late 1934, Joe Labovsky, the Russian immigrant, whose tailor–father made the suits and coats worn by the duPonts, began working for Carothers in his laboratory in the old Purity Hall. The first of Charles Stine's research buildings was expanded now; Carothers had an office and adjacent laboratory. Labovsky carried out experimental work that Carothers wanted done, work he would not assign to the scientists in his group. Labovsky worked alone in the laboratory but within earshot of Carothers. It was his place to carry out the experiments, to speak only if he was spoken to, to be a patient assistant to DuPont's great scientist.

Joe Labovsky watched Carothers. He said Carothers was moody and seldom talked to anyone. Carothers had a cot in a room behind the office. When he worked late, he would sleep there. Labovsky was in the laboratory, working in a corner of the room one languid day in the summer of 1935. Carothers came into the laboratory. "Damn, it is hot here!" Labovsky remembered Carothers saying. Labovsky recalled this insignificant quote because Carothers was speaking to no one in particular, and Carothers stalked around the room like a "madman," while he and W. S. Carpenter III watched silently.[1]

■■

The Faraday Society of London planned to hold an international assembly of polymer scientists at Cambridge University in late September 1935. The event would be only the second of its kind; the first, titled The Colloid Aspects of Textile Materials and Related Topics and held at Cambridge in September 1932, focused on polymer

---

[1]Joseph Labovsky interview with Matthew E. Hermes, June 28, 1991.

chain flexibility.[2] By the 1932 meeting, most chemists believed polymers were long chains; they had dropped the idea of colloidal massing as the critical element of polymer molecular structure. This change of belief came in large manner from reading of the synthetic products of Carothers work. This new 1935 conference would now bring together scores of academic and industrial chemists from around the world to recognize and discuss the emerging field of high molecular weight polymers. The theoreticians and experimentalists from Germany would meet their counterparts from the United States, and they would hear for the first time in person the details of the proof of the molecular chain theory. The title for the symposium was "Phenomena of Polymerization and Condensation."

In this title lies obvious recognition of the stature of Wallace Carothers. He was known in England and on the continent by his publications in the *Journal of the American Chemical Society* and by the reports of the occasional European visitor to the United States. But clearly, Carothers' synthesis of high molecular weight polymers with properties of natural materials—synthetic fibers and rubbers—by using known and predictable chemistry gave final, synthetic proof of polymer structure. The Faraday Society targeted Carothers' body of science in the title of the forum. For Carothers was the inventor and the author of condensation polymerization—no other chemist could speak for this work. The title itself became a direct invitation for Carothers to move to a worldwide stage.

What achievement for Wallace Carothers in the summer of 1935! He had moved giant DuPont into still-secret, companywide mobilization for the development of a synthetic fiber, based on the lingering but direct path from his first polyesters in 1930 up to Berchet's tough and durable polyamide 6-6 of 1935. Elmer Bolton, the rayon fiber experts, the family duPont, all knew they had in their hands a remarkable synthetic. They knew the development road ahead would reveal pitfalls, but development was their trump card; they had the funds and the talent to push ahead—fast and first in the world.

Roger Adams was preparing Carothers' application for the National Academy of Sciences. Election could come in the spring of 1936.

And Carothers could go to Cambridge and stand in the long shadow of Isaac Newton and talk about seven years of understanding, seven unlikely years building the bricks and mortar of a new science.

---

[2]Morawetz, H. *Polymers: The Origins and Growth of a Science;* John Wiley and Sons: New York, 1985; p xv.

And he could do this while never mentioning his latest achievements in polyamides. His reputation was established solely on his basic science and fundamental understanding of the physical world.

Despite his reluctance to speak to any group, and the fear he knew he would experience in the weeks before the presentation, Carothers decided early he wanted to go to the Cambridge meeting. Elmer Bolton approved the trip. Carothers began to prepare an accompanying text to his oral presentation. The Faraday Society required a written paper, which would be published in its *Transactions* following the meeting. Carothers sent the paper to England in early July 1935 and began the laborious process of preparing the charts and lantern slides he would use in Europe.[3] Labovsky remembers taking much of the summer to prepare graphs and charts for Carothers.

■ ■

Perhaps Elmer Bolton and Wallace Carothers never truly walked the same path. Carothers came to DuPont for its promise of basic research. Carothers made that promise a reality and brought notoriety of the best kind to himself and DuPont. He met in grand fashion the justifying tenets of the original Stine proposal. And in the course of that work, in his seven years of active science, with its high moments and its damaging neurotic spells, Wallace Carothers brought forth rubber and fiber, neoprene and nylon, and after all, that was what Elmer Bolton really wanted.

But Elmer Bolton never deviated from a course that demanded new products for DuPont. At times, Carothers seemed to walk that same path. At one crucial moment, just past, he refused Robert Maynard Hutchins' invitation to go to the University of Chicago. But with the journey to Europe, Carothers demonstrated how far he had strayed from the course Bolton steered. For in that summer of 1935, the promise of a true synthetic silk had brought to Carothers' group Gerry Berchet and Don Coffman along with W. J. Merrill, Harry Dykstra, Paul Flory, W. R. Peterson, G. W. Rigby, E. W. Spanagel, F. C. Wagner, R. W. Maxwell, and V. R. Hardy, all assembled for an immediate frontal assault on the new material. But Wallace Carothers would leave them for more than a month and meet with Katz of Amsterdam and Meyer of Geneva and Staudinger of Freiburg and Mark of Vienna and Houwink of Eindhoven. They would talk the basic physical chemistry of polymers, not the development of useful new products.

[3]Carothers' paper was published in *Transactions of the Faraday Society* **1936**, *32*, 39. The text of questions and comments by participants in the oral presentation is given, along with Carothers' answers.

In truth, Carothers and Bolton had been walking two, diverting trails for years. At first, the measure of distance between them was small. Each day, as each trod his own road, the gap between them widened. As Carothers, wounded by the past, and battling the overwhelming demons that periodically disabled him, prepared for his trip to England, Bolton prepared a change. He directed that Dr. George Graves, another Adams Ph.D. from Illinois, who had graduated the year before Carothers, begin to help Carothers with the polyamide development. Carothers wrote later:

> It should be noted that Graves became associated with the direction of the polyamide work during the latter part of August, 1935, and he had full charge from September 17, 1935 to October 25, when the direction again became joint until early in November. Graves then took complete charge of the problem except for minor aspects of a rather theoretical or exploratory nature which were left with me in the hands of Berchet, Flory and Peterson.[4]

■ ■

Carothers stopped to say goodbye to Helen Sweetman in the library overlooking Brandywine River before leaving work early on September 17, 1935. He carefully packed his slides and charts and drove out the iron gate at the entrance to the Experimental Station, turned left onto the short bridge across the Creek, now languid here above the old mill dam in the warm stillness of late summer, and drove up Rising Sun Lane to Pennsylvania Avenue. Under the crisping leaves of the tall sycamores, Rising Sun was lined as ever with the stately properties of the duPonts. And it was lined with the nineteenth century company-built homes of the mill workers who had tended the dangerous grinding wheels driven by the Brandywine's water to mix carbon, sulfur, and nitrate into DuPont's black powder in the stone casements that could still be seen just below the research buildings of the station. Carothers left the Experimental Station as he had come, more than seven years ago.

Only with the left turn toward Wilmington was there a sense of being on a public road, and after a mile or so, Carothers turned right onto Greenhill then drove to his home in the Wawaset Park Apartments. He completed his packing for the trip. He carefully folded his

---

[4]Wallace Carothers to Arthur P. Tanberg, "Early History of Polyamide Fibers," February 19, 1936. (Hagley Museum and Library Collection, Wilmington, DE, 1784, Box 18.)

new tuxedo that John Miles had helped him purchase at Wilmington's Mansure and Prettyman's. Carothers wouldn't go to the clothiers alone. He was too shy and didn't know how what to buy. His friend Miles arranged for them to have a special fitting at the elegant men's store.[5] He summoned a cab for the short ride to the Pennsylvania Station in Wilmington and boarded the train alone for the two hours to New York.

Ocean service to Europe was near its peak in 1935, yet the schedule of fast and convenient passages was still limited. For the commercial traveler, North German Lloyd was the sensible choice because its nighttime departures allowed the businessman to work the day he left. However, a journey on German bottoms was becoming a political decision with the voice of Hitler and the National Socialists firmly entrenched in an ever more militant Germany.

But the Society page of the *New York Times* for Tuesday, September 17, 1935, lists Dr. Wallace Carothers, along with tennis players Bill Tilden and H. Ellsworth Vines and publisher Joseph Pulitzer, among a dozen or so prominent passengers embarking for Europe on North German Lloyd's *Bremen* at midnight that night.[6]

The liner pushed away with ritualistic fanfare at the first moment of a Wednesday morning. Searchlights illuminated the huge, block letters on the hull, naming the ship. The lights played on its new German flag, the Swastika. Just the day before, in Berlin, the Reichstag defiantly ordered the Swastika to be flown as the national emblem and specifically ordered its display on the German commercial fleet.[7] Tugboats edged the luxury liner away from its pier. The ship's orchestra played Wagner.

The *Bremen* and its sister ship, *Europa*, began four-day Atlantic service—*Die Ozean Express* it was called in 1929. On its maiden voyage the *Bremen* wrested the "Blue Riband," the ornate trophy given for the fastest crossing of the Atlantic, from the *Mauretania*. Now the liners, capable of carrying 2200 passengers, alternated midnight departures each week from West 46th Street bound for Cherbourg, Southampton, and their home port of Bremen.

At work in Wilmington, Carothers was often a study in inaction, save the smoke from his cigarette rising from his hands as he thought or wrote in his notebooks. The *Bremen*'s smoking room, shaped like a

---

[5]John and Lib Miles interview with Adeline C. Strange, July 1978. (Hagley Museum and Library Collection, Wilmington, DE.)

[6]*The New York Times*, September 17, 1935, p 21.

[7]Ibid., September 16, p 1.

rotunda, with wood panelling and unusual burl veneer on pillared fascia, was one of the more comfortable areas on the otherwise office-like interior of the *Bremen*. Carothers prepared for his talk there and read from the large selection of German classics from the ship's library, which supplanted the reading he had brought with him.

But Carothers certainly took advantage of the special invitation afforded first-class passengers to dine in the evening in *Bremen*'s Sun Deck restaurant. The room, with its shipboard Romanesque windows, elaborate sconces, and pastoral oils, resembled the familiar elegance of the Green Room of DuPont's dowager inn, the Hotel duPont, whose hallways ran through the middle of DuPont's headquarters building in Wilmington. We can anticipate Carothers was quiet and uncomfortable at first in this setting, eating at the small tables with passengers whose names, nationality, and occupation he knew not. But perhaps with wine came moments of ease. Carothers could use the German native tongue to quote snatches of Nietzsche or sing with them the ballads of their homeland. Carothers spent four days travel-ing alone in the institutional-like setting of the *Bremen*, with its utilitar-ian pool, its shooting gallery below, and the modernistic dance hall whose central fountain changed colors with the mood of the music.

The *Bremen* was at sea until the weekend and docked, on the morning of the fifth day, first at Cherbourg at 7 a.m. Sunday morning for passengers for France and the Continent, then at 11 a.m. at Southampton where the London-Northeastern Railroad boat-train stood ready to deliver London-bound passengers into the city by the middle of the afternoon.

The train reached Paddington Station in London on Sunday afternoon. Some time that day, or perhaps the next day, Carothers went to King's-Cross, the terminal of the Great Western, for the 50-mile ride to Cambridge.[8] He had plenty of time to reach the uni-versity town before the start of the meeting. The Faraday Society would not gather until Thursday, September 26. Carothers took advantage of the time to explore the ancient university.

Carothers toured Cambridge with Professors Mills of Jesus Col-lege and Sedgwick of Oxford. They went out from the new chemistry building, which was built in 1887, down Trinity Lane and through the main quadrangle of the sixteenth century college, past the ornate, authentically Romanesque 1602 fountain at its center, then out to the River Cam where the three scientists paused in the shade of the wil-lows at the corner of the Wren Library and watched Cambridge stu-

---

[8]John Tovey of Westport, Connecticut, outlined Carothers' likely onshore travel for me.

dents float slowly by in small canoes. The library reaches out to the bank of the Cam here, almost as if the scholastic building was here first and the river was fit around it.[9]

Chemists love to talk shop. Every year, and to this day, thousands of chemists travel in the summer to New England to descend for a week on any of a dozen of the lesser known of New Hampshire's small boarding schools. They gather, 100 in a group, to attend Gordon Research Conferences. The scientists endure the spartan lodgings, two to a room in student beds with arcuate mattresses resting on tired springs; they share communal facilities in nineteenth century dormitories and attend long conferences, morning and night, broken in mid-afternoon with leisure hours. Most of them stay right there on the campus and talk and talk. A small number will hike up the mountain trails on free afternoons, some will nap, and a few will play tennis, but chemists glow in the excitement of visualizing and sharing what they really cannot see—the processes by which the smallest of particles—electrons, atoms, and molecules—maneuver to give us our nature and its animation.

So Carothers no doubt revealed his thinking to Sedgwick and Mills, who were anxious to get a preview of the American's presentation. In 1928 Herman Mark and K. H. Meyer, two of their contemporaries working for the German chemical trust, I. G. Farben industrie in Ludwigshafen, reported X-rays of cellulose and rubber, and by using mathematical pictures of what they saw on their films they were able to draw how the atoms in the two substances were arranged in space. This was an outstanding achievement for the two young Germans, but it answered only part of the problem of the structure of the materials. It was as if Mark had held a hand lens to examine the weave of a suit of clothes. The magnified image could identify a twill or a worsted, but that examination could in no way tell them the size of the suit.

At that time, many scientists preferred to believe the arrangements Mark and Meyer had discovered—the crystal structures they had observed—completely described, and in fact proved, the dimensions of cellulose or rubber molecules. These chemists then believed the molecules were no larger than the small crystals Mark and Meyer had X-rayed. There was plenty of evidence from others that natural polymer molecules were much larger than Mark's crystals, but many

---

[9]Wallace Carothers to Mary Carothers, April 15, 1937. In this letter, written two weeks before Carothers killed himself, he describes briefly to his mother his 1935 tour of Trinity.

chemists took the position that no molecule could be larger than the crystals Mark observed. Therefore, if molecules *seemed* to be larger, they must be massing together in aggregates of crystal-sized particles.

Carothers told Sedgwick and Mills, as they sat for a time under the trees at the bank of the River Cam, that from 1928 he was in basic agreement with another German scientist, Hermann Staudinger of Freiburg. Staudinger believed the entire mystery of the arrangement of the atoms in the natural materials could be understood by thinking in terms of long chains of carbon atoms, perhaps interspersed with other atoms such as oxygen or nitrogen. Staudinger feuded openly with Mark and Meyer and held that the X-ray pictures of crystals Mark had obtained represented only small portions of very large chains of atoms. They had discovered the weave, Staudinger believed, but did not know the size of the suit.

Carothers had chosen for his research, he explained to the Englishmen, the designed synthesis—the actual making—of materials similar to the natural substances. He believed he could employ careful chemical strategy to build up chains of atoms from small fragments and that he could make chains long enough to begin to behave like the natural materials. If he achieved that goal, he would be able to describe exactly what he had done, and he would now know the essential characteristics required in these structures of nature. He made the new polymers by a procedure he invented called condensation, and as he got better at it these polymers began to be useful, just like rubber and cotton and silk. They were formable into articles such as fibers that could be envisioned as cloth, rope, garments, tents, and stockings. Carothers likely did not disclose that his work was rushing forward within the DuPont Company now to make a fiber that seemed to be commercially important.

But he was here, in Cambridge in the late of an English summer, to describe to the great German chemists he had never before met how these products of polymerization and condensation came to possess the mechanical properties of strength and toughness that made them useful. He had made these products by design and was about to describe that work and the precise mathematical basis for its success.

By the time of the Faraday Society meeting, Mark and Meyer were no longer at I. G. Farben, the place of their fruitful collaboration on the crystal structure of cellulose and rubber. Mark was now professor at Vienna and Meyer professor in Geneva. Mark had one Jewish parent and was an Austrian citizen. His managers at I. G. Farben warned him to leave Germany, to accept an academic position elsewhere, promising I. G. Farben financial support. But they made it clear Mark could no longer advance in Ludwigshafen. Mark left Germany for Vienna.[10] In 1935, Herman Staudinger was still at Freiburg but his attendance at Cambridge was encumbered with conditions by the German government. Mrs. Staudinger must stay in Germany while Staudinger was in England. And Staudinger must not comment in any negative way about his government's policies. This was the penalty for past sins: He had spoken out years before against the use of poison gas warfare in the Great War.

Most of the nearly 200 scientists who came to Cambridge stayed at Pembroke College, founded in 1347. However, the college rooms were taxed beyond their capacity, and the overflow was shunted to

---

[10]This discussion, along with much of the technical interpretation of early polymer science, is taken from *Polymers: The Origins and Growth of a Science* by Herbert Morawetz of Polytechnic of New York in Brooklyn (John Wiley and Sons, 1985).

In 1938 Herman Mark was arrested in Vienna, Austria, bribed his way to North America, and began his half-century career in Canada and the United States.

In the fall of 1988, I traveled to Greenville, South Carolina, to the kind of cold and cheerless convention center used for the trade and exhibition shows of the less glamorous of our American industries to hear a short symposium on the future of textile fibers. I went because the after-lunch speaker would be Herman Mark.

I am not given to fancy meals and was left with sufficient time to arrive first to the meeting room—an assembled room within the convention center with moveable walls, concrete floor, and unpadded metal chairs. I sat for a few moments alone, reading and thinking, and only looked up as I heard a soft shuffle.

Professor Mark, 93 years old, short, bent, but well dressed with a dark blue suit and steel-rimmed glasses, moved purposefully with sliding steps toward the front row of chairs and, without seeing me, sat down carefully facing the podium. He paused for a moment and then allowed himself to bend forward a bit and to nap. I remained seated toward the back of the room, alone with him and quiet. Perhaps 15 minutes passed; he raised his head and stood and took his notes to another chair next to the overhead projector and sat now facing the empty room. I introduced myself as an acquaintance of one of his colleagues at Polytechnic of New York, and we chatted briefly.

He gave his talk—an eclectic summary of fiber science from before the days of Carothers, appended with vital predictions on new areas of research he felt would be productive for the future. This was Herman Mark, born two years before Carothers and speaking more than 60 years after his own fundamental studies done as research director at I. G. Farben. Now, and in his tenth decade, Mark was still looking to the future. Dr. Herman Mark died in 1992 at the age of 96.

the Bull Hotel. But all the scientists ate together in the refectory of the college. The presentations would be held at the department of zoology; tea would be served there morning and evening.

At the opening session of the Faraday meeting, held in the Zoology Hall, the vice-chancellor and master of Caius and Gonville Colleges, J. F. Cameron, welcomed the attendees. He asked the 34 foreign visitors to stand, each in turn, as he read their names. Only four other scientists had traveled from the United States: three industrial men from Union Carbide and Carbon and Standard Oil of New Jersey, and Thiokol and Dr. H. B. Weiser of Rice University. Perhaps J. R. Katz, the tall, blond former psychoanalyst now living in Amsterdam, whose appearance surprised Carothers in Wilmington, was the only person Wallace Carothers knew.

More than two dozen papers were given over three long days at the meeting. The society assumed the participants would have read the written papers beforehand. Each participant was to give a short summary, and then the floor was opened for long discussions, which would be recorded by scribes and published in the *Transactions* with the author's paper. These discussions were the heart of the Faraday Society format, providing an open forum for developing new scientific thought.[11] Carothers spoke on the first morning, on Thursday. He took the time to detail the nomenclature of condensation polymerization. He digressed to consider polymers in living molecules and then launched into a description of methods to calculate the length of polymer chains. The work was clear, convincing, but this last topic was not his. The calculation work was the product of his associate, Paul Flory, the young Ohio State scientist now doing theoretical science. Perhaps Carothers presented Flory's work to align his talk closely to the objectives of the Faraday Society. The society was not a home for study of organic chemistry but existed "to promote the study of electrochemistry, electrometallurgy, physical chemistry...."[12] It was this physical understanding of polymer structure which brought the synthetic chemist, Carothers, to Cambridge.

Professors Mark, Staudinger, Meyer, and Katz spoke. All now aligned themselves publicly with long-chain structures for polymers. But each struggled to place himself, in this decade of history of the science, along an enlightened pathway of understanding of the issue.

---

[11]Whiffen, David H. *The Royal Society of Chemistry: The First Hundred Years;* The Royal Society of London: London, 1991; p 36.

[12]These words are found on the title page of the *Transactions of the Faraday Society* **1936**, *32*.

The summaries of the responses in the *Transactions of the Faraday Society* show that Mark, Meyer, and Staudinger parried regularly after the talks. The *Transactions* record that Mark and Staudinger each made long comments after seven of the presentations. Meyer seemed to be on the stage all the time, challenging the speakers and tilting with Staudinger. Meyer made observations on 11 of the talks.

Wallace Carothers made but a single comment. After Katz presented extensive X-ray analysis in the second talk after his own, Carothers suggested the aliphatic polyesters he had made five years before would be appropriate objects for study. He did not analyze or propose or argue. If Carothers was there for all the presentations, he sat quietly, not mixing into the public debate. The participants organized themselves in long rows for a photograph during the meeting. Several years ago, Speed Marvel looked at the photograph carefully. He could not find Carothers among the 162 faces.[13]

A long series of scientific papers in a darkened room and the lively discussion that follows can be a tonic for the passionately engaged. But a tired man, hung over perhaps, fighting a blinding headache from the passage of a night's alcohol, twisting off to doze in the dim hall, confused by his need to be there and the inability to remain alert, hates to have to leave the room but exits with the appearance of intention, with dignity, summoning his last heave of energy to look busy, purposeful. But he is fleeing. Did Wallace Carothers, winding away from his science at the precise moment his science elevated him, disappear from the contact of his peers as they debated the very science he had brought them?

On Friday night at Cambridge at the formal guest dinner, Professor Donnan, Sam Lenher's old mentor at University College, proposed toasts for the overseas guests. Staudinger, Meyer, Katz, Mark, and Wallace Carothers, dressed in his exquisite tuxedo, responded.[14]

Do we see Wallace Carothers at Cambridge, surrounded, guarded, protected by the ancient tradition of knowledge and science seeping forth from the wood and stain, the plaster and glass of the old rooms? Do we see him standing, glass held high, in the halo of applause from his world's chemical peers? Can we see Carothers walk alone, retracing his steps across Trinity College, imagining nights long past, with the windows of the buildings showing only the faintest light as the most learned

---

[13]Carl S. Marvel to Robin Penfold, October 11, 1977. (Chemical Heritage Foundation, Philadelphia, PA.) I have examined this photograph and I, too, can not pick out Carothers.

[14]*Transactions of the Faraday Society* **1936**, *32*. The papers, and discussions afterward, appear. A brief summary of the meeting appears on pp 1 and 2.

of the scientists of the seventeenth century bent over their books. Trinity, after all, was the college of Isaac Newton.

How can Wallace Carothers be so unwell? He will leave Cambridge. He will go to Paris alone. "He went back to the old haunts in Paris and he didn't enjoy himself at all. Definitely depressed there. He was alone and he didn't have a good time—he told us this."[15] He returned home, sailing another 100 hours across the North Atlantic. He is alone as before, now less than 500 days until he feels the fearful pain of the cyanide.

■ ■

As the second of three topics of his September 1935 talk, Wallace Carothers made one final polymer proposition. Carothers, his friends maintain, believed he had conceived but one worthwhile idea. Carothers believed that all else, all his other work, flowed in triviality from the concept of synthesis as proof of polymer structure.

His last scientific conception is startling in its clarity of focus. Carothers predicted the basis for our catalyst and our biotechnology revolution. He wondered about the natural synthesis of proteins. Why did nature's amino acid monomers form these polymers called peptides? In the laboratory, these monomers would smoothly cyclize, two at a time, to form simple compounds called diketopiperazines. But organisms, in contrast, made polymers that served as structure and function.

> How can this be? Temperature and moderate dilution have small control over such matters, and perhaps then in the wrong direction. But if reaction is preceded by adsorption at an interface, as it might be biologically, the molecule is no longer free to assume its spatially probable configuration. Its head, tail and middle are fastened to a surface and the only terminal approaches possible may be intermolecular. This picture is not proposed as a solution of the mechanism of protein synthesis: it is introduced rather as an interestingly conceivable possibility...and in any event, the surfaces on bifunctional reactions present an almost completely unexplored field.[16]

This picture was not a solution. But just as Carothers' one great idea, his first glimpse of xABy monomers condensing to form long

---

[15]John and Lib Miles interview with Adeline C. Strange, July 1978. (Hagley Museum and Library Collection, Wilmington, DE, 1985.)

[16]Carothers, W. H. *Transactions of the Faraday Society* **1936**, *32*, 43.

chains, led through an agony of chance and change to the development of nylon, Carothers simple realization of shape and surface as a template driving force in chemistry could have led him to leadership in the great advances in catalysts and molecular biology. For we now know it is shape that drives living systems. The greatest catalysts, the enzymes, work because they fit and mold and propel chemical reactions. The proteins are made, not by the fitting of the amino acids themselves, but by the bonded match of DNA strands and RNA containing the amino acids in proper orientation. If the chemicals in a living system do not fit, they do not perform. In a sense, we no longer read journals. As the century turns, we look at the pictures, the shapes of enzymes and nucleic acids and saccharides, interpreting their chemistry by their ability to bond and fit in the infinite complexity of surface topology. In 1935, Carothers, like few other men, had a second superb idea. Give Carothers the lifespan of his peers, make him once-born with an affirmed faith, give him the hope that he would have another great thought, give him the resources of DuPont, and at the age of 55, he might well have been working in parallel with Linus Pauling when, in 1951, Pauling correctly postulated the helical shapes of proteins, or the next year, at the age of 56, with James Watson and Francis Crick, when the two young men first made their double helix, the model of nature's nucleic acids, which, in an instant, told the first chapter of nature's story of replication.[17, 18]

[17]Pauling, L.; Corey, R. B.; Branson, H. R. *Proceedings of the National Academy of Sciences* **1951**, *37*, 205.

[18]Watson, J. D.; Crick, F. H. C. *Nature (London)* **1953**, *171*, 737.

Chapter 17

Carothers Marries

After the Faraday Society gathering Carothers remained in Europe. He went to Paris, alone.[1] He went to the Black Forest to walk and walk; he went to figure out himself and his life.[2] His friends in Wilmington, his co-workers, had no idea when he would come back.[3] Wallace Carothers returned from Europe in late October to rejoin his group at the Experimental Station.

The polyamide work moved ahead quickly under George Graves during Carothers' absence. Bolton left Carothers in an awkward position for a time, with only a shared responsibility for the new fiber. He then removed Carothers altogether from responsibility for the synthetic polyamide. At the end of 1935, Wallace Carothers retained a skeleton crew of researchers: Flory, who was on his own, beginning his physical chemistry of the polymers, which would lead to the Nobel Prize in Chemistry; Berchet, who would do as directed to find yet another, superior candidate fiber; and Peterson, who continued work to find better ways of molecular weight control.[4]

---

[1]John and Lib Miles interview with Adeline C. Strange. (Hagley Museum and Library Collection, Wilmington, DE, 1985.)

[2]Adeline C. Strange taped recollections of her 1978 interview with Bill Mapel. Mapel referred to Carothers in the Black Forest before his marriage, according to the tape. That puts the walk in 1935, not as part of Carothers'1936 European trip. Also, in the summer of 1936, Mapel had left Wilmington.

[3]Julian Hill to Arthur B. Lamb, October 9, 1935. (Hagley Museum and Library Collection, Wilmington, DE, 1784.)

[4]Wallace Carothers to Arthur P. Tanberg, "Early History of Polyamide Fibers," February 19, 1936. (Hagley Museum and Library Collection, Wilmington, DE, 1784.)

From outside of DuPont, Carothers could be seen as a still rising star. He had dozens of publications and the credit for DuPrene and the early condensation polymers. His reputation, passed around among the academics, had brought him the Chicago invitation. Now there was the petition circulating among his peers to elect him to the National Academy of Sciences. And within DuPont, dozens of chemists and engineers now mobilized to make polyamide 6-6 a commercial product. They were grateful to Wallace Carothers, the inventor, whose skill was the source from which their very jobs had come.

Carothers was not yet 40. But Elmer Bolton recognized that the man who stepped off the boat from Europe could not stay with the pace and spirit of the fiber development. DuPont's Carothers and DuPont's fiber work went in disparate directions; one dazed and drifting, the other demanding and driving. Someone, Bolton or Tanberg, asked Carothers to write a chronicle of the polyamide work. He dug back to the records and on February 19, 1936, published a long, clear, and modest summary of his contributions. Carothers wrote his *Early History of Polyamide Fibers* with a backward reflection. He started with Harvard in 1927 and brought his "ideas first grasped as possibilities, which by slow growth became firm convictions" up through the beginning of the aggressive development of polyamide 6-6. Carothers' memorandum was written as history. It begins, "Factors that led to the development of polyamide fibers were..." The document shows no present and plans no future. "It will be observed," Carothers wrote in his final paragraph, "that this report does not attempt any critical review of work subsequent to the preliminary successful polyamides of Coffman and Peterson."[5] Wallace Carothers' career in research on high polymers was over.

■ ■

Two days later, on Friday, February 21, 1936, Wallace Carothers married Helen Sweetman in New York City. The East and New York were gripped by bitter, record-breaking chill that last week of February. On the evening before his wedding, Wallace Carothers hosted a dinner in his honor at Wilmington's University Club. At a small table in the comfortable old club, Carothers and Bill Mapel and Julian Hill, along with Sam Lenher and Hans Svanoe from Whiskey Acres and the physicist John Miles and Mapel's associate and junior editor Martin Klaver,

[5]Ibid.

ate and drank in quiet celebration. A few blocks away that night, Mrs. Sam Lenher held a small bridal shower for Helen Sweetman.[6]

On the morning of February 21, Julian Hill, Bill and Evelyn Mapel, Wallace Carothers, and Helen Sweetman, bundled against the cold, gathered in the steamy waiting room of the Pennsylvania Station in Wilmington for the train to New York.[7] With them were Helen's parents, Mr. and Mrs. Willard Sweetman. As the train approached, Julian Hill and the Mapels went outside. Helen and Wallace lingered a moment and said goodbye to the Sweetmans. As Julian Hill and the two young couples boarded the train for the trip to New York, Bill Mapel's photographer took a photograph of the bride and groom-to-be. Helen looked somber in a cloth coat with a huge fur collar and a dark beret. Carothers, in a double-breasted coat and light-colored, misshapen fedora, gave the photographer a truly bright smile. Mr. and Mrs. Sweetman stood beside the coaches, waving as the electric locomotive and its string of cars pulled silently away.

That afternoon, Hill, the Mapels, Wallace, and Helen went to Mrs. Mapel's sister's apartment in Tudor City in Manhattan on East 41st Street. There, the Rev. Dr. Phillip C. Jones, associate pastor of the Madison Avenue Presbyterian Church, married Helen and Wallace in the presence of the Mapels and Mr. and Mrs. Sam Browne, Evelyn's sister and brother-in-law. Julian Hill was Wallace's best man. Evelyn was matron of honor. "The bride wore an ensemble in brown," noted the Wilmington paper.[8] Bill Mapel said the group then went to the Plaza Hotel and left the newlyweds for a time. They picked Wallace and Helen up at 7:30 p.m. and "took them to a large night-club for dinner and an elaborate floor show. Very posh. End of wedding celebration."[9] The Carothers' honeymooned at the Savoy-Plaza for a week and returned to Wilmington.[10]

---

[6] *Wilmington Journal-Every Evening*, February 21, 1936.

[7] Julian and Polly Hill believe Polly did not go to New York for the wedding. In 1990, Polly said, "I don't remember, 'cause I just wasn't there." However, Bill Mapel listed both Julian and Polly as attending the wedding in the *Wilmington Journal-Every Evening*, February 22, 1936. This wedding announcement also indicates Mrs. Sam Lenher made the trip. *The New York Times* of February 22, 1936, p 18, records the scene slightly differently. It lists only Julian, the Mapels, and the Brownes as guests.

[8] *Wilmington Journal-Every Evening*, February 22, 1936.

[9] William C. Mapel interview with Adeline C. Strange, July 1978. (Hagley Museum and Library Collection, Wilmington, DE.)

[10] *The New York Times*, February 22, 1936, p 18.

Wallace Carothers and Helen Sweetman board train. (Reproduced with permission from Wilmington Journal-Every Evening, *Feb. 21, 1936.)*

"You wanted to know why the wedding was in New York?" Julian Hill said during my 1990 interview with Julian and Polly. "Well, Bill Mapel's wife had a sister whose name I've forgotten and her husband, Sam Browne, and for some reason (as I say this whole thing was very strange) they decided to have the wedding at Sam Browne's in New York. Bill Mapel took charge of the wedding."

Polly Hill said to Julian, "He insisted you be the best man. I do remember you said, 'Oh, you've done all this. Why should I be it?' No, he insisted [you] be the best man because he did have that sense of fitness."

"I don't think Wallace was consulted," said Julian Hill.

"It was out of his line to get married," said Polly Hill.

"As we said," Julian continued, "Bill Mapel was this charismatic, overmobilized character, and you either did what he said, or you didn't have anything to do with him. I don't think there was any middle ground, was there Polly?"[11]

■■

Bill Mapel said he loved Wallace Carothers like a brother. He remembered that Wallace was desperately lonely. Mapel said that Carothers told him that when he was in Europe, after the Faraday Society meeting, he traveled to Germany and walked the Black Forest for days, alone. He was trying to figure himself out, Mapel remembered.

When Wallace was planning to marry Helen, Mapel was unhappy with his choice. Helen was a "nonentity." Mapel said there was a general feeling Carothers could have married anybody. "Everyone said, 'Who is Helen Sweetman.'"

Mapel said Carothers was "very shy" about going to a jewelers for a wedding band. Mapel had the jeweler come out to his house with a tray of rings. Carothers made a choice and borrowed the ring sizer so he could get Helen's size.

Bill Mapel confirmed his wife arranged the wedding in New York.[12] A few days after the wedding, Jack Johnson received an announcement:[13]

> Mr. and Mrs. Willard Sweetman
> announce
> the marriage of their daughter
> Helen Everett
> to
> Mr. Wallace Hume Carothers
> on Friday, February twenty-first
> nineteen hundred and thirty-six
> New York City

---

[11]Julian and Polly Hill interview with Matthew E. Hermes, May 29, 1990.

[12]William C. Mapel interview with Adeline C. Strange, July 1978. (Hagley Museum and Library Collection, Wilmington, DE.)

[13]John R. Johnson file. (Hagley Museum and Library Collection, Wilmington, DE, 1842.)

and a small card:

At Home
after March first
Wawaset Apartments
Wilmington

■ ■

It was hardly unusual for a depression marriage to take place quietly, with just the participants and a few guests. It was such a spare time, with a developed habit of thrift, with no expectation of family financial contribution to the wedding ceremony. Carothers' sister, Elizabeth, married Robert Kyle, quietly, in New York in 1931. The parents, the Carothers and the Kyles, did not come east from Iowa and Ohio. The young painter, Marcus Rothkowitz, who would become expressionist Mark Rothko, married Edith Sachar at his apartment on 75th Street in New York in 1932. A rabbi and two witnesses were present.[14] My own parents married in the anteroom of a small Brooklyn church in 1934. Only the attendants and the officiating priest were there.

But the whirlwind trip to New York, the wedding in the Eastside apartment of strangers, all under the orchestration of Bill Mapel, reflects a powerlessness that both Wallace and Helen Carothers must have felt. In the grip of the Svengali-like Mapel, Carothers was compelled to the New York ceremony because he lacked the energy and vigor to resist. And Helen, who stands essentially silent to this day on the matter of her marriage, was not about to interrupt the sweep of events triggered by the powerful editor. It made no sense for Helen and Wallace, two single adults who never before had married, to marry in New York, but there it is.

The marriage ceremony makes clear distinction that two steps are necessary to conclude the union. "Are you willing...?" it asks. If an affirmative is received, it asks again, "Do you take...?" If three frogs sit on a log and one decides to jump, goes the test, how many remain? Answer: three. Willingness and action are separate and distinct. Beaudelaire wrote his mother six letters within 24 hours, threatening to punch Ancelle. His threats trailed off to a whimper. He could not act. Baudelaire "was an 'aboulic' who was incapable of settling down to regular work. His poetry is full of 'invitations to travel'; he clamored for escape from his surroundings, dreamed of undiscovered countries, but he hes-

---

[14]Breslin, James E. B. *Mark Rothko, A Biography;* University of Chicago Press: Chicago, IL, 1993; p 81.

itated for six months before making up his mind to got to Honfleur; and his one and only voyage seems to have been a long torment."[15]

Wallace Carothers could not act on his own desire to marry. If he and Helen wished to wed, they had first to ford the barrier of his own abulia. The desire and the action were separate and distinct. In chemical terms, Bill Mapel served as a powerful catalyst, lowering the activation energy required for the process. Wallace Carothers and Helen Sweetman married under the directive influence of the "overmobilized" Bill Mapel.

Many of Carothers friends report that after the wedding, they had very little to do with the newly married couple. The invitation implicit in the small card accompanying the wedding announcement went unanswered. A dislike for Helen as Carothers' new mate crystallized smoothly to action. Carothers' friends ignored the newlyweds. The Friday night gatherings for drink and song at Wawaset Park ended. Julian Hill said, "We didn't see him as much...after he was married."[16] Even though his wife hosted a shower for Helen Carothers, Sam Lenher said, "Neither my wife nor I were interested in his wife. I didn't think his choice of a wife was suitable."[17] Lib Miles said, "It was the most tragic thing; he never should have married her."[18] Gerard Berchet said, "Wallace was sick when he married Helen. I felt sorry for her. He was not happy."[19]

Dr. Luther Arnold, who came to DuPont on the recommendation of Sam Lenher in 1933 and knew Carothers through Lenher, and who now lived at Whiskey Acres, analyzed the situation. "That was an 'other side of the tracks' thing," he said.[20]

■ ■

Bill Mapel, the fresh wind blowing through the stodgy *News-Journal* papers of Wilmington, exhausted himself in less than two years.[21] But

[15]Sartre, Jean-Paul. *Baudelaire;* Turnell, M., Trans.; New Directions: New York, 1950; pp 15, 34.

[16]Julian Hill interview with Matthew E. Hermes, May 29, 1990.

[17]Samuel Lenher interview with Matthew E. Hermes, August 24, 1990.

[18]John and Lib Miles interview with Adeline C. Strange, July 1978. (Hagley Museum and Library Collection, Wilmington, DE, 1985.)

[19]Gerard Berchet interview with Adeline C. Strange, 1978. (Hagley Museum and Library Collection, Wilmington, DE, 1985.)

[20]Luther Arnold interview with Matthew E. Hermes, April 19, 1991.

[21]Reese, Charles L., Jr. "Christiana Farewell: A Newspaperman's Memoirs of the New-Journal Company, 1927–1977," *Delaware History* **1988**, *23*, 72.

while he was there, he promoted his friend Carothers as best he could. Mapel knew Roger Adams because over the previous two years Adams had joined the evening musicals and drinking at Wawaset Park when he came to Wilmington. Roger Adams told Bill Mapel that Wallace Carothers was likely to be elected to the National Academy of Sciences on April 30, 1936. Mapel arranged to have Adams call him with the news when the election was official. Mapel stationed himself in the press room of the *News-Journal* to insert Wallace Carothers' new distinction in the newspaper as soon as he could. The *Morning News* for April 30 announced the honor. The academy now had 293 members, elected after nomination by at least five of the scientists. Thirty-three chemists were members. Carothers was the first industrial organic chemist to be honored. Mapel quoted Roger Adams in his front-page story: "This is the highest attainment possible for a man in our science, and it is one Wallace Carothers richly merits."[22]

Wallace Carothers immediately thanked Adams for his help over the last 15 years.

> May 1, 1936
>
> Dear Roger
>
> Please accept my very sincere thanks for your telegram, and for more than that—for all that you have done to provide the occasion for the congratulations. I feel very highly honored to join so distinguished a group and am indeed very grateful to you.
>
> It was very nice to see you here. I enjoyed our visit a lot and look forward to another meeting before you go abroad.
>
> Sincerely
> Wallace Carothers[23]

Carothers' election came just three days after his fortieth birthday.[24]

■ ■

One month later, in June 1936, Arthur Tanberg began the difficult job of answering Wallace Carothers' mail for him. On June 16,

---

[22] *Wilmington Morning News*, April 30, 1936.

[23] Wallace Carothers to Roger Adams, May 1, 1936. (Adams Papers, University of Illinois Archives, Urbana, IL.)

[24] William C. Mapel interview with Adeline C. Strange, July 1978. (Hagley Museum and Library Collection, Wilmington, DE.) Press release by DuPont, undated but received by the chemical department on May 13, 1936. (Hagley Museum and Library Collection, Wilmington, DE, 1784.)

W. H. CAROTHERS WINS HIGHEST U. S. SCIENTIFIC HONOR

DuPont Research Chemist Elected Member of National Academy

Wilmingtonian First in Industrial Field to Be Admitted; 14 Others Chosen

Carothers elected to the National Academy of Sciences. (Reproduced with permission from Wilmington Journal-Every Evening, *April 30, 1936.)*

Tanberg wrote Prof. Arthur Lamb, for whom Carothers still served as associate editor of the *Journal of the American Chemical Society*, and returned two papers and a book for review that Lamb had sent Carothers. Arthur Tanberg wrote:

> I am sorry to inform you that Dr. Carothers, because of the need of treatment for a nervous condition, is likely to be absent from his office for some time. The situation is such that the doctors have decided it would be best for the time being if his associates do not undertake to communicate with him in connection with such responsibilities as he has had in the past. It is impossible to tell at present how long this situation will last, and we are telling his many friends that he has gone away to take a much needed rest. In view of the situation, it seems best to Mrs. Carothers and me that we return to you several things which you have sent him and suggest that until further notice nothing more be sent him in the way of papers and books for review.[25]

Tanberg wrote that Mrs. Carothers had found the text of the book, with an incomplete review written in Carothers' hand. It was

---

[25]Arthur B. Lamb to Arthur P. Tanberg, June 17, 1936. (Hagley Museum and Library Collection, Wilmington, DE, 1784, Box 18.)

evident Carothers had held the book for a long while. Tanberg suggested Carothers' notes be used as a guide for subsequent reviewers but not be published. Carothers had not given his consent, and Tanberg wrote he was not about to try to get it. It was clear that Helen Carothers was working with Arthur Tanberg, whom she worked for at the Experimental Station, to sort through the current scientific affairs of her husband. It was as if Wallace Carothers were now dead.

Arthur Lamb wrote a brief comment, nearly an obituary for Carothers, in his reply from Harvard to Tanberg the next day. He noted recovery from "nervous disorders" is "always distressingly slow." He continued, "I am deeply distressed at this news about Dr. Carothers. Our science in America can ill afford to lose the services of so brilliant an investigator." Lamb then recalled with some hope, Carothers' Harvard days, "Dr. Carothers, when he was here, had what I suppose were somewhat similar attacks from which he, however, recovered completely. I am trusting that an equally rapid recovery will occur in this instance."[26]

---

[26]Arthur P. Tanberg to Arthur B. Lamb, June 16, 1936. (Hagley Museum and Library Collection, Wilmington, DE, 1784, Box 18.)

Chapter 18

At the Philadelphia Institute

Sometime in the spring of 1936, in Wilmington's finest season, with banks of azaleas bursting red and white along the streets of the town, along the lanes and curving paths of Rockford Park and the parks along the run of the Brandywine, Wallace Carothers' behavior became fearful and intolerable. Perhaps now that he was married, and not living alone, his behavior could be truly appraised by Helen. We do not know what he did or said. He may have been frozen into inaction, unable to respond to Helen and his friends. He may have drunk to insensibility, gone into deep blackouts, unable to remember his days and nights. He may have threatened to kill himself, quietly and without remorse or loudly as an intimidating response to something Helen or someone else said or did. We do not know, but it would be in character for Carothers not to act rather than to act, to be overwhelmed and dazed by his life rather than to be dominating and threatening. Helen Carothers says very little about her life with Wallace Carothers. What we can expect was she quickly saw life with her new husband was not taking an expected course. His behavior, whether silent or bizarre, was frightening. Perhaps she asked him what was in the small capsule attached to his key chain.

What we know is that Wallace Carothers was absent from DuPont from Thursday, June 4, 1936, until Monday, September 14, 1936.[1] And we know that on about June 4, Wallace Carothers, in the custody

---

[1] Willard Sweetman to J. A. Horty, January 27, 1937. (Hagley Museum and Library Collection, Wilmington, DE, 1784.) Willard Sweetman, the Experimental Station's chief clerk, was Wallace Carothers' father-in-law.

of his wife and his DuPont employers, was admitted, against his will, to the Philadelphia Institute of the Pennsylvania Hospital on Forty-ninth Street in that city.

■ ■

This institute was a special place at the Pennsylvania Hospital. The Pennsylvania Hospital, the oldest hospital in the United States, founded with the energy of Benjamin Franklin in 1751, from its first charter was designed to give care to the insane. The cells of the old hospital held chains and leg locks and iron staples and handcuffs, forged for the care of the mentally ill by blacksmith John Cresson. Dr. Benjamin Rush, signer of the Declaration of Independence, pioneered in treatment of the insane for 30 years, tranquilizing excited patients by lashing them to a stout chair and enclosing their heads in a tight box. Rush's contemporary, Dr. Philipe Pinel, in Paris, began a humane approach to treatment of the insane. He believed the insane were medically ill people, suffering from a disease. He gave them work and a place to bathe and treated them with dignity.

The Pennsylvania Hospital for the Insane, a long, stone structure built on a 100-acre tract of land west of the Schuylkill River in 1841, continued this new tradition of treatment for the mentally troubled. But the hospital always maintained a desire for a facility for the treatment of "mild disorders." "Is not the fundamental need the treatment of those every-day people who want help in their adjustments to their families and their work?"

With exquisitely poor timing, the Philadelphia Institute opened its doors in 1930. The institute was located in its own new building and focused on the treatment of mild nervous disorders. The first floor contained a series of private consultation rooms for the psychiatrists and an area for out-patient treatment. Up the stairs, on the second and third floors, the patients were housed in fine accommodations, managed as a private club or small hotel. The patients would find outdoor gardens on the fourth floor and an area for occupational therapy. Hidden in the basement were the physiotherapy areas: the pool and certain special bathing rooms.

> From the first, the patients coming to the Institute were above expectations in the normalness of their problems and in abilities. Some of the commoner problems have been sleeplessness, fatigue, depressions, feelings of inferiority, and difficulties in getting on with other people. The patients have been decidedly above average in abilities and in other gifts.
>
> The number of patients has been small. While this is somewhat due to the business depression and to lack of

knowledge about the Institute's facilities, it is chiefly due to lack of funds. Such patients as we receive need interviews with a psychiatrist which are at least an hour long; one physician can only see eight patients a day and the physician must be a man of experience.[2]

Carothers called this place "an especially elegant, large, and elaborate semi-bug house."[3] He was brought to the institute after an extended series of events. The admission of a patient to an institution, an asylum, the hatch, the college, the bug-house, was not an easy matter in the 1930s, because once a person was there it was not a trivial matter to get out. William Seabrook, the author who, four years earlier, hid himself away in a cabin on the *Europa* as he fled from Paris and from alcohol, was stalled for more than a month before he was admitted to the plush-carpeted Bloomingdale's in Westchester County, New York. Even though Seabrook wanted a padded room, alone and in isolation, the process of getting a court order for admission consumed the month of November 1933. Finally a physician certified him as suffering from chronic alcoholism and neurasthenic symptoms and Bloomingdale's took him for a visit that lasted seven months.[4]

Wallace Carothers appeared at Philadelphia with Helen and with Elmer Bolton's assistant, Hamilton Bradshaw, and with a diagnosis from a Dr. Gehrmann of "neurocirculatory asthenia." We would call these attacks anxiety neuroses today—anxiety so deep it seemed that Carothers could not draw a next breath. The very thought of minutes and hours, a succession of future time with no relief from the consummate effort to breathe, made the attacks even more terrifying. He was fatigued; he was afraid of making any effort. Walking, even talking, any exercise, made him uncomfortable. He could not move.[5]

---

[2]Packard, Frances R. *Some Account of the Pennsylvania Hospital from Its First Rise to the Beginning of the Year 1938;* Engle: Philadelphia, PA, 1938; pp 1–5, 45–51, 119–127.

[3]Wallace Carothers to John R. Johnson, July 9, 1936. (Hagley Museum and Library Collection, Wilmington, DE, 1842.)

[4]Seabrook, William *Asylum;* Harcourt, Brace & Company: New York, 1935; pp vii, 15.

[5]Willard Sweetman to J. A. Horty, January 27, 1937. Willard Sweetman's letter footnotes Dr. Gehrmann's diagnosis as neurocirculatory asthma. There was no such condition; neurocirculatory asthenia is described in Dorland's *The American Illustrated Medical Dictionary*, 16th ed., 1935 and 18th ed., 1942. The 27th edition of the dictionary, published in 1988, suggests neurocirculatory asthenia is now considered a particular manifestation of anxiety neuroses. (Hagley Museum and Library Collection, Wilmington, DE, 1784.)

Carothers was now the patient of Dr. Kenneth Ellmaker Appel, born in 1896, the same year as Carothers, Harvard educated, a psychiatrist dedicated to the value of psychoanalysis. Appel had himself been analyzed in Europe by Otto Rank.

Appel held idealistic, optimistic values. If he showed interest and patience and friendliness, his own attitude would heal his patients. "Failure is inevitable, yet incidental, not final," Kenneth Appel wrote. "The present is the thing. Life lies in the pursuit, not always in attainment. It is in the unflagging striving that healing comes. The psychiatrist's presence and attitude keeps effort alive."[6] He believed psychiatry would bring great healing; he visited his patients nights, weekends, and holidays.

But before Wallace Carothers met with Dr. Appel he faced the standard routine of admission to the institute. While his friends at DuPont that summer came to work to sweat in the small laboratories and preproduction semiworks, attempting to extrude miles of continuous filaments of polyamide, carrying out their daily science and engineering, acting on their own, Wallace Carothers met a degrading, dehumanizing routine. He was reduced to nakedness; his shoes and pants and belt and all his clothing were taken from him. His razor and watch and matches and cigarettes were gone. If he brought the cyanide, it too was confiscated. The clublike atmosphere of the institute meant little at first, as Carothers was dressed only in the institutional robe and slippers. The robe had no tasseled cord. Wallace Carothers, the newest inductee into the National Academy of Sciences, was carefully bathed and shaved by an attendant; he was allowed to smoke only in the common areas where he could be watched by the staff.[7]

Dr. Appel knew little about Carothers beforehand. As with any new patient, the critical first days at the institute provided a cooling off period for the patient in crisis; an opportunity to distance himself from whatever incident triggered the hospitalization. And for the staff, for Dr. Appel, it would take days to understand the newly admitted man.

Dr. Appel liked to be the first one to talk to his patients. He felt he could put them at ease, help in this acclimatization to the new condition of treatment. He often began with a physical examination, probing for the organic cause for the hospitalization. With Wallace Carothers, who had a history of apprehension over his physical well-being, this physical examination was a necessity.

[6]Braceland, Francis J. "Memoriam, Kenneth Ellmaker Appel, 1896–1979," *American Journal of Psychiatry* **1980** *137(4)*, 501–503.

[7]Seabrook, William. *Asylum;* Harcourt, Brace & Company: New York, 1935; p 13.

Appel's examination was lengthy, going beyond routine measurement of pulse, respiration, and temperature. He had a two-page list to cover. He would know everything about the physical body of Wallace Carothers before he began a mental examination. He would list Carothers' apparent age and cite his surgical scar from the removal of his goiter. He would probe and test; he would note the texture of Carothers' skin and the color of his eyes. He would observe his breathing and his gait. He would look for tremors and spasms and test his reflexes at 12 different locations on his body. Dr. Appel was careful to note any lack of cooperation with his study. If Wallace Carothers slumped in his chair or muttered his answers or rolled his eyeballs in frustration or reacted to any of Dr. Appel's small talk, the psychiatrist would note it for his records.[8]

Dr. Appel was pursuing the "complex," the driving force that brought Carothers to him. Dr. Appel first asked Wallace Carothers, "Why have you come here?" And when Carothers told him it was not his idea, Dr. Appel asked, "But do you need help?"

"How do you feel?" he asked Carothers. "Are you blue or depressed? Do things seem hopeless? Does life not seem worth living? Are you afraid of anyone? Are you being watched? Is your body healthy? How are your finances? Do you ever become confused? Have you thought someone was speaking to you and there was no one there?" Dr. Appel noted whether Carothers talked rapidly or slowly, whether he rambled or jumped from one thought to another, joked, swore, gave irrelevant answers, was coherent. All of this to bring his patient somehow to the release of telling him what he felt and how it was making his life fail.

We know the questions Dr. Appel asked Wallace Carothers while they sat in his quiet study off the first floor hallway at the institute. We do not know Carothers' answers. Helen Carothers once told me she did not know why Wallace Carothers killed himself. Years later I suggested we could ask a contemporary psychiatrist to look through her husband's records at the institute and then talk with us. Then we might both get a better picture of her husband in his lonely weeks in Philadelphia. Helen Carothers agreed to consider my request but

---

[8]Appel, K. E.; Strecker, E. A. *Practical Examination of Personality and Behavior Disorders;* Macmillan: New York, 1936. In the year Dr. Appel examined Wallace Carothers, he published extensive guidelines for the physical and mental examination of patients, from which he would draw the important psychiatric evaluation. This process of Carothers' examination is drawn from Appel's contemporary writing.

suggested she would need to talk to her daughter, Jane, first. She refused permission to have Wallace Carothers' records opened.[9]

Dr. Appel needed to classify Wallace Carothers. If the psychiatrist could name the condition, his diagnosis would lead to treatment. Did Carothers suffer from a psychosis related to an infectious disease or hyperthyroidism or syphilis? Or was there a toxin involved—drugs or alcohol? Or an affective psychosis: mania or depression? Was he schizophrenic, paranoid or suffering a hysterical neuroses, anxiety attacks?

Wallace Carothers was in fact inert and barely responsive. But he had spent years of self-examination, talking to himself about the fears and anxieties and depressions, fighting them with his new routines, hopping boulders along the creeks near his home, walking alone in the forests of Europe. He was tired when he reached Philadelphia. He was weary of the fight for a brighter outlook. And he was impatient with the doctor who didn't know him at all and who must draw from him all the past clutter of which Carothers had become so indifferent.

Psychiatrist Adolf Meyer, Dr. Hohman's mentor, gave us the name depression. William Styron claims Adolf Meyer suffered from a "tin ear for the finer rhythms of English" when he named this suffocating wave of infirmity.[10] Depression overwhelms. It covers like a blanket, deceptively soft and inviting; compelling, not harsh. It lures its victims, tugging them with an undertow, sliding them to the attractive coolness at the bottom of a still pond. Depression attacks in the morning. Its victims come awake tranquilly, but in an instant they are alerted to the dread of the day. Any task comes magnified in the morning to an insurmountable obstacle. A pain grows in the gut and spreads up and down and across. Legs will not move, body cannot rise, eyes stay fixed on the room, tears roll and burn.

A day finally arrives when the depressed man knows his illness will kill him. He will choose one day that the terror of death has receded to a point that the terror of life is larger, more vivid. He is confused and lost. The clearest of summer days lays dark across his horizon. He dreams of death, organizing its aftermath as a triumph of his final action.

---

[9]I had hoped to ask Dr. Martin Orne of the institute to look at the records. Diane Middlebrook, with whom Dr. Orne cooperated on a biography of Anne Sexton, suggested that the only way to the documents would be with Mrs. Carothers' permission.

[10]Styron, William. *Darkness Visible;* Vintage Books: New York, 1992; p 37.

The depressed man can barely talk. His voice comes as a labored wheeze. He is confused, with his thinking poisoned by the run of fear and pain coursing through his body. He can but stammer. The ball of string his mind has generated has too many broken ends. If he begins a thought to attempt to unravel the mass, the thought trails off in an unconnected snippet. He must begin again, down a new thread, to a new and incomplete end. He never reaches the source. The depressed man never forgets the pain. You never can get to the bottom of it, make sense of it. So depression seems to have no solution but death.

■ ■

A patient can participate in therapy or observe it. There are but these two alternatives. It comes to the intellectual among us to view the process of hospitalization as a sequence to be recorded and judged, even as it is being applied for our own benefit. We arrive needing help, drained of the ability to function independently, yet we still fight for the controls in the foolish expectation we still can manage our own affairs. Our own best thinking has gotten us where we are, institutionalized, stripped of every dignity, yet we remain reluctant to turn over the feeble power we still retain.

William Seabrook at Bloomingdale's in 1933 struggled to take charge. But the naked man, hosed down at the showers by laughing strangers, is reduced to submission. Seabrook was immersed in quiet baths they told him were called hydrotherapy. He was wrapped in warm, wet sheets and left for hours, totally immobilized, in darkness. His nurses called that the ''Pack.'' As his confinement went on, Seabrook had no choice but to become part of his own recovery. His keepers cut away Seabrook's cords to the outside world, and he began to think and dream more freely, without the sensations of his frantic world, without its alcohol and drugs.

And one evening in his sterile room, while he was listening to Wagner's Siegfried coming across his small radio from Carnegie Hall, William Seabrook received more than he expected; he in fact got about all that he could get.

As he neared sleep, Seabrook experienced a quiet spiritual adventure. He was not alone. He had a visitor, Mr. Duval, his keeper. But he saw the man in a transparent and mystical aura, a guest for his soul, a light suffusing across the man, making him a kind protector, a parent, both parents. William Seabrook was secure and screened from harm. It was a wonderful, benevolent glow. He knew a gentle fact. He was safe, but it was not of his own doing. He was led, but nevertheless he was safe, saved. He would be all right.

Seabrook lived in that glow for a time as he continued his treatment. But the spiritual moments faded, and Seabrook took to analyzing them. They were the traces of memory of the best of drunken times, surfacing in one last fling after three months of sobriety. These glowing moments were the best of hard times on the difficult withdrawal from alcohol.[11]

■ ■

Whatever quiet visions entered the imprisoned soul of Wallace Carothers on the warm summer nights of June and July 1936, he took away with him. Carothers resisted engagement in his own recovery. Carothers was at the institute for more than a month when he wrote Jack Johnson a warm friendly letter, scrawling with a blunt pencil on yellow, lined paper. He crabbed about the course of treatment at the institute:

> 111 North 49th Street
> Philadelphia Pa
>
> Dear Jack;
> A long time ago you sent me a very nice and very generous letter, to which no reply has yet been offered. Perhaps it is too late to make an acceptable reply, and I am certainly in no condition to offer an appropriate one just at present. I merely wanted at this first opportunity to express my appreciation and gratitude.
> The above address is that of an especially elegant, large, and elaborate semi-bug house. It is so large that one only occasionally encounters a fellow patient. Most of them seem to be the bright and cheery, very well to do kind of professional hypochondriacs. Meeting a doctor is a rare event. Treatment consists in conversation, rambling, inconsequential, pointless and sometimes so repetitious and puerile as to be the source of laughter, amazement or anger. On the other hand there are tennis courts, bowling, a pool of sorts and a badminton court. Also hydrotherapy which is a very elaborate and impressive method of taking supervised baths. It must be terribly expensive, but they don't send bills until the patient's resistance has been fully built up (or broken down?). I am probably unconsciously bankrupt now financially and certainly am so morally. Just too lazy to move.

---

[11]Seabrook, William. *Asylum;* Harcourt, Brace & Company: New York, 1935; pp 141–150.

This is an involuntary business so far as I am concerned. It was sprung on me suddenly about 5 weeks ago, and here for the most part I have been ever since. It's an all right place to loaf and since the diplomatic axe fell and I was transported up here I haven't been capable of anything but loafing anyway. The future is pretty obscure. At the moment this assignment seems permanent unless I simply walk out. They can't very well prohibit that here. So it is quite within the range of possibility that you may before long discover a disreputable caller at your front (or perhaps better back) door step. If so don't shoot. It will only be me.

How are Hope and Keith? I would like to see you all again. If you take a notion to drop me a note, please address it to Wilmington. I am scarcely in a position to solicit anything of the kind but would appreciate it. In any event, I send my fondest regards to you all. (Also for Speed who was here yesterday.)

Auf Wiedersehen
Wallace[12]

■■

Speed visited Carothers at the institute, so did John and Lib Miles. Carothers was not held strictly in place under Dr. Appel's care. Crawford Greenewalt remembered that Carothers came to Wilmington in the company of a nurse and his wife Helen to attend a party at Greenewalt's house that summer. Greenewalt recalled the occasion because Wallace Carothers called Helen "Sylvia" several times that afternoon.[13]

Ten days after Carothers wrote Jack Johnson, he was allowed to go to Ithaca for a visit. He and Helen talked with the Johnsons at Cornell; they talked about Roger Adams. Carothers knew Adams was in Europe combining an August visit to a conference in Switzerland with a long-planned hiking trip in the Alps.[14] Wallace Carothers wished he could be in the Alps with Adams. Jack Johnson knew Professor James Flack Norris of Massachusetts Institute of Technology was leaving in a

---

[12]Wallace Carothers to John R. Johnson, July 9, 1936. (Hagley Museum and Library Collection, Wilmington, DE, 1842.)

[13]Crawford Greenewalt interview with Adeline C. Strange, 1978. (Hagley Museum and Library Collection, Wilmington, DE.)

[14]Tarbell, D. S.; Tarbell, A. T. *Roger Adams: Scientist and Statesman;* American Chemical Society: Washington, DC, 1981; p 117.

few days to tramp with his Illinois friend, Adams. Johnson saw the possibilities. Perhaps Carothers could arrange to leave with Norris to meet Adams in Europe. Perhaps he could have his wish for the European vacation. The Carothers hurried to Boston to catch Professor Norris before he left. And Carothers cabled Adams in Munich, trying to determine his mentor's schedule.[15] A few days later, Carothers wrote Jack Johnson, outlining his plans:

> Wednesday
> Midnight
> Philadelphia
>
> Dear Jack and Hope,
> Many thanks for your kind hospitality. Not only kind but lavish. I was sorry the visit was so brief, but in fact we just barely got to Boston in time, and then to Philadelphia rather late Tuesday when the decision was made of sailing Thursday. This naturally took some scurrying. The only possible available boat is the Europa and it is pack-jammed. I haven't even a space assigned—only a guarantee of space which may turn out to be between the boilers. Everyone is catching the last sailings for the Olympics. The whole thing is rather arbitrary and my state of bankruptcy is too appalling [sic] to contemplate. I am not very optimistic concerning the sudden healing qualities of Bavarian air, but Bavarian beer will undoubtedly provide at least a temporary solace. Perhaps temporary solaces are all one should expect. In any event this will surely be adventure of a sort. I wish you were both going along. Please give my kindest thoughts to Keith. And again many thanks for your very generous consideration.
>
> As ever,
> Wallace[16]

Wallace and Helen had rushed from Boston to Philadelphia, not Wilmington. He was still a patient. He was a man for whom, in his own words, "the decision was made." Dr. Appel, a man who would stretch to be flexible, would allow him to travel across the ocean to meet with Roger Adams and Professor Norris to hike for two weeks in the Austrian Alps.

---

[15]Telegram from Roger Adams to Wallace Carothers, July 18, 1936, responding to Carothers' question. (Hagley Museum and Library Collection, Wilmington, DE, 1784.)

[16]Wallace Carothers to John R. Johnson, July 23, 1936. (Hagley Museum and Library Collection, Wilmington, DE, 1842.)

And on Thursday, July 23, 1936, Wallace Carothers boarded the *Bremen*'s sister ship, *Europa*, bound for the second year in a row for Europe. This time *The New York Times* took no note of his passage. He was not cured. He was not optimistic of a cure. In 1935 the Black Forest had done nothing for him. But what is clear is that the cloud of inertia Wallace Carothers brought with him to the institute in June was now lifting.

Once again Wallace Carothers traveled alone. Helen went home to Wilmington.

Chapter 19

In the Alps with Carothers

Roger Adams last saw Europe in his postdoctoral year spent with Otto Diels and Richard Willstätter in 1912–1913. So much had happened: the war, the 1920s, resurgent German science imbedded in Hitler's regime. Adams wanted to return to Europe. He planned a two-month vacation first, then a visit to the Twelfth International Conference on Chemistry in Lucerne and Zurich between August 16 and 22, 1936.[1]

Roger Adams left the United States in June, soon after Wallace Carothers was placed in the Philadelphia Institute. He was unaware of Carothers' new troubles. Adams went to Ireland first, then joined tours of the English countryside, visiting Bath, Stonehenge, and Oxford.

On Sunday, the 28th of June, he met an old friend at breakfast at the Savoy Hotel in London. They chatted a while and, his friend, Elmer Bolton, whom Adams still called "Keis," told him a story that Bolton wished to be kept confidential. Wallace Carothers was now in an insane asylum in Philadelphia, Bolton revealed. Bolton told his friend that he and other executives of the chemical department had arranged the confinement. There are times that Carothers is not normal, Bolton told Adams. Bolton expected Carothers would soon be transferred to a new institution where Bolton thought the inventor would stay for six months. He hoped this long hospitalization would cure Carothers completely.

While Adams was touring in Europe, he wrote regularly to his wife, Lucile, who remained in Urbana, Illinois. Although Bolton

---

[1]Tarbell, D. S.; Tarbell, A. T. *Roger Adams: Scientist and Statesman;* American Chemical Society: Washington, DC, 1981; p 117.

wished this news about Carothers to be kept secret, Adams was not about to withhold the news from Lucile for long. He wrote her and included Bolton's story on July 1. But Adams just communicated the facts as he knew them. If he had any feelings about Carothers' mental slide, he did not share them with his wife at this time.[2]

In the next week, Adams visited Chester, near Liverpool, and walked along the completely preserved Roman walls, dating from the first century, that surround the town. He returned to London and ran into Elmer Bolton again. Naturally, they talked about Carothers. The latest news from Bolton was that Carothers was much better. Then Roger Adams and Elmer Bolton fell into a long-distance analysis of their friend's marriage, "...K & I deduced it might be a Platonic marriage but we of course don't know."[3]

Later in July, Adams visited Vienna, then took a scheduled airplane flight from Vienna to Munich. He settled down in the Hotel Bayerischer Hof to wait for Norris. On the 18th of July he got a cable from Carothers asking for Adams' itinerary. Adams responded he would be in Munich until August 1 and that Carothers should contact American Express to trace him after that.[4] On Wednesday, July 29, Adams wrote Lucile that he expected James Norris to meet him there in Munich within a day or two. And by now, he had heard speculation that Carothers was actually on his way to Europe to meet him. Apparently Carothers would come right there to Munich. But Adams did not know for certain whether Carothers had even left the United States.

In fact, Carothers should have been there by the 29th. He sailed from New York on Thursday, July 23. The *Europa* would have reached Cherbourg on the 27th. If Carothers had not yet arrived in Munich, he could be expected within hours.

Adams went to the theater that Wednesday night and watched a German vaudeville company: acrobats, singers, and a chorus of girls. He wrote he was exhausted after three hours of nonstop entertainment by the cast and its chorus made up of about half "tall lanky blonds—not selected for leg-form, most of them homely—some with reddish skin, some white & some with knees that looked as though they hadn't been washed for weeks" and half dark girls, one "certainly an ordinary mulatto—gold teeth—rather clever at dancing but

[2]Roger Adams to Lucile Adams, July 1, 1936. (Adams Papers, University of Illinois Archives, Urbana, IL.)

[3]Ibid., July 8.

[4]Telegram from Roger Adams to Wallace Carothers, July 18, 1936. (Hagley Museum and Library Collection, Wilmington, DE, 1784.)

repulsive-looking...." Adams stopped by at Willstätter's house the next day, but his old mentor was not home and would not be back for a couple of weeks. Adams was enjoying Europe and Munich. He wrote Lucile he had "forgotten all the chemistry I know on this trip...;" he hoped he would return to serious study when he came home.[5]

On Friday morning, July 31, Wallace Carothers appeared at the Bayerischer Hof. Adams wrote Lucile that when Carothers arrived, "He looked rather dejected—had been drinking heavily but I think the walk with Norris and me will help him."[6]

So Wallace Carothers had come across the ocean and into southern Germany on a liquid tide. He was drinking enough so that his friend knew that morning, immediately, from the stale odor of the previous night's alcohol, now transpiring from Carothers' every pore, from the cloudy cast to his reddish eyes, that Carothers was still in deepest trouble. Here is this man, his best student, his friend, joining Adams on his life adventure in the Alps, arriving after more than a month in the asylum, drunk and dejected. Adams was a graceful man. And for his friend, Wallace Carothers, he did the very best he could do.

Adams and Carothers walked a bit, talked a bit that day. They each took a nap in the afternoon. Norris showed up on Saturday, August 1. Norris and Carothers shopped for a while. Norris was an experienced hiker. He picked out for Carothers proper leather hiking boots and warm clothing for the chill mornings in the mountains. The three men ate a long lunch, which didn't end until past four o'clock, and then lingered at their dinner until midnight. Carothers translated as they talked to a German factory manager. Sunday passed the same way, with eating and conversation consuming the hours.

It was a strange combination: three fine scientists, men who saw each other only on special occasions and one of them now there in Munich after the most distressing of sequences—a long, forced hospitalization for his deep and despairing depression, and likely, his threatened suicide. Certainly Adams and Norris watched Carothers each minute, measuring his mood, hoping for some sign of progress, some signal of recovery. Adams and Norris hoped most fervently that Carothers would not repeat whatever unmanageable behavior had caused his hospitalization while he was there in Europe, literally in their care.

---

[5]Roger Adams to Lucile Adams, July 30, 1936. (Adams Papers, University of Illinois Archives, Urbana, IL.)

[6]Ibid., August 2.

Adams and Norris had planned to take the train to Innsbruck, Austria, in the Province of Tyrol on Sunday. But Carothers lacked a visa. Carothers visited the consulate Monday, August 3, got the necessary papers, and Adams, Norris, and Carothers boarded the train at 2 p.m. for the trip across the border. Adams reported hastily to Lucile on Monday that, "Carothers appeared more optimistic last evening and even more so this morning so I hope by a couple of weeks he will be feeling more optimistic."[7]

The train south from Munich to Innsbruck picks up the Inn River at Rosenheim, then rises with the river toward Innsbruck. It is not a long ride. Past Rosenheim, the Eastern Alps rise up on both sides. This is the Tyrol, the alpine country of deeply cut valleys flanked by high, glaciated peaks. Innsbruck lies on a wide flood plain at just under 2000 feet elevation, where the Inn and the Sill rivers meet. Past Innsbruck, further south, the railroad crosses Brenner Pass, the critical route to northern Italy.

The Alps surround the city of Innsbruck. Even in summertime, their north-facing peaks, above the tree line are still dotted with snow fields. Along the narrow, old streets of Innsbruck, Adams and Carothers and Norris saw closely built seventeenth century stuccoed houses with shingled roofs. The Germans—most of Innsbruck's 60,000 residents spoke High German—cycled along the streets to the university with its huge library and fine museum and its several dozen American students.

Adams wrote that he and Carothers and Norris would use the Europische Hof at Innsbruck as a base for two five-day hikes in the Tyrolean Alps. Adams wrote a brief message to Lucile on August 5. They had stayed at a small village called Bärenbad, where they were the first Americans that year. They walked through the tiny Alpine village, then climbed into the higher mountains, above the trees, up to the barren stone huts that served as way stations for the hikers in these mountains. The Nürnberger Hütte clings to the slope of a broad talus, above the tree line. It is a huge, four-story building with · sloping roof on each side and gables on every slope. No roads now lead to the structure. Adams and his two friends ate lunch there on August 7. Outside, a brief summer snow squall raged. But the skies cleared and Adams and Carothers climbed up into the glaciers. It was a strenuous climb, particularly for Carothers, whose lack of vigor had been a critical element in his hospitalization. Adams suggested they had gone up more than 3000 feet. They returned for a long dinner at

---

[7]Ibid. Adams completed this letter on the morning of August 3.

the Hütte. Adams wrote, "Carothers is a changed man—quite normal & enjoying it."[8] On Sunday, August 9, almost a week into the trip, Adams, Carothers, and Norris stayed at a village called Gries. They slept at another Alpine inn, the Grieser-Hof, this one located in a small valley surrounded by pastures and tall spruce trees. Roger Adams wrote that his party would walk 10 to 15 miles a day and would take two days to go over a pass to a village called Ötztal, where, on Thursday, August 13, they would return to Innsbruck and he would go to Lucerne for his meeting.

By Tuesday night, they had reached the pass, a saddle called Kühtai, at nearly 7000 feet above the tree line. Adams posted a card from Kühtai's sprawling inn on the night of the 11th. It was cold in the inn, no more than 50 °F, and the temperature outside was near freezing. Adams did not mention Carothers.[9] On Wednesday they walked past the walled church in Öbergurgz, the highest church in Austria, then made their way back to a bus that took them to Ötztal. It was only a few miles from there, back down the Inn River to Innsbruck.

By the end of the second week in August, by Friday the 14th, Roger Adams was preparing to leave Innsbruck for Switzerland; he finally left on the 16th. Carothers remained behind at the Europische Hof.

More than two weeks later, on September 2, Roger Adams returned to the United States on the *Bremen*. Early the next week, on September 7, Adams received a long-distance telephone call from Merlin Brubaker, one of his former students at Illinois. Brubaker worked in Wilmington at the chemical department with Wallace Carothers. Helen Carothers probably worked directly for Brubaker at the time. Brubaker's call was frightening. Where was Carothers? Did Adams know what had become of him after their hike? Had he heard from him? No one in Delaware, his DuPont managers or his wife, knew where Carothers was. They had not heard from him in nearly a month.

Did he lay injured at the foot of a steep glacier in the Alps? Had he drunk to insensibility and stumbled off a trail? Was he dead? Had he crouched down beside a cool stream tumbling from the mountains and chosen the cyanide? Dr. Appel may have wondered, perhaps wished, that Wallace Carothers had slipped off to Switzerland, to Zurich, to see Carl Jung. He was in Vienna for a time. Did he visit old Freud?

Brubaker telegraphed Arthur Tanberg on Monday, September 7. Tanberg was in Pittsburgh at the time. Brubaker informed Tanberg

---

[8]Ibid., August 7. Roger Adams sent three postcards to Lucile: Two show photographs of the Nürnberg Hütte, and the third shows the valley village where they stayed the night before.

[9]Ibid., August 11.

that he had spoken with Adams and knew all that Adams knew. But that was very little. Adams had last communicated with Carothers on August 22. Adams told Brubaker that Carothers was in "good spirits" during the last week of their hike, up until about the 16th, but not before. On the 22nd, Carothers was still hiking. Hiking alone. On September 8, Adams talked to Brubaker by telephone again. Adams said he had left Carothers on August 16 but knew Carothers was still in Innsbruck at the Europische Hof on the 22nd. But more than two weeks had passed, and Adams knew nothing further.[10]

One week later Tanberg, Brubaker, Helen Carothers, and Dr. Appel were still searching frantically for Wallace Carothers. On the morning of September 14, Helen Carothers looked up from the work on her desk and saw Wallace Carothers walk into her office at the Experimental Station.[11]

■ ■

Time ceased to flow for a few moments that morning. One mystery was immediately solved, of course. Carothers was alive. But for all the parties in the drama—Carothers himself, Helen, the working staff at the Experimental Station who knew little beyond the obvious that Carothers had been gone for months, his managers who were frantically tracing the passenger lists of the incoming ocean liners looking for him, and Dr. Appel under whose care Carothers was still situated—Carothers' reappearance caused a radical readjustment.

The behavior of a sick person monopolizes the energies of the healthy. Carothers likely arrived acting as if all was well and normal. As if he was ready to pick up where he had left things in June. And the temptation among his friends and co-workers was certainly to play along. To ask him how he was, to ask how his vacation in Europe had gone, to be pleasant to his face and listen attentively to his suggestions.

But in true fact, everything had changed for Carothers. First, Helen, Dr. Appel, and the DuPont managers huddled to assess their patient and try to make sense of his return. What was Carothers' condition? He had left the institute; did he need now to return? What was the next step?

---

[10]Telegram from Merlin Brubaker to Arthur Tanberg, September 7, 1936 and September 8, 1936. (Hagley Museum and Library Collection, Wilmington, DE, 1784.)

[11]Telegram from Roger Adams to Arthur Tanberg, September 14, 1936. (Hagley Museum and Library Collection, Wilmington, DE, 1784.) This telegram provides Tanberg with Carothers' Innsbruck address and shows that on the day Carothers returned, his friends and family were still looking for him.

Carothers no longer had real responsibility at DuPont. It quickly dawned on Bolton, Tanberg, and Ernest Benger that the best they could hope was that intermittent visits to the Experimental Station by Wallace Carothers served to ease his mind and maintain some contact with the work.[12] Carothers was not recommitted to the institute in Philadelphia. But for a time he stayed there at night. He saw Dr. Appel regularly; by day he drove to Wilmington to visit at the Experimental Station, and each evening he returned to Philadelphia.

One recent tenant at Whiskey Acres, Dr. Luther Arnold, related that Carothers began to sleep there at night, too. Dr. Arnold remembered that Helen and Dr. Appel had agreed that she was not strong enough to watch over Wallace Carothers. After Dr. Appel no longer required Carothers to stay overnight at the hospital, Helen called Leigh Williams and suggested Wallace be allowed to stay with Williams, Hans Svanoe, and Dr. Arnold at the bachelor house in Fairville where Carothers had lived until mid-1933. Carothers took up residence in their empty room.

Dr. Arnold remembered Wallace Carothers at Whiskey Acres as "limp, lackadaisical." Arnold said he had played a lot of squash with Carothers in the first couple of years he had been in Wilmington; Carothers always won easily. Dr. Arnold played him again after Carothers returned to Whiskey Acres. "He was terrible," Arnold said. Dr. Arnold remembered that at this time Carothers was obsessed with his physical health, his "internal plumbing," and he made several trips to Johns Hopkins. He and Leigh Williams paid little attention to his "moods." Carothers' physical condition seemed to be more important to him.

Carothers shrank from their guests. "He was terrified of strangers," Dr. Arnold said. Leigh Williams was an outgoing, personable man who entertained many visitors. "Often strangers would come at 5:30 or 6 and sit around and have some drinks. Wallace was afraid and sat in a corner in a cold sweat. If they were strangers or their conversations didn't fit, he was uncomfortable. He would disappear."

Wallace Carothers was not always found at Whiskey Acres. Dr. Arnold said, "...he disappeared from time to time, to go back to Helen I think. She was a lovely girl and I think they got along well."[13]

---

[12]Willard Sweetman to J. A. Horty, January 27, 1937. (Hagley Museum and Library Collection, Wilmington, DE, 1784.)

[13]Luther Arnold interview with Matthew E. Hermes, April 19, 1991.

Chapter 20

Isobel Carothers Dies

Sometime in the Christmas season of 1936, in the cold Chicago winter, Isobel Carothers became ill and she dropped out of her radio broadcasts. She soon developed a streptococcal infection. In this last decade before Alexander Fleming's invention of penicillin, such infections could threaten life. After two weeks, Isobel Carothers contracted pneumonia, and on Friday, January 8, 1937, she died in an Evanston, Illinois, hospital. Her husband, Dr. Howard Berolzheimer, a professor at the Northwestern University School of Speech was at her side. She left her husband and an adopted son, David.[1]

Wallace and probably Helen Carothers rushed to Chicago and they mourned the death of his sister, who was but 36 when she died. Wallace and Helen travelled back to Des Moines with his parents for Isobel's burial in that city but returned quickly to Wilmington.

It is likely Helen met Wallace's parents for the first time at this sad gathering. Ira and Mary Carothers were unaware of the string of dramatic events in Wallace's life over the past few years: his disappearance to Baltimore in 1934 soon after they moved back to Des Moines, his long hospitalization in Philadelphia the previous summer, and the period of weeks in which Carothers' whereabouts were unknown except to himself. Ira Carothers maintained at his son's death that he was taken by surprise and had no hint of any problems.[2]

At some point during Carothers' unexplained disappearance, Helen must have considered letting Wallace's parents know Wallace was missing and that she had a growing apprehension over his

---

[1] *The New York Times,* January 9, 1937.

[2] Ira Carothers to Roger Adams, December 2, 1937. (Adams Papers, University of Illinois Archives, Urbana, IL.)

Isobel Carothers. (Helen Carothers.)

absence. She must have considered revealing the whole stream of events. But she did not. Now the ceremony and mourning of Isobel in Chicago and her burial in Des Moines would deflect attention from Wallace and Helen. On the long train ride to Chicago they certainly must have prepared carefully what they would and would not reveal of the turmoil of their new marriage, of strange behavior, of talk of suicide, of hospitalizations, of a sudden departure for Europe, of unaccounted weeks, of the virtual end of a professional career, of separation, and of the impending doom that the death of Isobel seemed most grimly to foreshadow.

DuPont records show Carothers came back to Wilmington and immediately went on a short leave of absence from January 14 to 24, suffering a repeat of the "neurocirculatory asthenia" of the previous summer.[3] Several of Carothers' friends alluded to a second hospitalization at the institute for their friend. They remember he loved his sister dearly and spoke about her all the time. Memories have faded, but it is likely Carothers was recommitted for a short stay under the care of Dr. Appel.

In the shadows of winter the depressing effect of Isobel's death remained with Wallace. The nonsense of "Clara, Lu 'n' Em" no longer came across the airwaves. Instead, Wallace Carothers could hear Hal Kemp's dispiriting recording of "Gloomy Sunday," a Hungarian lament whose lyrics it was said had driven more than a dozen impressionable Hungarians to their death:

> Sunday is gloomy, my hours are slumberless.
> Dearest the shadows I live with are numberless.
> Little white flowers will never awaken you,
> Not where the black coach of sorrow has taken you.
> Angels have no thought of ever returning you.
> Would they be angry if I thought of joining you?
> Gloomy Sunday.[4]

Wallace Carothers returned to work on Monday, January 25. On the night of January 28, he wrote his mother from Wilmington. He was, as ever, responding to and not initiating the correspondence. He gave his mother all the support he could muster in the difficult time. His letter is addressed to his mother, not to both his parents, and he leaves only a perfunctory salute to his father, who has just lost his daughter to sudden and unexpected illness:

Wilmington Jan 28

Dear Mother:
    I have been reading over again your lovely letter about Isobel. It says so beautifully what everybody who knew her

---

[3]Willard Sweetman to J. A. Horty, January 27, 1937. (Hagley Museum and Library Collection, Wilmington, DE, 1784.)

[4]*Time*, March 30, 1936, pp 66, 67, quotes the lyrics of the "Famous Hungarian Suicide Song." The magazine claims more than a dozen "impressionable Magyars" had been led to suicide through the lyrics. Brunswick, Decca, and Victor recorded the ballad in the United States.

at all must have felt. Her going away was a terrible shock to
all of us and most of all to you who was so dear to her and
appreciated more than anyone else the fineness of her
qualities. There is no good trying to argue away the sorrow
of her going, but for us and for you especially, there is the
consolation that she is your child, and she added so much
to the happiness of so many lives. How terribly proud you
must feel. And though her task here is finished, her spirit
of joy and courage is still with us.

I am sorry there was not time for a longer visit with
you and dad. Also that I missed out on seeing old friends
at D. M. Please give Ray my best regards. Let me know how
you and dad and Helen are getting along.

Love
Wallace.[5]

■ ■

For a couple of weeks Carothers seemed to return to some peace
and some normality. He still visited Dr. Appel in Philadelphia and split
his time between home and Whiskey Acres. But on February 17
Carothers wrote Arthur Lamb and resigned as associate editor of the
Journal of the American Chemical Society. He cited "current personal circumstances" in giving up the position he had treasured.[6] In early spring
his old associate, James Conant, now nearly four years into his Harvard
presidency, asked for some advice on the chemical stability of two exotic
ring structures, cyclooctatetraene (*see* 11 in Appendix A) and cyclopentadiene (*see* 12 in Appendix A). He wanted, too, catalogs for molecular
models that he knew Carothers used. Carothers responded quickly.[7]

Carothers' reputation brought him a new proposal that spring.
England's Rubber Producers Research Association (RPRA), the prestigious organization that collaborated in the research and development of natural rubber, came looking for Carothers to join them.
Professor Eric Rideal of Cambridge began a correspondence with

---

[5]Wallace Carothers to Mary Carothers, January 28, 1937. (Terrill.) The Helen
mentioned by Carothers is not his wife, but an aunt.

[6]Wallace Carothers to Arthur B. Lamb, February 17, 1937. (Hagley Museum
and Library Collection, Wilmington, DE, 1784.)

[7]James B. Conant to Wallace Carothers, April 6, 1937. Wallace Carothers to
James B. Conant, April 9, 1937. (Hagley Museum and Library Collection,
Wilmington, DE, 1784.)

Carothers. Carothers talked to some of his friends about the offer and wrote back to Professor Rideal some time in April asking for further clarification.[8]

■ ■

But all was not going well. In fact Carothers situation was more desperate than ever. In early April, Dr. Kenneth Appel called Sam Lenher on the telephone. Dr. Lenher knew Dr. Appel was taking care of Wallace Carothers, but the telephone call came to Lenher as a surprise.

Dr. Appel said, "Mr. Lenher, I'd like you to come and call on me at my office, for reasons that I won't give, but you can assume they are related to your friend Carothers." So Lenher drove to Philadelphia, to the institute, and sat in front of the psychiatrist.

"Friends of Carothers have told me that you are a good friend of his." Dr. Appel said. "It is on that basis that I am telling you what I have not told anyone related to the Carothers' case." Dr. Appel blinked and swallowed. He said, "I have full charge of Carothers. I can't do anything for him. I can't help Carothers. I am sure—that he's going to kill himself." Lenher gripped his chair. Appel said, "I think I owe that to you as a good friend."

Sam Lenher, the most precise and organized and focused of the young DuPont scientists who came to Wilmington to make their fortunes during the depths of the Great Depression, went to his car. He tried to start the car but he could not turn the key. He sat there in the parking lot of the Philadelphia Institute and cried until he had no more tears.[9]

---

[8]Eric Rideal to Wallace Carothers, May 10, 1937. Arthur P. Tanberg to Eric Rideal, May 25, 1937. (Hagley Museum and Library Collection, Wilmington, DE, 1784.) Details of the exact nature of RPRA's offer have disappeared. All that remains is a letter from Rideal to Carothers thanking him for his comments, which would be taken up with the board of the association. But Dr. Rideal's letter was posted after Carothers was dead. Arthur Tanberg answered Rideal and revealed that Carothers had been telling people about this new opportunity.

[9]Samuel Lenher interview with Matthew E. Hermes, August 24, 1990.

Chapter 21

Wallace Carothers Swallows Cyanide and Dies

Dr. Appel's call to Sam Lenher came a few weeks before Carothers died. In the score of days before Carothers swallowed his poison, he wrote two notes that, in themselves, reveal no threat of imminent danger. But analyzed in the context of his life and death, the two letters reveal a serenity of choice coupled with a plea for companionship.

On Thursday, April 15, 1937, Wallace Carothers finished something very unusual. He wrote his mother. He was not responding to her letter. He did not use his Wawaset Apartments stationary, but a simple, lined, institutional yellow pad.

Wilmington
April 15th

Dear Mother:

Spring is arriving now and I wish very often that you and dad could be with us here to see the beauty of the Delaware and Pennsylvania hills. The star magnolias are out and the dogwoods are just starting to bloom. But spring is with you too and I hope it will be a very lovely and fine one.

Wilmington is booming at a great rate now. New York, Cleveland and Buffalo offices are being moved in by the dozen, so that we should have hundreds of new families. Building is going on at a great rate but it can't keep up with the demand.

Roger Adams is here, and as usual he inquired about

you. He is just back from Chapel Hill N. C. for the A. C. S.
meeting but goes to Urbana and will return to Washington
next week for the Academy meeting. With all this dashing
about, he seems to get a great deal done.

For me at the laboratory things are rather quiet—just
coasting for the present. One of the old problems that was
reborn during the Arden days is now bursting into real
flower. We have about 60 patents and about 40 engineers
are working on the thing now, so that we should have a
real development before long. Meanwhile however it is all
very confidential and secret. Besides I haven't very much
to do with it now and am really looking for something
quite new. Meanwhile the synthetic rubber is getting on its
feet financially; and it is amazing to read now how the Rus-
sians and the Germans are retracing every one of our
experiments.

On Thursday I had a call from Mills of Jesus College at
Cambridge. When I was in Cambridge last year he and
Sedgewick of Oxford took me for a walk and showed me
Trinity where Newton worked. Jesus isn't quite as hand-
some but it is much older.

I had a note lately from Howard. Not much news but
apparently some progress is being made in getting affairs
straightened.

This isn't much of a letter. I just wanted to give you my
love. Also to dad and Helen.

Wallace.[1]

If a descendant of Wallace Carothers comes across this letter and
knows nothing about the man, a quick inquiry would show his "old
problem that was reborn during the Arden days" became nylon, the
synthetic rubber was neoprene, and the letter is a sensitive and modest recounting of an astonishing career of accomplishment, a series of
successes now becoming the focus of worldwide attention. The reader
must study further to date the letter accurately. The yearless date
Carothers wrote, April 15, and the reference to "Cambridge last
year" might suggest 1936. But the reference to Isobel's husband
Howard, and the winding up of affairs, pins the date as April 15, 1937,
after Isobel's death and just 14 days before Carothers' own end.

The reader might wonder at Carothers' feeling for the landscape
of the fruitful spring. But the low hills rolling north of Wilmington
and up to Kennett Square and West Chester are truly rendered in a

---

[1]Wallace Carothers to Mary Carothers, April 15, 1937. (Terrill.)

background green so powerful that the eyes are not to be believed. Against this lush tone, the flowers do not simply open, they burst at center stage into bloom and color.

The reader might puzzle, too, at the unusual penmanship of the document. Was it customary at the time to complete letters near the first and last letter of a word, but to run the vowels and consonants in the middle in a low, straight line, hugging the ruled line, daring the reader to decipher the code? Could they unravel h—dreds, de—d, Pe—sylvania, look—g, us—l, boo—g, so—th—g?

■ ■

It becomes the biographer's task to attempt a more complete analysis. The first two pages of the letter repeat this strange penmanship. Carothers had shown this before, in some of the letters he wrote to Frances Spencer—written when he said he was drinking. However, the third page of this April 15 letter, the section beginning, "On Thursday...," is quite different in appearance. Carothers wrote this page in a more representative hand. April 15, the day Carothers dated the letter, was a Thursday. So it is likely Carothers was writing this third page, writing about a call from Mills on Thursday, a day or two later and in a sober state.

More to the point is the reflection that Wallace Carothers wrote his mother just two weeks before he took his life. The sense of peace Carothers transmitted suggests he may have come to the decision, he had become willing to die, he had taken the important first step, the willingness that must precede the action.

Wallace Carothers listed his accomplishments in the letter, guiding us deftly to neoprene, nylon, the National Academy, and the worldwide scientific stage. But this is not a self-serving obituary: the listing of accomplishments by a man knowing he is about to die, fashioning for his biographers the facts of his life in the best possible light.

Such an accounting would not fit Wallace Carothers. Dr. Appel wrote that Carothers' illness was that he was obsessed with his own failure. Carothers believed that he was finished and could do nothing more that would be worthwhile.[2]

So the actual writing of a letter to his mother, unprompted by a letter from her, is a puzzle. In addition, the peacefulness of the letter

---

[2]Ira Carothers to Roger Adams, December 2, 1937. (Adams Papers, University of Illinois Archives, Urbana, IL.)

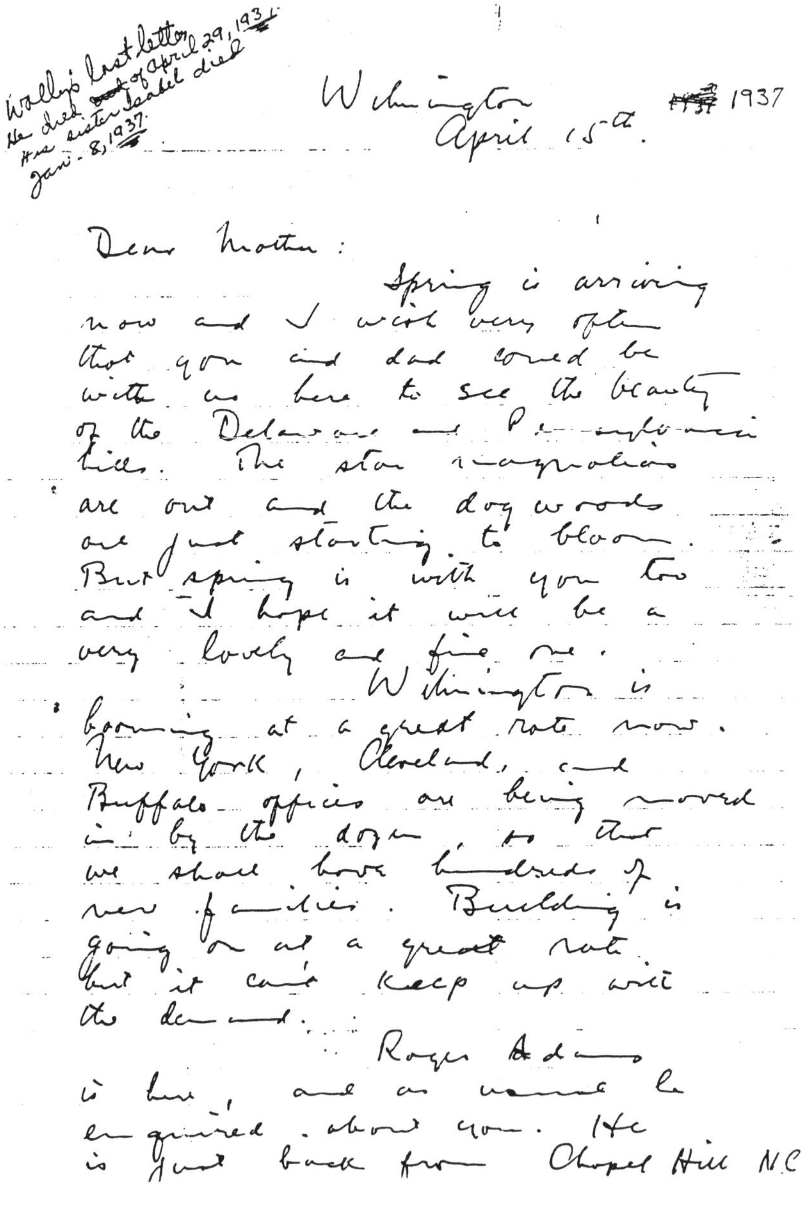

Wally's last letter, April 29, 1937.
He died of April 29, 1937.
His sister Isabel died
Jan. 8, 1937.

Wilmington
April 15th.   1937

Dear Mother:

Spring is arriving
now and I wish very often
that you and dad could be
with us here to see the beauty
of the Delaware and Pennsylvania
hills. The star magnolias
are out and the dogwoods
are just starting to bloom.
But spring is with you too
and I hope it will be a
very lovely and fine one.
Wilmington is
booming at a great rate now.
New York, Cleveland, and
Buffalo offices are being moved
in by the dozen, so that
we shall have hundreds of
new families. Building is
going on at a great rate
but it can't keep up with
the demand.
Roger Adams
is here, and as usual he
enquired about you. He
is just back from Chapel Hill N.C

Wallace Carothers letter to Mother, April 15, 1937. (Suzie Terrill.)

Wallace Carothers letter to Mother, April 15, 1937. (Suzie Terrill.) (Continued)

and its recitation of accomplishment clash. If Carothers rested as easily with his new directions and with the worldwide interest in his work, as his words indicate, he would not be looking forward in calmness to now die.

I believe Dr. Appel had a hand in this letter. Carothers' psychiatrist is frustrated, he is unable to help Carothers, but he never gives up. I hear the echo of Appel's voice, "You are not a failure. Take this pad of paper. Just once write down what you have done in proper form and look at it! Do you ever let anyone know what you have accomplished? This once, write it down and let someone see it!" And Carothers took the pad away and over a drink and then a few drinks he wrote his mother with a reflective calmness. He fell asleep after writing the first two pages. Perhaps he spilled something on an incomplete third sheet. In any event, the last page was a new letter on a new date written with even wider margins than Carothers usually used. By this letter, Carothers was able to please Dr. Appel, but at least for this day he had reserved to himself the right to die.

In a letter to Jack Johnson dated Thursday, April 22, 1937, just one week later, Carothers wrote that "A prospect of a visit from you is the best news I have heard in a long time." Carothers made a precise invitation to his friend. "You are hereby specifically and categorically invited to stop at Whiskey Acres. The accommodations are not elegant but we can make a bed for you."

Carothers' added that the countryside was a "gorgeous retrospective" and that everyone there would be glad to see him. Carothers' wish for the visit was that he and Johnson could go down to the shore for a few days of "loafing and sailing."[3]

This final known letter by Wallace Carothers discloses a sorrowful tale of the state of his soul. "My own circumstances are nothing to brag about," he wrote Johnson that day.

Wallace Carothers was living at Whiskey Acres; his invitation to Johnson included his friends but not his wife. He anticipated a trip to the Delaware shore in late April, where the wind blusters off the cold ocean. Nothing makes sense. On one side the letter offers the action of driving to Rehoboth Beach but then suggests rest. It was April. Cornell was in session as was DuPont. But work, research, and the science that Carothers and Johnson had shared for 15 years were distant from Wallace Carothers' mind.

■■

A few days before he died, Carothers went home from work to Wawaset Park, to Helen. But Carothers was not planning to spend an evening at their apartment in Wawaset. He began to get ready to go out. Luther Arnold remembered them as his "disappearing acts." But this night, as Arnold related the story, as Wallace was shaving, Helen Carothers looked into the mirror so he could see her. She said, "I am pregnant."[4] Lib Miles said that Helen's message came as a betrayal to Wallace. Wallace did not want children because he feared

---

[3]Wallace Carothers to John R. Johnson. Johnson read this letter and read the date to Adeline Strange in 1978. Its contents are on tapes held in the Hagley collection of Mrs. Strange's notes. I believe the date was added by Johnson when he received it. He had observed that practive with other Carothers letters. Carothers wrote the letter between the 20th and the 22nd. (Hagley Museum and Library Collection, Wilmington, DE, 1985.)

[4]Luther Arnold interview with Matthew E. Hermes, April 19, 1991.

he would pass along to a child the illnesses he suffered. And Helen had agreed. Mrs. Miles cited Dr. Margaret Handy as a source for her story. Dr. Handy was a highly respected pediatrician in the Wilmington area: a distinct and respected woman. Andrew Wyeth painted Dr. Handy. His portrait is called "Country Doctor."[5]

■ ■

The last 30 hours of Wallace Carothers' life come together from newspaper reports and from these fragments of memories, some of them second- and third-hand recollections. Carothers went to work at the Experimental Station on the morning of Wednesday, April 28, 1937. He had celebrated his 41st birthday the day before.

That evening, he called at the home of John and Lib Miles on Bucks Lane. But they were not at home; they were attending a movie. A young woman, a live-in maid, told Carothers the Miles would be out for a time. Carothers, who was a frequent visitor there, let himself in and read a book, sitting on the couch for an hour or an hour and a half. Then he left before the Miles returned.[6]

Carothers then drove to Philadelphia. Arnold said Carothers was a frequenter of the bars and clubs in the city and he probably drank long into the night. It was nearly 5 a.m. on April 29 when Carothers parked his car in the garage of the Philadelphia Hotel on the west side of the city and rented a room.

Almost 12 hours later, guests who were staying in a nearby room heard anguished groans from somewhere on their floor. One of the guests called Daniel Crawford, the manager of the hotel. Crawford let himself into Wallace Carothers' hotel room and found him lying on the floor, now dead. Next to his body was a squeezed lemon and crystals of a substance, which proved to be a cyanide salt. Crawford called the Philadelphia police. Detective Sergeant Bernard O'Donnell found cards identifying Wallace Hume Carothers, SSN 221-07-6944, and his wallet containing $52. The Philadelphia detective found no note.[7]

---

[5]This story comes from the Miles' interview. If true, it must have come from Helen herself. The Miles date the story as the very night Carothers went to Philadelphia. If Dr. Arnold heard the story from Carothers, the Miles have the date wrong. If Arnold heard it from a source other than Carothers, then Miles could have the timing correct.

[6]John and Lib Miles interview with Adeline C. Strange, July 1978. (Hagley Museum and Library Collection, Wilmington, DE.)

[7]*Wilmington Morning News*, April 30, 1937, p 1. *The New York Times*, April 30, 1937.

DR. CAROTHERS POISON VICTIM; NOTED CHEMIST

Death of Member Of National Academy Of Science Suicide, Say Police in Philadelphia

Found Dead in Hotel Room There; Wilmingtonian Had Major Part in Development Of Synthetic Rubber

Dr. Wallace Hume Carothers, 41-year-old DuPont research chemist who won fame through his part in the development of synthetic rubber, was found dead late yesterday afternoon in a room in the Hotel Philadelphia, Philadelphia. He lived in the Wawaset Apartments, this city.

Guests in an adjoining room at the hotel heard groans and called Daniel Crawford, the manager, who entered the room. He found Dr. Carothers lying on the floor. Crystals of poison and a squeezed lemon were nearby.

Carothers' death reported in Wilmington.(Reproduced with permission from Wilmington, Morning News, *April 30, 1937.)*

Helen Carothers was looking for Wallace on the 29th. She called Leigh Williams at Whiskey Acres, expecting to find him there as she often did. But Leigh had no idea where her husband had gone.

■ ■

Leigh Williams went to Philadelphia and escorted Wallace Carothers' body back to Wilmington. Carothers was cremated in Wilmington with two of his friends observing, as was the law. On Monday, May 3, the Rev. Charles Clash of the Immanuel Protestant Episcopal Church led a small, private service for Wallace at the Chandler Funeral Home in Wilmington.[8]

■ ■

The DuPont Company held a memorial service for Wallace Carothers at the Playhouse Theater, which is located within the DuPont building. The service echoed the editorial reflections of the Wilmington papers:[9]

> In the untimely death of Dr. Wallace H. Carothers the world of scientific research loses a diligent and able worker who had achieved signal distinction. His skill, knowledge and perseverance had given him an important part in the perfection of synthetic rubber for commercial uses. He also achieved other chemical triumphs.
> Dr. Carothers was one of a group of worthy members of our community who are devoting their lives to efforts to promote the well-being and happiness of mankind. As they are working quietly behind the scenes, they make no display. Yet they comprise one of the most important elements in this specialized phase of industry which depends for its success upon the study of possibilities of development of the secrets of nature's storehouse.

■ ■

Helen Carothers said she has no idea why Wallace killed himself. After his death, she immersed herself in preparation for the birth of their child. Wallace and Helen Carothers' daughter, Jane, was born on November 26, 1937.[10]

---

[8] *Wilmington Journal Every Evening,* May 1, 1937.

[9] *Wilmington Morning News,* May 1, 1937.

[10] Helen Carothers interview with Matthew E. Hermes, February 19, 1990.

■ ■

Perhaps, as Helen Carothers sifted through her husband's papers in the months after his death, she found a few dog-eared pages torn from *The Forum* from February 1933. "A Man's Life is his Own Affair" was the title of the article. Frank Cross wrote:

> The majority of arguments against self-destruction are of a very sentimental and fatuous nature. It is maintained that most suicides are either psychotic or neurotic, and hence, by some strange process of reasoning, deplorable. Why is it deplorable for an insane person to make away with himself? The commonest mental disturbance to lead to suicide is melancholia, or manic depressive insanity. It is an intermittent affliction, in its usual form, from which permanent recovery is extremely rare. At first the attacks of despondency which characterize it may be separated by hopefully long periods of normal behavior, but in the typical case these periods become progressively shorter until the victim enters a state of chronic derangement.

Cross continued:

> If recovery from any type of insanity which frequently leads to suicide were at all common, there would be some reason to deplore the self-inflicted death of a victim. Almost without exception, however, he is doomed to be a lifelong burden to himself and upon those who care for him. Death is the only escape. The popular solicitude for preserving his life through years of mental anguish can have no possible justification, it seems to me, aside from sheer sentimentalism. It is fatuous to argue that he may be wrong about it—that his ailment has been incorrectly diagnosed, or that a new cure might be discovered to-morrow. Such a forlorn hope has little appeal to a mind of intelligence.[11]

---

[11]Cross, Frank, C. *The Forum*, February 1933, pp 76–78.

Chapter 22

DuPont and Ira Carothers

The fiber Wallace Carothers invented was called Fiber 66 for two years after Elmer Bolton specified poly(hexamethylene adipamide) would be chosen as DuPont's new synthetic. But the product needed a name. Hale Charch, the rayon research director at Buffalo once suggested playfully "Duparooh" for "DuPont Pulls a Rabbit Out of the Hat," but Charch's tongue-in-cheek recommendation didn't take. A committee of three formed in 1938 collected a list of 400 names. None of them met everyone's approval. One of the committee members, Dr. E. K. Gladding, suggested "norun" but changed it to "nuron" when he was reminded that stockings of the new fiber would run. "Nuron" sounded medical; a change to "nulon" presented the difficulty of the redundant phrase, "new nulon"; and a change to "nilon" offered three confusing pronunciations. Nylon, with an unambiguous spelling and pronunciation became the name of the new fiber.[1]

The invention of nylon was a triumph because DuPont was poised to capitalize on Wallace Carothers' research. Elmer Bolton forced his scientists to develop efficient manufacture of the critical adipic acid and hexamethylenediamine (*see* Reaction 15 in Appendix A). There are times when science cannot be forced, but Bolton intuitively believed the two monomers could be teased from the same compound, benzene, which he believed would become readily available. Just as Bolton believed they would, the chemists at DuPont soon found that diacid and diamine could be produced efficiently from benzene. DuPont's managers poured money into manufacturing

---

[1]Hounshell, David; Smith, John Kenly. *Science & Corporate Strategy, DuPont R&D 1902–1980;* Cambridge University Press: Cambridge, England, 1988; pp 268, 269.

plants at Belle, West Virginia, to make the intermediate chemical for fiber preparation, Peterson's nylon salt. In 1939, in a field on the flat farmlands of southern Delaware near the town of Seaford, with nearly $10 million, they erected the world's first nylon plant to convert Peterson's salt to fiber. By the end of that year, nylon filaments, forced through rows of spinnerettes, began to be twisted into fibers and converted at the nation's hosiery mills into production quantities of fashion hosiery. The whole affair went so fast and so well that by mid-1940, *Fortune* predicted, "The Giant Molecule is a greater fact of history than Adolf Hitler, although it may take vision to believe it."[2]

By 1945, DuPont was shipping 25 million pounds of nylon a year into the war effort. With war's end, the company expanded its capacity dramatically and by late 1949, peacetime sales neared 100 million pounds a year. Wallace Carothers' nylon brought enormous profits to the DuPont Company and to the duPonts who still held majority control. And one man who led in the development of the fiber now ran the company. Crawford Greenewalt, Carothers' associate and friend, now in 1949 was just 45. But he was the president of DuPont. As a young engineer he had supervised the semiworks for nylon-fiber production. Greenewalt combined technical competence with management brilliance. The elders of the duPont family wanted the family to continue in charge of the company. In 1948, they chose Greenewalt, who long ago married the daughter of Irénée duPont, to lead them.[3]

At the chemical department, the memory of Wallace Carothers faded. Elmer Bolton remained in charge, with a portfolio of research success set against a collection of long-held research principles. Principle and achievement intertwined in an inseparable knot. The chemical department with its Illinois tradition, with Roger Adams and Speed Marvel and Jack Johnson as its ever present consultants feeding new scientists from the heartland into the company, was a center for basic research once again as the scientists searched ahead for the next success, the next nylon.

■ ■

On Wednesday, November 2, 1949, in Des Moines, Iowa, Wallace Carothers' father, Ira Carothers, sat at his desk in the simple home he and Mary Carothers now owned. He took up his pen and began writing a very difficult letter. It was more than 12 years since Wallace had

---

[2] *Fortune* **1940**, *22*, 58.

[3] Hounshell, David; Smith, John Kenly. *Science & Corporate Strategy, DuPont R&D 1902–1980;* Cambridge University Press: Cambridge, England, 1988; p 358.

Ira Carothers. (Helen Carothers.)

killed himself, but his father had carried the fragment of a very practical memory forward for more than a decade.

2206-33d St.
Des Moines 10, Iowa, Nov. 2, 1949

Dr. E. K. Bolton
Wilmington, Delaware

   Dear Dr. Bolton,- I am writing you on a rather delicate and very confidential matter.

   Following the death of our son Wallace Carothers, we were told in a round-about way that the Company intended to make us, his parents, a financial gift. I had a feeling at the time that our informant was not well posted.

   I do not know what the Company's policy is in matters of this nature, and this is my reason for writing you personally, rather than to some official connected directly with their finances.

   I do not want Helen, Wallace's wife to have any hint that I have written you on this matter.

   We are not in straitened circumstances, but present conditions are drawing on our capital.

   I shall appreciate your writing me very frankly on this matter.

   With kind regards I am

Very sincerely yours,
I H Carothers[4]

   Elmer Bolton acted immediately. He likely called Crawford Greenewalt on the telephone and told him what he had received from Ira Carothers. Greenewalt would indicate the next step. Bolton immediately sent Ira Carothers' letter to A. E. Buchanan, assistant general manager of the rayon department—the department now cashing in on the nylon success. Buchanan showed the letter from Carothers' father to Bob Richards, the general manger of rayon. On Monday, November 7, 1949, just the third working day since Ira Carothers mailed his letter, Buchanan responded to Elmer Bolton with some surprising information and a generous recommendation:

   I am returning Mr. I. H. Carothers' letter. Bonus payments to the widow have aggregated $372,727; an additional $11,139 was paid to Wilmington Trust Co. (as executor).

---

[4]Ira Carothers to Elmer K. Bolton, November 2, 1949. (Hagley Museum and Library Collection, Wilmington, DE, 1784, Box 18.)

All things considered the Company has treated the widow quite generously and it would be most unfortunate from the public relations standpoint, if some crusading journalist should uncork a story on the deteriorating circumstances of the inventor's parents against the background of our spectacular success with nylon. I feel sure the Carothers are not the type of people who would voluntarily have any part in such a story, but the success of nylon is so conspicuous that a muck-raking feature-writer could hardly resist the story if he found any basis for it whatever.

Bob Richards and I feel that Rayon Department could afford cheerfully to accept its share of the cost of providing an income of say $5000 per year to each of the parents for life. The amount suggested might be modified after, as we would recommend, an investigation of Mr. & Mrs. Carothers' circumstances is conducted by the Treasurer's Department. We think, however, that the amount should be sufficient to permit the Carothers to maintain their accustomed standard of living and to be reasonably free from financial worry arising from the infirmities of old age.[5]

The bonus money that had already flowed to Helen Carothers was astonishing and extraordinary. Buchanan's letter implies the awards were continuing. The size of the awards showed the complete confidence among the DuPont hierarchy that nylon's success was unprecedented and would continue. The fact that Helen continued to receive money was unheard of within DuPont.

DuPont had a long-established policy of granting bonuses on an annual basis to its most important employees. The stock awards came in two classes. The "B" bonus was the most common and consisted of small awards, generally a few shares in the company, which competent professional employees—sales personnel, manufacturing managers, and researchers—could come to count on as an annual ritual and reward. But the real prize to be won was an "A" bonus award, which the company masters granted only for heroic, specific contributions. These bonuses were far fewer in number and existed in a kind of mystical place. The mystery was enhanced by the practice among DuPont employees to remain closed-mouthed about such financial and compensation matters. So the hint that an employee had received an "A" bonus came to mark that person as someone quite special. Often, of

---

[5]A. E. Buchanan to Elmer K. Bolton, November 7, 1949. (Hagley Museum and Library Collection, Wilmington, DE, 1784, Box 18.)

course, the granting of an "A" award caused controversy. Two of DuPont's most successful research leaders, Herman Schroeder and Bob Joyce, recall the "A" bonus as "a very complicated system to award a pittance to the wrong person."[6]

Wallace Carothers received bonuses throughout his years at DuPont. At his death, he held stock worth about $38,000—presumably mostly DuPont shares. And this stock comprised the bulk of his estate.[7] Sam Lenher said that after Carothers died, the "A" bonus committee made a very special arrangement to provide his widow with continuing funds. Dr. Lenher gave no date for this decision, but the action must have been taken soon after Carothers' death and before his will was probated, because the initial money went not to Helen but to his executor.[8]

The receipt of more than a third of a million dollars over the dozen years since Wallace Carothers died made Helen and her daughter a substantially well-to-do family. These were days of 9-cent bread, $900 automobiles, and fully furnished and ample $9000 homes. Regardless of whether DuPont made the special awards to Helen Carothers out of gratitude over Wallace Carothers' contributions or out of an overriding sense of guilt after the daydream that they could prevent his suicide had turned to a nightmare, their support for Wallace's widow was remarkable.

■■

Buchanan of rayon had suggested Elmer Bolton might investigate the actual status of Ira and Mary Carothers before they determined a final level of support for them. Bolton convinced Buchanan to do the investigation. Buchanan had no trouble obtaining the contents of Carothers' will from his executor, Wilmington Trust. And he sent an investigator to Des Moines to dig into the financial status and the lifestyle of the old couple, who lived in a tiny, frame house on Thirty-third Street.

---

[6]Hounshell, David; Smith, John Kenly. *Science & Corporate Strategy, DuPont R&D 1902–1980;* Cambridge University Press: Cambridge, England, 1988; pp 305 and 306. I worked for both Schroeder and Joyce. Both men had the habit of telling things as they were, with little held back.

[7]Carl E. Geuther, November 15, 1949. Geuther worked for Wilmington Trust, the executor of Carothers' estate. (Hagley Museum and Library Collection, Wilmington, DE.)

[8]Samuel Lenher interview with Matthew E. Hermes, August 24, 1990.

With the investigation underway, Elmer Bolton gave Ira Carothers the frank answer he had requested. He wrote on November 15:

> In response to your letter of November 2, I regret that you have been misinformed regarding the intention of the Company to make you a financial gift. It may have come to your attention that the Company has an "A" Bonus Plan for rewarding employes [*sic*] who have rendered conspicuous service. One of the forms of conspicuous service is an invention resulting in profit to the Company. A bonus award can only be authorized by the Executive Committee which, as you may know, consists of the President and eight Vice-Presidents. No individual has any authority to commit the Company respecting an "A" bonus award. I believe you are aware that during Wallace's life time, and subsequently, he and his wife received very substantial "A" bonuses, which were awarded in recognition of his inventions and from which the Company has derived profit.
>
> I regret to advise you that the Company does not have any plan that would provide financial assistance to parents of a deceased employee. Moreover it does not have any policy in this respect inasmuch as no similar case has arisen so far as I can determine.[9]

But having given the carefully worded bureaucratic response to Ira Carothers' bureaucratically phrased question, Bolton wrote that he wished to explore the Carothers' request further. He assured Ira Carothers that he would do his best to keep the matter from coming to the attention of Helen Carothers.

On the day he wrote Ira Carothers, Elmer Bolton learned that Wallace had made a new will in January 1937, while Isobel was gravely ill and just days before she died. In the document, Wallace made a specific cash bequest of $5000 each to his father and mother and left the bulk of his estate, including all his personal belongings, to Helen. The estate totaled just under $45,000 at Carothers' death. Bolton knew Ira and Mary Carothers had once gotten this reasonable sum, but there was no provision in Wallace's will that would direct income gained after his death toward his parents.[10]

---

[9]Elmer K. Bolton to Ira Carothers, November 15, 1949. (Hagley Museum and Library Collection, Wilmington, DE, 1784.)

[10]Carl E. Geuther, November 15, 1949. (Hagley Museum and Library Collection, Wilmington, DE.)

Buchanan's investigator was Dr. Ernest Benger. He had been Bolton's assistant director in the chemical department but now ran research on fibers in Buchanan's rayon department.[11] Dr. Benger had known Wallace Carothers well and had taken part in his 1936 hospitalization at the Philadelphia Institute.

Dr. Benger arrived in Des Moines on the morning of November 25, 1949. Within hours he determined that the elderly couple was not then in Des Moines and was visiting their daughter, Elizabeth, her husband, Robert Kyle, and their three small children, in Cuyahoga Falls, near Cleveland. Benger met with the president and vice-president of the Central National Bank and Trust, with a friend of Ira's who worshiped at the Westminster Church, with the president of the Commercial College from which Ira was retired, and finally with a neighbor.

Dr. Benger reported the Carothers lived in a new postwar development, in a 750 square foot house on a 50 x 75 foot lot. The house was so simple that it lacked an electric door bell or a mechanical knocker. It was assessed at $3960, which led Benger to conclude the house was worth $9000. Benger reported the Carothers' old Dodge automobile was parked in the garage.[12]

He reported Ira Carothers was highly respected, though the bankers considered him somewhat straight-laced. Dr. Benger learned Ira had been sick during much of the summer of 1949 and had incurred about $200 in medical expenses.

Benger went so far as proposing a monthly budget for the Carothers' household. Amortization and interest, $30.36; taxes, $25.16; gas heat, $49.10; and groceries, $45.00. He came to a total of $174.62 and rounded it to $175. Benger assumed the Carothers had a combined Social Security income of $60 a month and calculated assets of $40,000 would be required to provide the income the old couple needed. From his talks with the bankers, he doubted the Carothers had that kind of money. Dr. Benger suggested $150 a month would be generous and would protect the Carothers against all but catastrophic medical expenses.

And so it became. Elmer Bolton went to Crawford Greenewalt with a proposal to send the Carothers $150 a month. On December 20, 1949, Dr. Bolton wrote to Ira Carothers:

---

[11]A two-page letter sent from Des Moines on November 25, 1949, by Benger to A. E. Buchanan describes the investigation. (Hagley Museum and Library Collection, Wilmington, DE, 1784.)

[12]The Carothers actually owned a similar-looking Plymouth.

In my letter of November 15, I advised you that the Company does not have any plan that provides financial assistance to parents of a deceased employee. After bringing to the attention of several of the top executives of the Company your letter of November 2, in which you indicated that it was necessary under present conditions to use some of your capital, the decision was reached that the Company should make an exception in your case in view of the outstanding scientific work which was performed by Wallace during the period of his employment.

I have accordingly been authorized by Mr. C. H. Greenewalt, President, to advise you that beginning January 1, 1950, the Treasurer's Department will forward to you, on a monthly basis, a check for $150 as a gift to be used in defraying some of your expenses. The Company desires to continue making these payments on a monthly basis so long as you are in need of financial assistance.

Elmer Bolton assured Ira Carothers the company would keep the payments confidential and asked him to do the same. Finally, Elmer Bolton wrote:

As the Company is interested in the welfare of you and Mrs. Carothers, will you kindly write to me about every four months in order to let me know whether you are both enjoying good health and how you are getting along so that I may report to the President.[13]

■ ■

Ira Carothers never forgot his instructions, and as the monthly checks came to Ira and Mary in Des Moines, and later, after Mary died in 1956, to Ira, living with Elizabeth and Robert Kyle and their teenage daughters in Cuyahoga Falls, Ira wrote every four months. It was an obligation.

For 11 years, the directors of the chemical department—Elmer Bolton for a year until he retired in 1951, Cole Coolidge until his early and sudden death in 1953, and Paul Salzberg, who as a student of Carothers once snuck into Carothers' laboratory at Illinois to work over the Thanksgiving recess—kept an ever-thickening file of Ira Carothers' letters to DuPont.

Ira Carothers' first letter, sent on the eve of Christmas, 1949, flowed with his gratitude. He acknowledged he understood the money was a

[13]Elmer K. Bolton to Ira Carothers, December 20, 1949. (Hagley Museum and Library Collection, Wilmington, DE, 1784, Box 18.)

gift. As he was in his 81st year, and with his wife almost as old, he predicted "our tenure will probably not be very much longer." He promised to follow Dr. Bolton's suggestion and send "occasional reports."[14]

For the next 11 years, Ira Carothers never wavered from his task. He wrote 45 letters to Bolton, Coolidge, and Salzberg, never less than three a year. That was what he must do. In the difficult time when Mary Carothers, whom he called Molly, was dying, Ira's letters came a bit late. He apologized for his overdue reports.

The chemical department directors always replied immediately. Bolton and Coolidge and particularly Salzberg avoided any perfunctory or obligatory sense. Their responses were warm and chatty. They wrote of seeing Helen about town and of cold winters with ice choking the Chesapeake and Delaware Canal. Salzberg reminisced with Ira Carothers, sending the old man copies of anything in the public press that dealt with nylon or any of the work Wallace had initiated with his fundamental polymer discoveries.[15]

The correspondence for both Ira Carothers and for Elmer Bolton and Cole Coolidge and Paul Salzberg was a means to mourn Wallace Carothers in a private and healing way. For Ira Carothers, who certainly displayed the same distance from his elder son as Wallace held from his father, this last decade of remembrance buried the hurt and anger over Wallace's detached life and sudden and unexplained death. Time after time, Ira Carothers picked an obscure item from the news that referred to nylon or any of DuPont's polymer achievements and asked the chemical department director how this news related to the uniqueness of his son's inventions. The answer from Wilmington was always the same. The inventions of Wallace Carothers formed the base and plain on which the success of the revolution in high polymers was established. Wallace had been their leader, the founder of the chemical feast. The three directors sent Ira Carothers the *DuPont Magazine* and audiotapes of DuPont's "Cavalcade of America" radio broadcasts—anything that had reference to nylon or Wallace. For the directors, who chose not to memorialize the achievements of Carothers to the scores of chemists now inventing in his place at the chemical department and who chose to silence any open discussion of Wallace Carothers' end, the correspondence with the inventor's father was a small and holy and private

---

[14]Ira Carothers to Elmer K. Bolton, December 24, 1949. (Hagley Museum and Library Collection, Wilmington, DE, 1784.)

[15]The Hagley Museum and Library Collection, Wilmington, DE, 1784, Box 5, contains the complete record of Ira Carothers' letters and the DuPont responses.

task, a votive candle lit periodically for all the men and women in the laboratory, to honor the man upon whom they all owed their success.

Late in the summer of 1959, I joined that laboratory, just after its name was changed to the central research department. I was young, just 23, with a new Ph.D. degree, and DuPont expected me to contribute in the research tradition of the department. I worked far down the ladder, but I reported ultimately to Paul Salzberg and remember him as regal and polite. But I had no sense of the deep roots that fed my department and, of course, no idea of the long and loyal attention Dr. Salzberg was still paying to its tradition.

Near the end of his days, in October 1960, Ira Carothers wrote Dr. Salzberg that he had passed his 91st birthday, "with a feeling I have lived too long."[16] Four months later, Elizabeth wrote Paul Salzberg and told him her father had died peacefully at her home in Ohio. She remembered that Julian Hill once worked with her brother in Wilmington and that he had often stopped in Des Moines to visit with her parents. She wanted Dr. Salzberg to make certain Julian knew of Ira Carothers' death. Paul Salzberg was away from DuPont when Elizabeth's letter arrived at DuPont.

Merlin Brubaker, now Salzberg's assistant director, read the message. He was puzzled by the letter until Salzberg's secretary, Viola Kipe, remembered that a payment was being made to Ira Carothers. Dr. Brubaker took care to notify the treasurer so that payments would stop.

The small, quiet but important gift of a few dollars a month was being made for more than a decade, and even those closest to the top of the chemical department were unaware of its existence. Elmer Bolton had promised secrecy to Ira Carothers, of course. But there are times when the finest thing one person can do for another is to give them something—and not tell anyone else about it. And a few men at DuPont—Bolton, Coolidge, and Salzberg—unobtrusively gave the elderly couple not just the money, but a second gift, a trail of meaningful correspondence that became a greater offering, separate and distinct from the freely given financial contribution.

■■

Julian Hill gave the Carothers' time and interest. As Elizabeth Kyle remembered, Julian Hill had visited Wallace's parents in Des Moines several times, first in 1952, then again in 1953, in 1956, and

---

[16]Ira Carothers to Paul Salzberg, October 18, 1960. (Hagley Museum and Library Collection, Wilmington, DE, 1784.)

finally after Molly died, in 1958. But the visits were not merely social. Julian became a guardian for his friend's parents in this second decade after Wallace Carothers' death. Julian's initial visit, in 1952, was designed to quietly reinvestigate the Carothers' financial status. Vice-President Roger Williams initiated the visit, concerned that the monthly gift to the Carothers might need to be increased because of inflation.

In the summer of 1952, Julian Hill wrote Ira Carothers a personal letter and asked to visit. He wrote, "As you know, I was devoted to Wallace and I remember well the pleasant times we had together when you were with him here."[17] Julian carried with him to Des Moines letters of introduction to two bankers and a physician he would visit after seeing Wallace's parents. He visited in their home on Thirty-third Street, drove them on some errands, noting they had 39,000 miles on their Plymouth four-door sedan, and took them to lunch. The bankers cheerfully disclosed the Carothers' financial status in detail. The Carothers did indeed have a $40 per month mortgage payment, a bit higher than Dr. Benger had estimated. Ira Carothers was "ultraprompt" in his payments.

Although Ira Carothers' physician, Dr. Sones, was wary at first, after Julian explained the nature of his mission, the Des Moines doctor gave fragments of Ira's physical history to Julian. Dr. Sones stressed that the Carothers should be left alone. He believed that the old couple would fare best if left independent, without domestic or other help.

Julian Hill wrote a lengthy report to Cole Coolidge, recommending no change in the monthly gift. He stressed it was his personal experience in his own family that elderly people should be allowed to live their own lives. And he wrote, "The best thing of all is attention." But in relating his conversations with Ira and Molly, he noted she said they had considered purchasing a television set. But since there was then only a single channel in Des Moines, they had decided against it.[18]

Cole Coolidge, nevertheless recommended an increase of $50 per month to the gift. He recorded conversations with the treasurer, Vice-President Roger Williams, and a long discussion among several executives over whether President Greenewalt needed to approve the $600 per year change.[19] The fact that Mrs. Carothers had rejected

---

[17]Julian Hill to Ira Carothers, June 4, 1952. (Hagley Museum and Library Collection, Wilmington, DE, 1784.)

[18]Julian Hill to Cole Coolidge, July 14, 1952. (Hagley Museum and Library Collection, Wilmington, DE, 1784, Box 18.)

[19]Cole Coolidge, August 7, 1952, and September 2, 1952. (Hagley Museum and Library Collection, Wilmington, DE.)

purchase of a television set prompted Coolidge to suggest the company buy one for them for Christmas and have Julian Hill present it. But Julian Hill thought such a gift would be gratuitous and patronizing if given by the company and lavish and unexplainable if given in his name.[20] The DuPont executives thought about this matter until November, whereupon the idea was finally abandoned. Julian Hill returned to Des Moines the next year and visited the Carothers. He stopped to see Dr. Sones who reported the old couple's health remained good considering their advanced age.[21] Julian next visited Des Moines in 1956, just one month before Molly died. He took Ira and Molly out to lunch—or rather Ira and Molly took him out, for Ira, at 87, drove his 1941 Plymouth, still with less than 50,000 miles on it, carefully through the Des Moines streets. Julian wrote the Carothers seemed familiar with the city's restaurants and they reminisced about Wallace and the tragedy of the family's children. With Wallace and Isobel dead in 1937, their second son, John, had died in 1943, now only Elizabeth remained alive.[22] In 1958, Julian visited Ira Carothers one last time in Des Moines. The old man had just given up driving after surgery. He was impatient with his recovery, but he seemed to Julian to be in good spirits.[23]

■ ■

Wallace's sister, Elizabeth Kyle wrote Paul Salzberg in 1961 that, "it meant a great deal to Dad to have continued contact with you..."[24]

It was nearly 30 years later that I began research for this book, and I reopened contact with the Kyles to talk about Wallace Carothers, the inventor of nylon.

---

[20]Julian Hill to Cole Coolidge, November 4, 1952. (Hagley Museum and Library Collection, Wilmington, DE, 1784.)

[21]Julian Hill to Paul Salzberg, July 14, 1953. (Hagley Museum and Library Collection, Wilmington, DE, 1784.)

[22]Julian Hill to Paul Salzberg, July 24, 1956. (Hagley Museum and Library Collection, Wilmington, DE, 1784.)

[23]Julian Hill to Paul Salzberg, July 2, 1958. (Hagley Museum and Library Collection, Wilmington, DE, 1784.)

[24]Elizabeth Kyle to Paul Salzberg, February 25, 1961. (Hagley Museum and Library Collection, Wilmington, DE, 1784, Box 18.)

Epilogue
What Is the Family Illness?

When I interviewed Helen Carothers in 1990, in the early stages of my research on her husband's life, she offered that I should talk to her niece, Suzie Wallace Kyle Terrill, who lived in Kent, Ohio. Suzie, Helen told me, was the youngest of the three daughters of Elizabeth Carothers Kyle. She had once planned to write a biography of Wallace Carothers, Helen remembered. Suzie's middle name was Wallace and she had long been interested in her inventor uncle. But Helen said Suzie was dissuaded from writing the biography by Julian Hill, who recommended that task should be left to someone with scientific training.

■■

Suzie Terrill was 45 years old when I met her. She was tall and graceful, with smooth, unlined skin and large, green eyes under strong eyebrows. Suzie has a beautifully framed face with long hair, curled only at the bottom. She wore no makeup save a trace of lipstick. On the telephone she had a soft, placid, even consoling voice, but now in person she was gracious but seemed taut and perplexed. Suzie told me she had located, in a box of her mother's things, a large number of letters from Wallace and from Isobel, written to her grandmother Carothers years ago. But as quickly as she had found them, she had misplaced them. They were, to all appearances, gone.[1]

Suzie wanted to talk, not about Wallace, whom she knew only by her mother's and father's and grandfather's stories, but about her sister, Nancy. Nancy was born in 1940 and was five years older than

---

[1]Later in 1990, Suzie Kyle found letters and photographs and made them available to me.

Suzie. The sisters were close as children, closer as adults because both women searched for an elusive mental peace. Nancy married, had three sons, and worked as a physician's assistant. But for 20 years Nancy seemed to concentrate on her own death. She repeatedly threatened suicide. She was hospitalized many times. Nancy and Suzie searched together over the years for the "family illness," some explanation linking them to Wallace and to their mother, Elizabeth, Wallace's sister, whom they both knew brought to them the turbulent mental states they suffered.

When their mother, Elizabeth Kyle, died in 1983, Nancy missed her terribly. She could no longer work; she became drug-dependent. Nancy lost her last job in Columbus, Ohio, and drove to Cincinnati. There, on her mother's birthday, February 25, 1987, Nancy took a precisely titrated drug overdose and died. She wrote a note before she died, reminding her family how much she had wanted this death for so many years. Nancy's death came as Suzie was arranging for one final hospitalization for her sister at the Hazelden Institute in Minnesota.

Nancy was most likely manic–depressive, Suzie said. That was the most reasonable of many diagnoses. And she believed her uncle Wallace had been manic–depressive also.

■ ■

In the process of biography, the biographer, the imperfect scribe, holds certain fragmentary keys to the life of his subject. Repeated references to Wallace Carothers' drinking formed one of these critical elements. Again and again in my research Wallace's friends remarked about his drinking. Each separate story was only a bit of information, isolated and by itself of little importance. But as the stories accumulated, a pattern emerged. Only with Wallace Carothers' letters to Frances Spencer was the extent of his daily use of alcohol and its deadening effect truly confirmed. Whatever else Wallace Carothers suffered, he carried the message of alcoholism in his genetic makeup.

I did not have Wallace's letters to Frances when I met with Suzie in 1990. But I said to her, "Many of Wallace's friends report he drank quite heavily." Suzie was stunned. She was silent. She whispered, finally, "My mother was an alcoholic."[2]

■ ■

Robert Kyle said that he had repeatedly tried to get his wife to stop her drinking. She had been in treatment of some kind once, but

---

[2]Susan Wallace Kyle Terrill interview with Matthew E. Hermes, May 2, 1990.

it did not change her behavior. Finally, and without Robert Kyle's prompting, Elizabeth stopped drinking alcohol. She was attending the meetings of the group called Alcoholics Anonymous (AA). That was about 1973. Elizabeth Kyle never drank again, and she continued her affiliation with AA until she died 10 years later. But Suzie Terrill and Robert Kyle both noted that the end of Elizabeth's drinking did not entirely solve her problems. Of Elizabeth they spoke no more.[3]

■ ■

On the warm May afternoon I traveled to visit Suzie Kyle Terrill and Robert Kyle, a bright sun colored the new and pale green spirit of a Midwestern spring. I rolled the window down on the rental car I drove on Ohio Route 43, entering Kent from the north. The campus radio station of the state university had preempted its classical music that warm afternoon. Instead it replayed tapes made 20 years ago that day—tapes reporting the progress of the National Guard down this same road on May 2, 1970, enroute to the killing field in front of the Kent State University Library.

It was difficult for me to identify this softly shaded Ohio landscape and the sounds of its spring with those freshly armed Ohio Guardsmen and the noises of their half-tracks. But I knew the sun painted the white and the brown skin and the new fatigues and the shiny boots of the troopers rolling anxiously south on this road in 1970 as truly as it tinted the bright faces and starched dresses of the school children skipping innocently across the highway in 1990 under the direction of the uniformed town policeman who controlled the crossing.

History is not drab and bleached—but my memory once quietly faded the past. I remember looking at some old home movies recently—eight-millimeter movies taken at Yankee Stadium soon after color film became available. It may have been 1939. The infield was emerald green, the basepaths that special mustard brown, and the Yankees, Gordon and Keller, wore uniforms whose whiteness was almost a painful contrast to the vivid colors of the stadium. I was shocked by the bright hues of the old movie. I forgot that the past was distinct, exotic, and emotional, and the replacement I have for the past— books, letters, and memories— strain to express that fullness, that color, that life. Unhappily, the greyness of my own vision of the past originates from periods of depression. I understand William Styron

---

[3]Robert Kyle interview with Matthew E. Hermes, May 2, 1990.

who wrote, "The weather of depression is unmodulated, its light is a brownout."[4] Happily, that time is long in the past, and things have come back to me, slowly, and in bright color.

My interest in Wallace Carothers, from the start was more than his magnificent science. Carothers' depressions attracted me to him. I identified with his sense of inadequacy, with his hopelessness. However the challenge is to write the life, including the drabness he surely felt, but also embracing invention, endurance, and vigor. For those who knew and loved Wallace Carothers were a strong and energetic group. And the pattern of their lives and their science had those moments of magic—sights and sounds and smells unique to the creation of chemistry.

■■

Arnold Collins teased the small rubbery mass of polychloroprene from his flask on an April morning in 1930, in the bright light of early spring in a laboratory filled with the tangy odor of the volatile chemicals fleeing from the opened flask. Julian Hill probably went to get Carothers. Wallace Carothers bounded into the laboratory. Can you see his face? Look past the three-piece suit, his hand darting about with the cigarette scribing trails of thin fire. See a quick smile. But the smile fades as fast as it appears, for Wallace Carothers' mind is racing on at the instant of a new challenge. So many questions have now been posed by the bit of rubber, questions born in Carothers' brain as suddenly as he has appeared in the room. What is the structure of the polymer Collins bounces across the slate laboratory bench? Where did it come from? Which of the imaginary images of carbon skeletons on the field of his mind actually conforms to this new matter? What will they do next?

Carothers never hesitates. His deep voice fills the room over the chatter of the excited young chemists. He tells Arnold Collins to make more of the volatile oil, which he instantly recognizes may be chloroprene.

Two weeks later, Collins summons Wallace Carothers as he returns from a trip to the DuPont building, downtown. And as Carothers enters the laboratory again, Julian Hill turns to him, eyes wide open, mouth agape, a gleeful child pulling yards of lacy fiber from a small tube. Carothers draws his own strands. Tube in one

---

[4]Styron, William. *Darkness Visible;* Vintage Books: New York, 1992; p 19.

hand, tweezers in the other, arms pulled wide now with fiber stretched between his hands, as far as he can reach. He laughs, he turns his head slowly, looking from one hand to the other at the glistening yarn. It is a wonderful moment.

Carothers has scrambled his way up the long grade of a gigantic amusement park coaster, looking up, pushed into his hard seat by the continent's gravity, and suddenly he is at the top, has arced over the top, his body falls away as the coaster drops, but his spirit soars, for a moment disassociated from his already plunging body. He has discovered man's first synthetic rubber and man's first synthetic fiber, surely sufficient magical creation for one lifetime.

Seven years later, nearly to the day, he sits, alone, lost, in a hard chair, in a corner at Whiskey Acres as the laughter and swirl of a party of young men and women surround him. His breath is shallow, he feels his heart flutter, and he cannot remove his attention from the stuttering organ. He is tired, and what's more, he is afraid to move because it seems the least effort at moving makes it nearly impossible to raise his chest to breathe against the column of the Earth's mantle of air. His friends all knew men they had called "shell shocked." Men who had suffered the terror of the trenches in Europe 20 years before and who never lost the vacant gaze in their eyes. Carothers, without the trauma of combat, and with a trail of success equaled by few men, is frightened into immobility. The plunge of the amusement park roller coaster, with its cargo of screaming souls heard faintly in his ears, had ended for his body. His body straining at its confines, at the bottom, under maximum pressure, can fall no further. But his spirit, taken through the material substance with the inertia of the dive, disassembles and parts from his body and keeps dropping away from him. Wallace Carothers' spirit is hopelessly gone.

■ ■

What took Wallace Carothers down? His own story leaves a few traces of his physical and mental and spiritual passage. He admitted abulia, suffered neurotic spells, confessed an angry, tortured memory of his father, was haunted with fear over ambiguous physical symptoms, wrote of physical and moral bankruptcy. The defining image of Wallace Carothers has him clinging to the edge of his desk on a normal work morning trying simply to remain upright in his chair.

He had vagotonia, one physician suggested, "hyperexcitability of the vagus nerve; a condition in which the vagus nerve dominates the general functioning of the body organs."[5] Vagotonia is marked by vaso-motor instability, constipation, sweating, and involuntary motor spasms with pain. From Wilmington, he would be sent to the hospital as suffering from "neurocirculatory asthenia," a package of symptoms now offered as a special presentation of anxiety neuroses. In his last winter, he was often in pain, hunched in agony, in an unremitting abdominal insult. Physicians struggle with this kind of pain. Anxiety and depression heighten the sense of pain. What is everyday and meaningless discomfort in a stable person, exhibits itself as a focus of constant attention for an anxious man, a man with his gaze fixed firmly on himself.[6]

■ ■

Biochemistry and the spirit are intertwined in Wallace Carothers, as surely as they are for all mankind. F. Scott Fitzgerald, Wallace Carothers' exact contemporary, came to the age of 39 and deemed he had lost his soul. He wrote of how "an exceptionally optimistic young man experienced a crack-up of all values...."[7] Wallace Carothers had the opportunity to read Fitzgerald describe his "crack-up" on the train to New York for his wedding, for the initial installment of the three-part essay appeared in the first-ever edition of *Esquire*, dated February 1936.

Fitzgerald wrote:

> Now a man can crack in many ways—can crack in the head—in which the power of decision is taken from you by others! or in the body, when one can but submit to the white hospital world; or in the nerves. William Seabrook in an unsympathetic book tells, with some pride and a movie ending how he became a public charge. What led to his alcoholism or was bound up with it, was a collapse of his nervous system. Though the present writer was not so entangled—having at the time not tasted so much as a glass of beer for six months—it was his nervous reflexes that were giving way—too much anger and too many tears.[8]

F. Scott Fitzgerald and William Seabrook, their stories, and their ultimate ends serve as defining parables for the silent and bankrupt

---

[5]Dorland, W. A. N. *The American Illustrated Medical Dictionary*, 19th ed.; W. B. Saunders: Philadelphia, PA, 1942.

[6]*The Merck Manual*, 15th ed.; Merck & Co.: Rahway, NJ, p 750.

[7]Fitzgerald, F. Scott. *Esquire*, April 1936. Fitzgerald's essays on his "crack up" appear in *Esquire*, February, March, and April 1936.

[8]Fitzgerald, F. Scott. *Esquire*, February 1936.

spirit of Wallace Carothers. Fitzgerald believed he resolved his "crack-up" with a long end-run around his own soul. He endeavored to remain a writer but to leave behind the man he had become, all of it. "There was to be no more giving of myself...." He wrote, "This is what I think now: that the natural state of the sentient adult is qualified unhappiness."[9]

The parable of Fitzgerald and that of Seabrook has a common bulletin. Neither man, left to his own devices could solve what became his torture. For both men, their own best thoughts left them at the bottom. Neither man had a "movie ending."

Fitzgerald's withdrawal gave no solace. His ready denial of his alcoholic drinking, based on a short period of abstinence, shows as a foolhardy inoculation of this new value, soon to be everted. By the next summer Fitzgerald was drinking heavily again and had hired a private nurse to watch over him to control his drinking. Of course, he found a way to elude her.[10] Fitzgerald was ashen, frail, distracted, evasive, a binge drinker, an old man, drained and wan when he died of heart failure at 45.

William Seabrook returned to drinking after his half-year at Bloomingdale's and moved to bizarre forms of ceremonial degradation. He made his estate in Rhinebeck, New York, a center for devil worship. His barn housed a series of under-age "research workers," young women who would eventually reappear and crawl into a hired taxi to be taken back to New York City. Seabrook's drinking became a compulsive nightmare. One night, drunk, enraged at the slavery of drink, he plunged his arms, up to the elbows, in boiling water as a self-punishment for his drinking.[11] Seabrook killed himself in 1943. His spiritual visitor of 1933 had come to have little impact.

■ ■

But recovery of the spirit alone is not sufficient. Carl Jung struggled with the suffering of "a human being who has not discovered what life means for him."[12] As a physician, he met men and women

---

[9]Fitzgerald, F. Scott. *Esquire*, April 1936.

[10]Turnbull, Andrew. *Scott Fitzgerald;* Charles Scribner's Sons: New York, 1962; p 275.

[11]Seabrook, William. *No Hiding Place;* Lippincott & Co.: Philadelphia, PA, 1942; pp 367-394

[12]Jung, C. G. *Modern Man in Search of a Soul;* Dell, W. S.; Baynes, C. F., Trans.; Harcourt, Brace & World: New York, 1933, p 225.

whose suffering came not from physical illness, but from a failure of reason, of a lack of sound personal judgment. But more often, as a philosopher and psychiatrist, he met men and women whose suffering came from a meaninglessness, an absence of the spirit. It was to that spirit Carl Jung returned as he wrote in 1933:

> During the last thirty years, people of all the civilized countries of the earth have consulted me. I have treated many hundreds of patients.... Among all my patients in the second half of life—that is to say, over thirty-five—there has not been one whose problem in the last resort was not finding a religious outlook on life.[13]

About 1930, an American industrialist, a man by the name of Rowland H., came to Carl Jung as a patient. He described his problems with alcohol, but he could not stop drinking, even under Jung's care. Jung described Rowland's condition as hopeless. He said that as a psychiatrist, he could do nothing further for Rowland's drinking. Only one thing could intervene to rescue the man: a religious event beyond Jung's power. Rowland H. left Zurich and joined a strict sect in New York, the Oxford Group, whose tenets held principles of absolute honesty. That religious focus helped Rowland H.; he became sober, and in doing so, he dragged several of his friends into temporary, reluctant sobriety.

One of them, a Vermonter named Ebby, visited a boyhood friend, William Wilson, a stockbroker and a drunk who was drying out at a small hospital on the upper west side of Manhattan. It was December 1934, just past a year since Seabrook's quiet visitor at Bloomingdale's. When Ebby left Wilson's hospital room, the stockbroker received a spiritual phenomenon as ephemeral yet as real as the vision William Seabrook recorded:

> Suddenly my room was blazed with an indescribably white light. I was seized with an ecstasy beyond description. Every joy I had known was pale by comparison.
>
> Then, seen in my mind's eye, there was a mountain. I stood upon its summit, where a great wind blew. A wind, not of air, but of spirit.... Then came the blazing thought 'You are a free man'.[14]

---

[13]Ibid., p 229.

[14]Anonymous. *Pass It On;* Alcoholics Anonymous World Services: New York, 1984; p 121.

Wilson felt that no matter how poorly things had gone for him, his world would be right; he felt peace. Wilson was comforted, closed in, taken care of.[15]

Bill Wilson's encounter with the spirit worked for him. He never drank again. But his recovery came from far more than the few instants when he was visited by the peaceful wind. He decided to tell his story to others—to stay sober by the constant reminder of the hell he had experienced. He learned alcohol was more than the spirit. It was a physical compulsion and a substance for which he had a terrible allergy. He could not drink, not even a dram, because even the most trifling return to alcohol would drive him to believe he could handle it once again. Drinking too was a constant mental obsession that faded only with his daily attention to not drinking. Bill Wilson's formula—not drinking, a dependence on the spirit and telling his story to others—became Alcoholics Anonymous, the group of men and women Elizabeth Kyle joined in 1973, more than 35 years after Bill Wilson and a few other sober people began AA, and more than 35 years after her brother Wallace died.

■ ■

Wallace Carothers was most assuredly alcoholic. He carried the peculiar allergy in his genes. The disease of alcoholism guaranteed he could not drink as a normal man, sipping a beer or testing a Scotch whiskey. For Carothers alcohol meant lots of alcohol, consumed until he was in a stuporous state. The physical pressure to drink, the mental preoccupation, and the loss of any spirit as a result of long years of drinking mark Wallace Carothers indelibly.

His physicians may have suggested the pain in his gut was the beginning of the agony of pancreatitis, the radiating pain that bores to the backbone and is unmodulated no matter how the afflicted man contorts to relieve the pain. They would know chronic pancreatitis is virtually certain to be initiated by heavy drinking. Dr. Appel frantically tried to understand Wallace Carothers' depression, unable through the talking cure to reconcile Carothers' success with the lethargic man he observed. Carothers became a paradigm of his generation's fallen warriors, the war veterans, exactly Carothers' age, who sat blank and motionless along the walls of institutional wards.

---

[15]Townsend Cann knew all these men. In the early 1930s he was engaged to Rowland H.'s daughter. For many years before he died in 1990, Townsend and I talked regularly about these men who were to be founders of the AA movement.

But Wallace Carothers was alcoholic and alcoholics kill themselves sometimes. There was no was way out of the drinking haze for Carothers because AA barely existed then. The peculiar convergence that led to AA and a successful, combined physical, mental, and spiritual approach to alcoholics was so fragile and new. It was so little known and so little understood, even by its own participants, that the movement by the time of Carothers' suicide, extended no further than a handful of smoky living rooms in New York, Akron, and Cleveland.

Wallace Carothers would have a complex diagnosis today. Not just alcoholism. Not just depression. But alcoholism and depression. That dual diagnosis is the family illness.

■ ■

Alcoholics kill themselves when suffering depression. Depressives kill themselves when suffering from alcoholism. Somehow most depressed people can keep barriers in place and not tumble across to the decision that all hope is lost; that they must kill themselves. But alcoholism teases away the barriers. The spirit alcohol washes down protective walls. Alcohol tells the alcoholic that what was once sane and sober no longer makes sense. Brains become scrambled. What seems logical and appropriate to an alcoholic, locked in his lonely shell, talking to no one, fails the test of rational exposure. An alcoholic can make it seem sensible to swallow poison, even if he is Wallace Carothers, even if he has invented a synthetic silk and a synthetic rubber, even if he suspects he is headed to Stockholm and the Nobel Prize.

■ ■

Carothers' biochemistry and his spirit were locked in a tight embrace. The biochemistry of alcoholism, damning Carothers' spirit as its last great gust of ill wind, worked on a body already driven by the biochemistry of depression. Alcohol itself is a central nervous system depressant. And when alcohol is removed, its depressive effects are lessened.

But many people run a second and truer depressive course. They cycle reliably in a biochemical rhythm between some height of action and passion, often exhibiting a compelling functional brilliance. Then they drop to a gloomy melancholy, distraught and inert. There is no single description encompassing all these people. Some exhibit cycles of passion approaching madness as if an internal rheostat was forced beyond its pin, its excess of energy sparking and spitting across its gaps. These depressives are called manic. Others seem to achieve an entirely workable mania, characterized as hypomania. Between cycles of depression, these men and women run to achievement, with heightened energy and crisp judgment.[16]

Somewhere in this scan of behavior lay Wallace Carothers. His cycles were long in duration. His depression of 1932 followed three years of astonishing activity. His friends called him a manic–depressive, but they described no mania. He worked hard and was enormously productive but did not exhibit any of the bizarre. In depression he foundered and found alcohol. The drink drove him deeper. Carothers rallied for a time in early 1934; then in the aftermath of his separation from Sylvia Moore, and with his parents packed off to Des Moines, Carothers packed off to Baltimore. He waged a desperate struggle for the last three years, incapable of putting down the alcohol so he could deal with the true depressive illness.

But Wallace Carothers was locked into a chemical imbalance for which there was no cure in 1937. Just as AA was in its anonymous infancy, the biochemical basis of depressive illness lay undiscovered. Not until 1969 were the simple chemical salts of lithium recognized widely for their control of bipolar depression. Advances in treatment since that time offer a real chemical balancing regime to the depressed person, controlling, but not curing, much of the natural incidence of depressive illness. The "forlorn hope" that the suicide's ailments might be diagnosed and cured should he but wait would have worked out truly for Wallace Carothers, but he would have had to wait for a long, long time.[17]

DuPont's Carothers suffered long "neurotic spells of diminished capacity," but in a brief, bright flame of his power, he brought us nylon and the synthetic revolution.

---

[16]Fieve, Ronald *Moodswing;* Bantam Books: New York, 1989; pp 1–12.

[17]Cross, Frank C. *The Forum,* February 1933, pp 76–78.

Appendix A: Chemistry Cited

$$N\equiv C-N=N-C\equiv N$$

Dicyanocarbodiimide

1

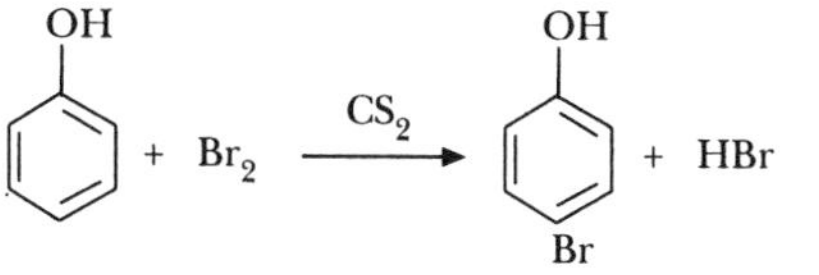

OH + Br$_2$ $\xrightarrow{\text{CS}_2}$ OH Br + HBr

Phenol p-Bromophenol

Reaction 1

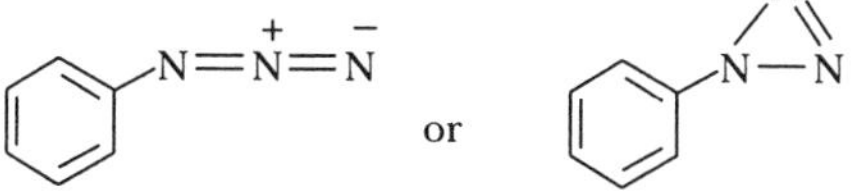

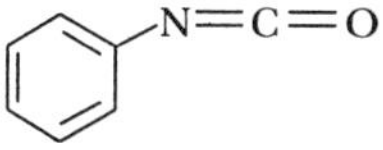

"Diazobenzene-imide"   Phenyl Isocyanate

2   or   **3**

$$RCH=CHCHO + H_2 \xrightarrow[\text{Fe}^{2+}]{\text{PtO}_2} RCH=CHCH_2OH$$

Reaction 2

OHCCH$_2$CH$_2$CHO

Succinaldehyde

4

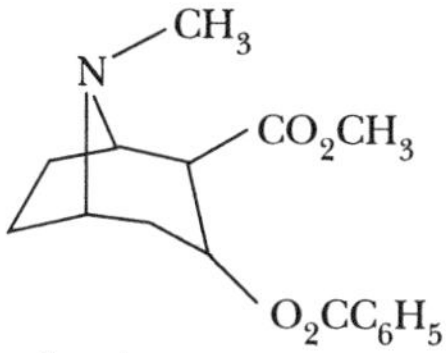

Cocaine

5

Van Natta:

$$HOCH_2(CH_2)_xCH_2OH + HCO_3H \longrightarrow \left(OCH_2(CH_2)_xCH_2OCO\right)_n$$

Reaction 3

Dorough:

$$HOCH_2CH_2OH + HO_2CCH_2CH_2CO_2H \longrightarrow \left(OCH_2CH_2OCOCH_2CH_2CO\right)_n$$

Reaction 4

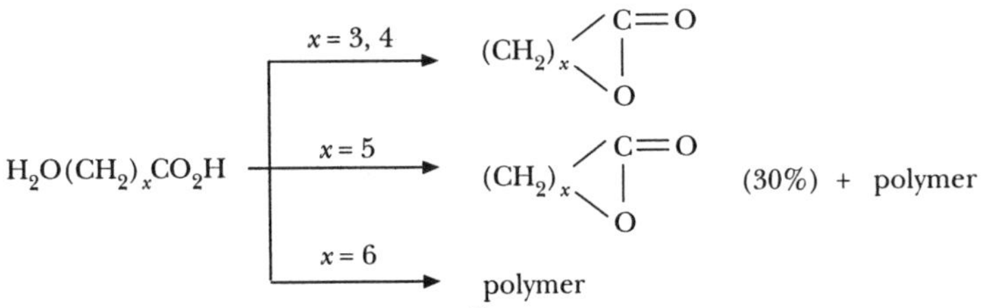

$$H_2O(CH_2)_xCO_2H$$

x = 3, 4 → $(CH_2)_x$ C=O / O

x = 5 → $(CH_2)_x$ C=O / O   (30%) + polymer

x = 6 → polymer

6

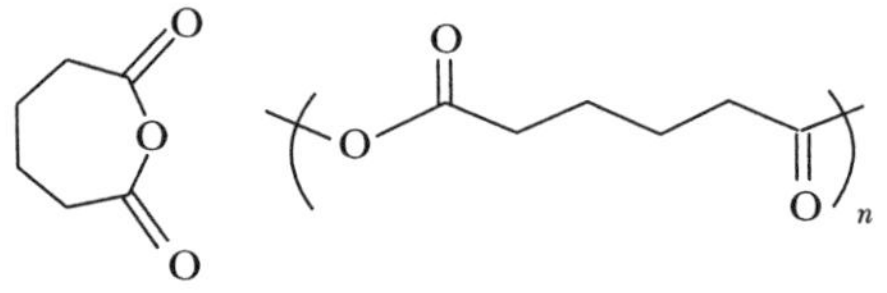

Adipic Anhydride

7

$$HC\equiv CH \xrightarrow{(CuCl)_2} CH_2=\overset{\overset{H}{|}}{C}-C\equiv CH \;+\; CH_2=\overset{\overset{H}{|}}{C}-C\equiv C-\overset{\overset{H}{|}}{C}=CH_2$$

Vinylacetylene   Divinylacetylene

$$CH_2=\overset{\overset{H}{|}}{C}-C\equiv CH + H_2 \longrightarrow CH_2=\overset{\overset{H}{|}}{C}-\overset{\overset{H}{|}}{C}=CH_2$$

Vinylacetylene   Butadiene

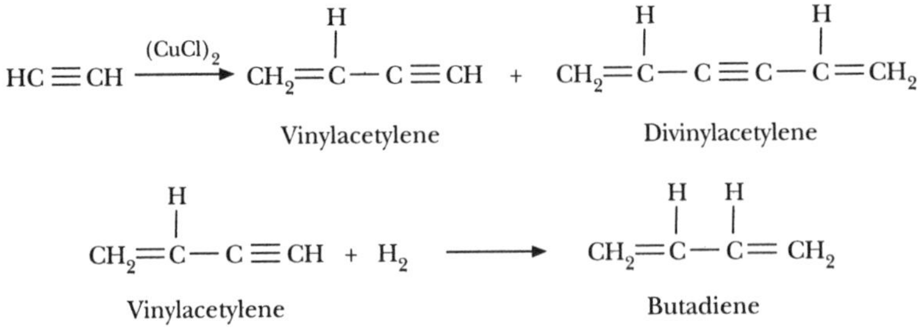

Reaction 5

$$H(CH_2)_{10}Na + H(CH_2)_{20}Br \longrightarrow H(CH_2)_{30}H + NaBr$$

Reaction 6

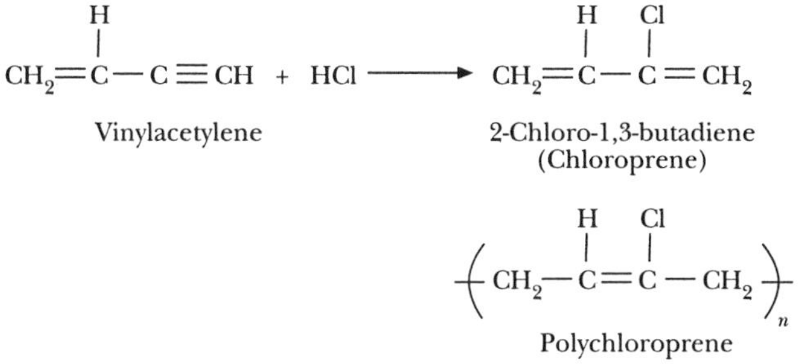

$$\underset{\text{Vinylacetylene}}{CH_2\!=\!\overset{\overset{\displaystyle H}{|}}{C}\!-\!C\!\equiv\!CH} + HCl \longrightarrow \underset{\substack{\text{2-Chloro-1,3-butadiene}\\\text{(Chloroprene)}}}{CH_2\!=\!\overset{\overset{\displaystyle H}{|}}{C}\!-\!\overset{\overset{\displaystyle Cl}{|}}{C}\!=\!CH_2}$$

$$\underset{\text{Polychloroprene}}{\left(\!CH_2\!-\!\overset{\overset{\displaystyle H}{|}}{C}\!=\!\overset{\overset{\displaystyle Cl}{|}}{C}\!-\!CH_2\!\right)_n}$$

Reaction 7

$$HO_2C(CH_2)_{14}CO_2H + HO(CH_2)_3OH \xrightarrow[\text{still}]{\text{molecular}} \left(\!OC(CH_2)_{14}CO_2(CH_2)_3O\!\right)_n$$

$$n \approx 37$$

Reaction 8

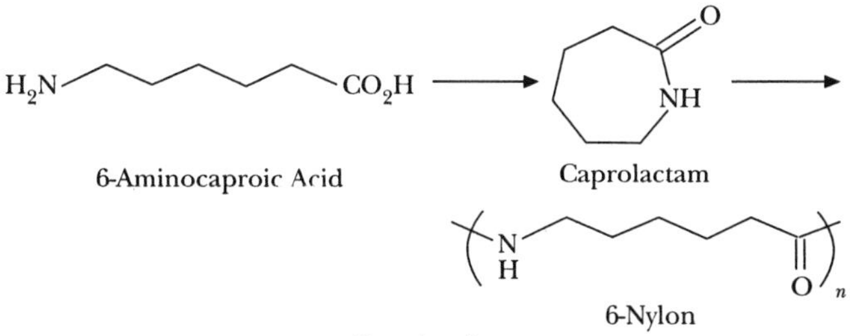

6-Aminocaproic Acid   Caprolactam

6-Nylon

Reaction 9

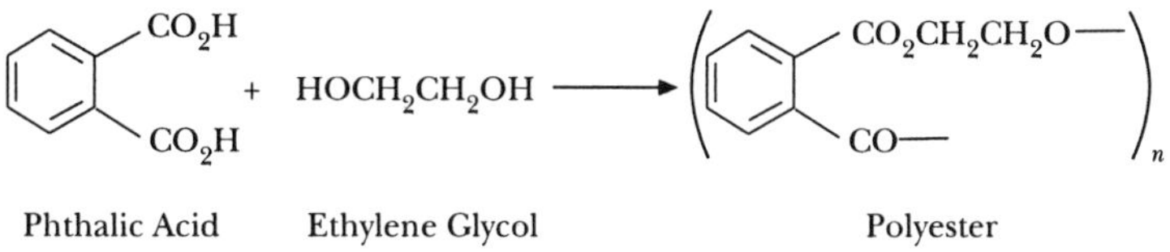

Phthalic Acid   Ethylene Glycol   Polyester

Reaction 10

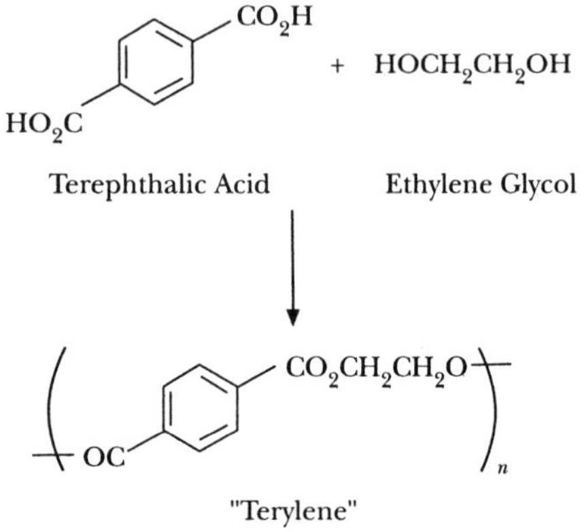
CO2H
+ HOCH2CH2OH
HO2C
Terephthalic Acid
Ethylene Glycol
CO2CH2CH2O
OC
n
"Terylene"

Reaction 11

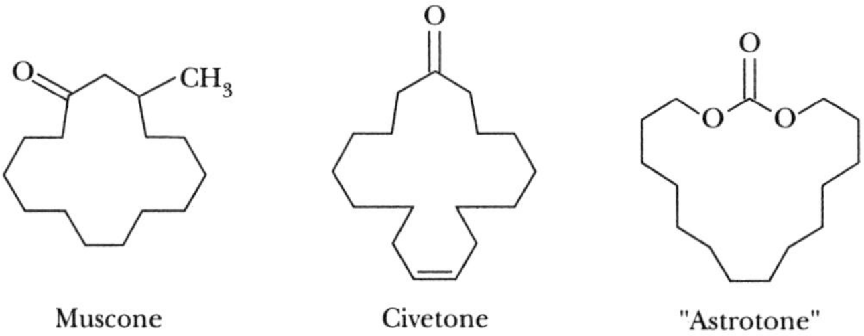
O
CH3
Muscone
O
Civetone
O
O
O
"Astrotone"

8

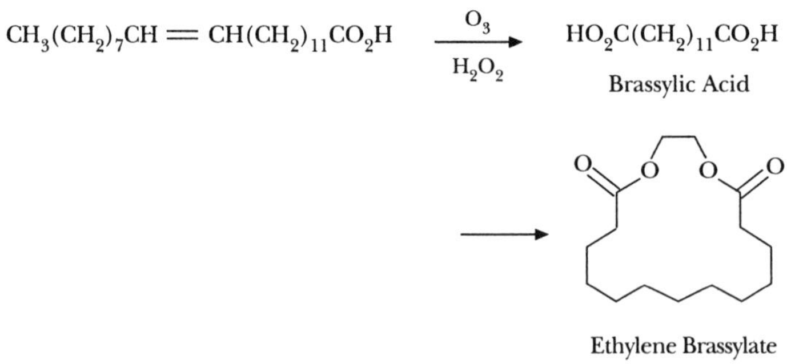
CH3(CH2)7CH = CH(CH2)11CO2H
O3
H2O2
HO2C(CH2)11CO2H
Brassylic Acid
O
O
O
O
Ethylene Brassylate

Reaction 12

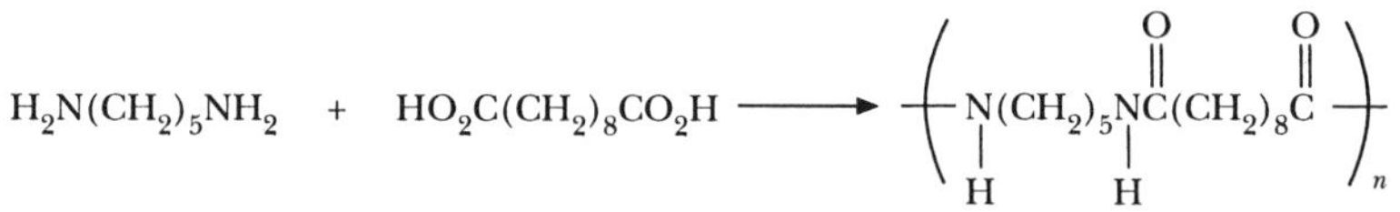

$$H_2N(CH_2)_5NH_2 \quad + \quad HO_2C(CH_2)_8CO_2H \longrightarrow \left(-N(CH_2)_5NC(CH_2)_8C- \right)_n$$

Pentamethylenediamine   Sebacic Acid   Nylon 5-10

Reaction 13

$$CH_3COCH = C = O$$

Acetylketene

9

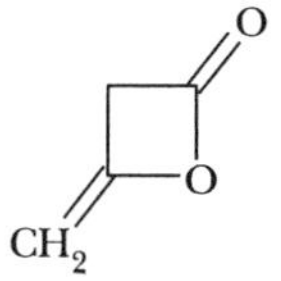

Methylenepropiolactone
(ketene dimer)

10

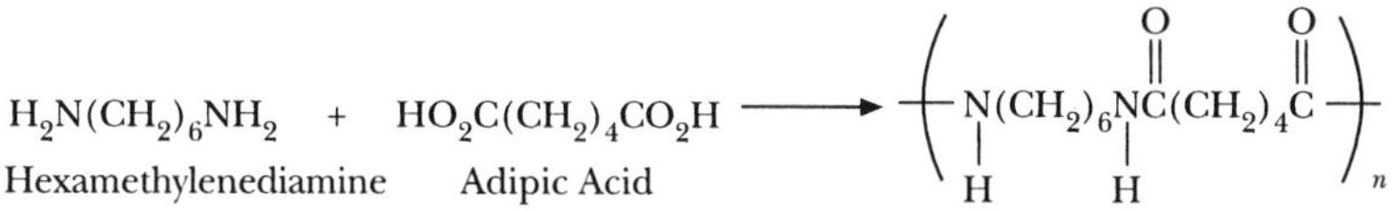

$$H_2N(CH_2)_6NH_2 \quad + \quad HO_2C(CH_2)_4CO_2H \longrightarrow \left(-N(CH_2)_6NC(CH_2)_4C- \right)_n$$

Hexamethylenediamine   Adipic Acid

Nylon 6-6

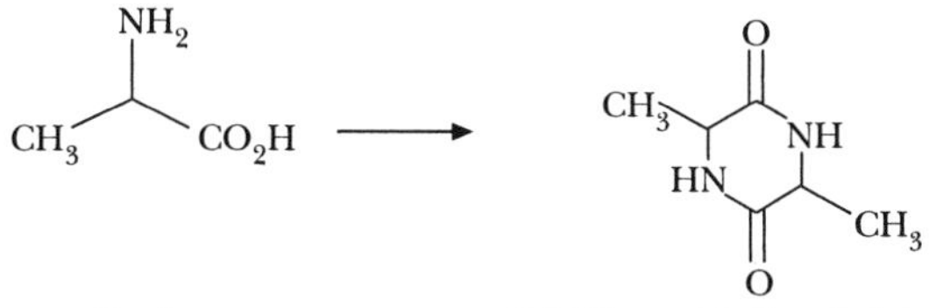

Alanine   2,5-Diketo-3,6-dimethylpiperazine

Reaction 14

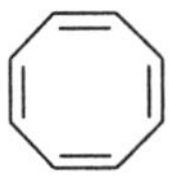

Cyclooctatetraene
11

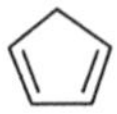

Cyclopentadiene
12

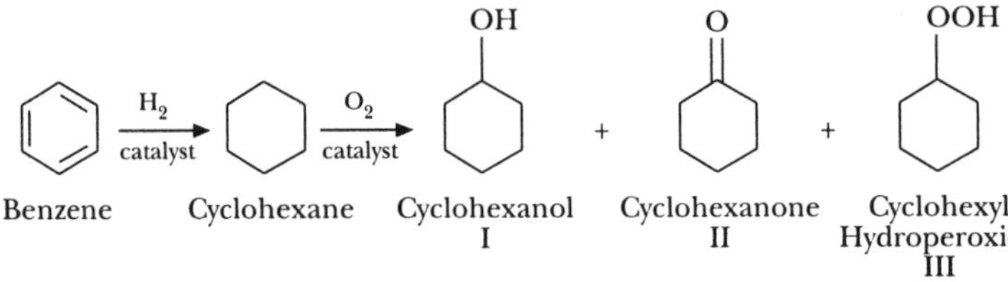
OH
O
OOH
H_2
catalyst
O_2
catalyst
+
+
Benzene
Cyclohexane
Cyclohexanol
I
Cyclohexanone
II
Cyclohexyl
Hydroperoxide
III

$$\text{III} \xrightarrow{\text{Heat}} \text{I} + \text{II} \xrightarrow{HNO_3} HO_2C(CH_2)_4CO_2H$$

Adipic Acid

$$HO_2C(CH_2)_4CO_2H \xrightarrow{NH_3} H_2NOC(CH_2)_4CONH_2$$

Adipic Acid — Adipamide

$$\xrightarrow{\text{dehydrate}} NC(CH_2)_4CN \xrightarrow[\text{catlayst}]{H_2} H_2N(CH_2)_6NH_2$$

Adiponitrile — Hexamethylenediamine

Reaction 15

Acknowledgments

I think my hours with Julian Hill, in the great, grey light of a Martha's Vineyard afternoon in May 1990, taught me the durability of the DuPont culture in which he and Wallace Carothers had prospered and from which I, too, had come. For the DuPont Company—no, more particularly its researchers: decent, talented, cultured—form the core of this book. Dr. Hill, at close to 90 years of age, sparked with a joy of life. He offered music, good food, and a discerning assessment of the beautiful young woman who served our lunch. He offered, too, the great fortune of his life in science, its grand achievements within DuPont, and the continuing pain of his loss of Carothers.

This biography of Carothers arises from my own experience at DuPont's chemical/central research department, where I was but a young and certainly minor player from 1959–1964. The research engine envisioned by Dr. Charles Stine came forward with little change for nearly 40 years through the period I was there. And that was a good thing. The department was stable, productive, driving for research contribution, and to a remarkable degree, unselfishly disengaged from profit motive.

Simply put, we lived off Wallace Carothers' brilliant acievements in a search for new, fundamental science that could throw off a "new nylon." But with tradition and culture now fallen before the necessary impact of business rationalization, all that has changed.

I offer this life of Wallace Carothers in gratitude to my science, my profession. For in a life in chemistry I have received far more from the giants of the science than I ever can repay. Three personal mentors stand out. William Bailey at Maryland taught me what he had learned from Carothers through Speed Marvel and tried to lead me

327

to maturity. I saw Howard Simmons' brilliance at DuPont close on. Burt Anderson, my occasional associate for 35 years, has taught me more about the process of scientific thinking than any other man.

■ ■

The words cooperative and forthcoming are not synonymous. Helen S. Carothers accepted me into her life, allowed me several visits, and cooperated with each and every request. Mrs. Carothers remains quietly proud of her husband's achievements. But she was reluctant to accompany me into the archives of the Philadelphia Institute where we might learn of the summer of 1936, of Dr. Appel and Dr. Carothers together. And without her acquiescence I could not make that visit.

But Mrs. Carothers sent me to her niece, Suzie, daughter of Wallace's sister Elizabeth. And she knew Suzie would be open with me. Suzie told me of her family and herself, leading me from the death of Wallace into the present.

Dr. Sam Lenher carefully managed his clear recollections of the early days at the chemical department, but he offered the contrasting view of that segment of the scientists who actually preferred working to the profit motive.

The elegant, literate Frances Spencer, who in her tenth decade wrote of her past so eloquently, but who was reluctant to converse or meet with me, was not forthcoming until near the end of her life, when the death of her only son prompted her to offer me a treasure of letters from Wallace. And in them, of course, Wallace Carothers defines himself for us at the critical period through his great creative burst.

■ ■

I thank all whom I met, especially Luther Arnold who placed Carothers back at Whiskey Acres at the time of his death, Carothers' brother-in-law Robert Kyle, Ora Machetanz and her son and daughter, Fred Machetanz and Barbara Osborn. Adeline Strange's interviews were of a quality that I appreciated more as this project lengthened; David Hounshell's and John K. Smith's *Science and Corporate Strategy* formed the background relief for the whole biography. In addition, Prof. Hounshell gave me the critical path through the Machetanz letters that led to Frances Spencer.

I ackowledge assistance from the Hagley Museum and Library in Wilmington, Delaware, and Michael Nash and Marjorie McNinch in their manuscripts and archives department. Arnold Thackray and Theodor Benfey of the Chemical Heritage Foundation encouraged this work and advised on its publication. Dr. Bob Joyce, for whom I worked briefly at DuPont, took the time to give my final manuscript a valuable detailed edit. A number of readers suggested a chemical appendix for the book. Bob prepared one for me, and I acknowledge that unanticipated and greatly appreciated assistance.

■ ■

One sultry afternoon in the summer of 1989 I sat with Cas Speare in the screened porch of his primitive camp on a rocky islet in Georgian Bay off Lake Huron. I opened *Science and Corporate Strategy* and told Cas how I wanted to write Carothers' life. He encouraged me. That afternoon, with Cas' help, I recognized I lacked the "necessaries" for authorship. I would have to return to school and learn to write. At Wesleyan University that fall I met Anne Greene, who led me through a new, midlife masters' degree and edited the early versions of this work.

To Cas, to whom I have turned for guidance in every difficult moment—personal and professional—since 1983, I owe my deepest thanks and gratitude. I want to thank Anne who taught and edited and who forced me to pay far more attention than I wished to a couple of rigorous reviewers at the Wesleyan Writer's Conference in 1993.

Bibliography

Adams, Roger. *Organic Syntheses;* John Wiley and Sons: New York, 1921; Vol. 1.

Anonymous. *Pass It On;* Alcoholics Anonymous World Services: New York, 1984.

Appel, Kenneth E.; Strecker, Edward A. *Practical Examination of Personality and Behavior Disorders;* Macmillan: New York, 1936.

Brandrup, J.; Immergut, E. H. *Polymer Handbook;* Intersciences: New York, 1967.

Brock, William H. *The Norton History of Chemistry;* W. W. Norton & Company: New York, 1992.

Burk, Robert F. *The Corporate State and the Broker State, The Du Ponts and American National Politics, 1925–1940;* Harvard University Press: Cambridge, MA, 1990.

Chandler, Alfred D.; Salsbury, Stephen. *Pierre S. duPont and the Making of the Modern American Corporation;* Harper & Row: New York, 1971.

Chandler, Alfred D. *The Visible Hand, The Management Revolution in American Business;* Belknap Press of Harvard University Press: Cambridge, MA, 1977.

Conant, James B. *My Several Lives;* Harper & Row: New York, 1970.

Conant, James B. *The Chemistry of Organic Compounds;* Macmillan: New York, 1939.

Cowley, Malcolm *Exile's Return;* Viking: New York, 1956.

Dorland, W. A. Newman. *The American Illustrated Medical Dictionary,* 19th ed.; W. B. Saunders: Philadelphia, PA, 1942.

Duncan, Robert Kennedy. *Some Chemical Problems of Today;* Harper & Brothers: New York, 1911.

Duncan, Robert Kennedy. *The Chemistry of Commerce;* Harper & Brothers: New York, 1907.

Duncan, Robert Kennedy. *The New Knowledge;* A. S. Barnes & Company: New York, 1906.

Dutton, William S. *DuPont, One Hundred and Forty Years;* Charles Scribner's Sons: New York, 1942.

Eliot, T. S. *Collected Poems, 1909–1962;* Harcourt, Brace & World: New York, 1970.

Encyclopedia of Psychology; Corsini, Raymond J., Ed.; John Wiley and Sons: New York, 1984.

Federal Writers Project. *Delaware, A Guide to the First State;* Viking: New York, 1938.

Fieve, Ronald. *Moodswing,* Revised ed.; Bantam Books: New York, 1989.

Flory, Paul J. *Statistical Mechanics of Chain Molecules;* Hanser: Munich, Germany, 1989.

Goodman, Lisl M. *Death and the Creative Life;* Springer: New York, 1981.

Haynes, William. *This Chemical Age, The Miracle of Man-Made Materials;* Alfred A. Knopf: New York, 1942.

Hemingway, Ernest. *A Moveable Feast;* Macmillan: New York, 1964.

Hemingway, Ernest. *The Sun Also Rises;* MacMillan: New York, 1926.

Hermes, Matthew E. "Synthetic Fibers from 'Pure Science': DuPont Hires Carothers;" In *Manmade Fibers: Their Origin and Development;* Seymour, Raymond B.; Porter, Roger S., Eds.; Elsevier Applied Science: London, 1993; pp 227–243.

Hounshell, David A.; Smith, John Kenly. *Science and Corporate Strategy, DuPont R&D 1902–1980;* Cambridge University Press: New York, 1988.

Hughes, Thomas. *American Genesis;* Viking: New York, 1989.

James, William. *The Varieties of Religious Experience;* Longmans Green and Company: New York, 1902.

Jung, C. G. *Modern Man in Search of a Soul;* Dell, W. S.; Baynes, C. F., Trans.; Harcourt, Brace & World: New York, 1933.

Lewis, Gilbert Newton. *Valence and the Structure of Atoms and Molecules;* American Chemical Society Monograph Series; American Chemical Society: Washington, DC, 1923. Available from University Microfilms International, 300 North Zeeb Road, Ann Arbor, MI 48106.

Manmade Fibers: Their Origin and Development; Seymour, Raymond; Porter, Roger, Eds.; Elsevier Applied Science: London, 1993.

Mark, H. "Polymer Chemistry in Europe and America—How It All Began," *J. Chem. Educ.* **1981**, *58*, 527.

Mark, H.; Whitby, G. S. *Collected Papers of W. H. Carothers on High Polymeric Substances;* Interscience: New York, 1940.

Marvel, Carl S. "The Development of Polymer Chemistry in America—The Early Days," *J. Chem. Educ.* **1981**, *58*, 535.

Milford, Nancy. *Zelda, A Biography;* Harper & Row: New York, 1970.

Millikan, Robert Andrews. *The Electron;* University of Chicago Press: Chicago, IL, 1924.

Morawetz, Herbert. *Polymers: The Origin and Growth of a Science;* John Wiley and Sons: New York, 1985.

Noble, David F. *America by Design;* Alfred A. Knopf: New York, 1979.

Packard, Frances R. *Some Account of the Pennsylvania Hospital from Its First Rise to the Beginning of the Year 1938;* Engle: Philadelphia, PA, 1938.

Pauling, Linus. *The Nature of the Chemical Bond;* Cornell University Press: Ithaca, NY, 1942.

Pensak, Robert J.; Williams, Dwight A. *Raising Lazarus;* G. P. Putnam's Sons: New York, 1994.

Perkin, W. H.; Kipping, F. Stanley. *Organic Chemistry;* W. & R. Chambers: London, 1911.

Pratt, Herbert T. "Textile Mill Scale-up of Nylon Hosiery: 1937–1938;" In *Manmade Fibers: Their Origin and Development;* Seymour, Raymond B.; Porter, Roger S., Eds.; Elsevier Applied Science: New York, 1993; pp 244–266.

Proceedings of the Robert A. Welch Foundation Conferences on Chemical Research, XX. American Chemistry—Bicentennial; Milligan, W. O., Ed.; Houston, TX, 1977.

Sartre, Jean-Paul. *Baudelaire;* Turnell, Martin, Trans.; New Directions: New York, 1950.

Seabrook, William. *Asylum;* Harcourt, Brace & Company: New York, 1935.

Seabrook, William. *No Hiding Place;* Lippincott & Co.: Philadelphia, PA, 1942.

Smith, John K. "The Ten-Year Invention: Neoprene and DuPont Research, 1930–1939," *Tech. Cult.* **1985**, *26*, 34.

Styron, William. *Darkness Visible, A Memoir of Madness;* Vintage Books: New York, 1992.

Tarbell, D. Stanley; Tarbell, Ann Tracy. *Roger Adams: Scientist and Statesman;* American Chemical Society: Washington, DC, 1981.

Thackray, Arnold; Sturchio, Jeffrey L.; Carroll, P. Thomas; Bud, Robert. *Chemistry in America, 1876–1976;* D. Reidel: Dordrecht, Holland, 1985.

Turnbull, Andrew. *Scott Fitzgerald;* Charles Scribner's Sons: New York, 1962.

Wolff, Geoffrey. *Black Sun;* Vintage Books: New York, 1977.

Document Sources

Hagley Museum and Library

Accession Number	Collection Subject
1662	Records of the E. I. duPont de Nemours & Co., administrative papers of the Office of the President
1689	Notes and interviews for Chandler & Salsbury, *Pierre S. duPont and the Making of the Modern Corporation*

Accession Number	*Collection Subject*
1784	Records of the DuPont Central Research and Development Department
1842	Correspondence of John R. Johnson and Wallace H. Carothers
1850	Papers relating to the preparation of Hounshell and Smith, *Science and Corporate Strategy*
1878	Hounshell and Smith interviews
1896	Wallace H. Carothers correspondence
1927	Pre-employment correspondence between DuPont and Wallace H. Carothers
1985	Mrs. A. B. C. Strange Collection of Carothers material

University of Illinois Archives
>Roger Adams Papers
>Board of Trustees Reports
>R. C. Fuson Papers

Chemical Heritage Foundation
>C. S. Marvel Papers

University of South Dakota
>Chemistry department records

Susan Wallace Kyle Terrill
>Letters, photographs, and Carothers' family records

Barbara Osborn
>Carothers letter, *The Tarkiana 1917*

Ora Machetanz
>Machetanz family photographs

Fred Machetanz
>Carothers' letters to Wilko Machetanz

Frances Spencer
>Fourteen letters between Mrs. Spencer and Wallace Carothers, 1929–1934.

Helen Carothers
>Carothers' family photographs

Anne Knepley
>Gerard Berchet letter and documents

Dennis Berchet
>Gerard Berchet documents and tape

Interviews

Adams, Roger. Interview at the Chemists Club, New York, 1964. (Adams Papers, University of Illinois Archives, Urbana, IL, Box 9.)

Alvarado, A. Interview with Adeline C. Strange, 1978. (Hagley Museum and Library Collection, Wilmington, DE.)

Anderson, Burton C. Interview with Matthew E. Hermes, August 20, 1994.

Arnold, Luther. Interview with Matthew E. Hermes, April 19, 1991.

Berchet, Dennis. Interview with Matthew E. Hermes, July 11, 1994.

Berchet, Gerard. Interview with John K. Smith, November 10, 1982. (Hagley Museum and Library Collection, Wilmington, DE, 1878.) Interview with Adeline C. Strange, 1978. (Hagley Museum and Library Collection, Wilmington, DE.)

Bolton, Elmer K. Interview with Alfred D. Chandler, Richmond D. Williams, and Norman Wilkinson, 1961. (Hagley Museum and Library Collection, Wilmington, DE, 1689.) Interview with Adeline C. Strange, 1978. (Hagley Museum and Library Collection, Wilmington, DE.)

Brubaker, Merlin. Interview with Adeline C. Strange, 1978. (Hagley Museum and Library Collection, Wilmington, DE.) Interview with John K. Smith, September 27, 1982. (Hagley Museum and Library Collection, Wilmington, DE, 1878.)

Cann, Townsend. Interviews with Matthew E. Hermes, 1985–1990.

Carothers, Helen. Interview with Matthew E. Hermes, February 19, 1990.

Cupery, Hal. Interview with Adeline C. Strange, August 2, 1978. (Hagley Museum and Library Collection, Wilmington, DE.)

Duncan, Virginia. Interview with Adeline C. Strange, June 1978. (Hagley Museum and Library Collection, Wilmington, DE.)

Dykstra, Harry. Interview with Adeline C. Strange, August 2, 1978. (Hagley Museum and Library Collection, Wilmington, DE.)

Greenewalt, Crawford. Interview with Adeline C. Strange, 1978. (Hagley Museum and Library Collection, Wilmington, DE.)

Hanford, William E. Interview with John K. Smith, October 29, 1985. (Hagley Museum and Library Collection, Wilmington, DE, 1878.) Interview with Matthew E. Hermes, July 9, 1990.

Hill, Julian. Interview with David A. Hounshell and John K. Smith, December 1, 1982. (Hagley Museum and Library Collection, Wilmington, DE, 1878.) Interview with Adeline C. Strange, 1978. (Hagley Museum and Library Collection, Wilmington, DE.) Interview with Matthew E. Hermes, May 29, 1990.

Hill, Polly. Interview with Adeline C. Strange, 1978. (Hagley Museum and Library Collection, Wilmington, DE.) Interview with Matthew E. Hermes, May 29, 1990.

Hounshell, David. Interview with Matthew E. Hermes, August 7, 1989.

Johnson, John R. Interview with Adeline C. Strange, July 1978. (Hagley Museum and Library Collection, Wilmington, DE.)

Joyce, Robert. Interview with Matthew E. Hermes, August 4, 1995.

Knepley, Anne. Interview with Matthew E. Hermes, July 11, 1994.

Kyle, Elizabeth. Interview with Adeline C. Strange, July 3, 1978. (Hagley Museum and Library Collection, Wilmington, DE.)

Kyle, Robert. Interview with Matthew E. Hermes, May 2, 1990.

Labovsky, Joseph. Interview with Matthew E. Hermes, June 28, 1991.

Lenher, Samuel. Interview with Matthew E. Hermes, August 24, 1990.

Machetanz, Ora. Interview with Matthew E. Hermes, December 9, 1989.

Machetanz, Fred. Interview with Matthew E. Hermes, September 25, 1989.

MacKenzie, Anne. Interview with Matthew E. Hermes, November 27, 1992.

Mapel, William C. Interview with Adeline C. Strange, July 1978. (Hagley Museum and Library Collection, Wilmington, DE.)

Marvel, Carl S. Interview with David A. Hounshell and John K. Smith, May 2, 1983. (Hagley Museum and Library Collection, Wilmington, DE, 1878.) Interview with J. E. Mulvaney. (Chemical Heritage Foundation, Philadelphia, PA.)

Miles, John and Lib. Interview with Adeline C. Strange, July 1978. (Hagley Museum and Library Collection, Wilmington, DE.)

Osborn, Barbara. Interview with Matthew E. Hermes, April 27, 1990.

Pratt, Herbert. Interview with Matthew E. Hermes, April 18, 1991.

Reese, Charles L. Interview with Adeline C. Strange, 1978. (Hagley Museum and Library Collection, Wilmington, DE.)

Salzberg, Paul L. Interview with John K. Smith, September 29, 1982. (Hagley Museum and Library Collection, Wilmington, DE, 1878.)

Schwartz, A. Truman. Interview with Matthew E. Hermes, June 1, 1993.

Stockmayer, Walter. Interview with Matthew E. Hermes, April 1, 1991.

Strange, Adeline Cook. Interview with Matthew E. Hermes, October 3, 1989.

Terrill, Susan Wallace Kyle. Interview with Matthew E. Hermes, May 2, 1990.

Whybrow, Peter. Interview with Matthew E. Hermes, May 1992.

Index

A

Abulia
  Spring 1923, 34
  Wallace Carothers, 255
Adams, Roger
  Alps hiking trip with Carothers, 271–276
  career, 232
  evening musicals and drinking at Wawaset Park, 256
  preparing Carothers' application for National Academy of Sciences, 236
  reaction to Carothers' hospitalization, 199–200
  refusal to move to DuPont, 75
  research consultant, 95
  stock market speculation, 97–98
  University of Illinois Department of Chemistry, 31
  University of Illinois Organic Chemistry Division, 24
Adams' platinum, hydrogenation, 31–32
Addition polymerization, synthetic rubber, 106
Adipic anhydride, Julian Hill, 99–100
Alcohol
  compounded mental problems, 153–154
  effect on Carothers, *iv*
  Elizabeth Carothers Kyle, 310–311
  F. Scott Fitzgerald, 315
  first drink, 12
  repeal of Prohibition, 193
  speech at organic symposium, 144
  trip to Alps, 273
  Wallace Carothers, 310
  weekend guests, 149
  William Seabrook, 315
Alcoholics Anonymous, foundation, 317
Alcoholism
  effect of, 318
  genetic source, 317
Aldehyde, platinum catalyst, 32
Alkanes, stable, long-chain, 108–109
Alps, hiking trip, 267
Amides, program to measure basicity, 52

Aminononanoic ester, superpolymer, 187–188
Andrew, Nancy (grandmother), move to Iowa, 2
Anxiety neuroses, Wallace Carothers, 261
Appel, Kenneth Ellmaker
  attempt to prevent suicide, 289
  Carothers' psychoanalyst, 262–264
  prediction of Carothers' suicide, 283
Arden, Delaware
  house shared with parents, 173–174
  parents returned to Des Moines, 196
Artificial silk
  DuPont, 62–63
  fiber-forming properties of polymers, 115
Arvin, James A.
  glycols and dicarboxylic acids, 95
  research staff, 94
Astrotone, fragrant cyclic carbonate sold as perfume ingredient, 164
AT&T, Bell Labs, 66

B

Backdoor behavior, instances of depression, 203
Bell Labs, fundamental research facility, 66
Belle, West Virginia, manufacture of nylon salt, 295–296
Benger, Ernest, investigation of situation of Ira and Mary Carothers, 301–302
Benzene, feedstock for nylon, 218–219, 295
Berchet, Gerard
  chemist at College de France, 47
  discovered nylon in study of polyamides, 216–217
  DuPont career, 229–230
  immigration, 48
  vinylacetylene, 107
Berolzheimer, Howard (brother-in law), husband of Isobel Carothers, 279
Bolton, Elmer K.
  arranged Carothers' confinement, 271–276

chose polyamide 6–6 as commercial
    candidate, 218
compared to Carothers, 237–238
DuPont director of dye research, 61–62
DuPont research and development, 181
elimination of fundamental research,
    158
head of the chemical department,
    121–134
research organization, 110
synthetic rubber, 105–112
Bonding, repeated multifold in polymers,
    56–57
Bonus money, Helen Carothers, 298–300
Bootlegger, near Whiskey Acres, 154
Bradshaw, Hamilton
    interest in polymerization, 55
    request from Carothers, 53
Bremen, ocean liner to Faraday Society
    Meeting, 239–240
Bridal shower, Helen Sweetman, 251
Brubaker, Merlin
    artificial musks, 164
    University of Illinois, 40
Budget cuts, research program, 158

C

Capital Cities Commercial College
    Ira Carothers, 2
    Wallace Carothers' attendance, 5
Caprolactam, publication and patent
    claims, 134
Carothers, Andrew (great grandfather),
    Pennsylvania farmer, 2
Carothers, Elizabeth (sister)
    wedding, 142
    See also Kyle, Elizabeth Carothers
Carothers, Helen (wife)
    bonus payments from DuPont, 298–299
    pregnancy, 290
    See also Sweetman, Helen
Carothers, Ira Hume (father)
    death, 305
    marriage, 2
    pension because of son's contribution
        to DuPont, 296–299
    photo, 297
    provided staff for Tarkio College, 1
    reports to DuPont, 303–304
    son's memories, 178
    unaware of Wallace's strange behavior,
        279
Carothers, Isobel (sister)
    comic radio performance, 175–176
    death, 279–282

photo, 280
    visit to University of Illinois, 35–36
    See also Berolzheimer, Howard
Carothers, Jane (daughter), birth, 293
Carothers, John (grandfather), move to
    Iowa, 2
Carothers, Mary (mother)
    death, 304
    last letter from Wallace, 285–286
    pension because of son's contribution
        to DuPont, 296–299
    unaware of Wallace's strange behavior,
        279
    See also McMullen, Mary
Carothers, Roger (great-great grandfa-
    ther), Pennsylvania farm, 2
Carothers, Wallace Hume
    arrival at Tarkio College, 1
    assessment of his own personality,
        86–87
    boarding train for New York wedding
        (photo), 252
    description of daily life, 148
    early interest in science, 3
    friends' dislike for Helen, 255
    inability to carry out desire to marry,
        255
    last letter to mother (photo), 288–289
    marriage, 250–255
    newspaper report of death (photo), 292
    pessimistic outlook on life, 7–9
    Philadelphia Institute, 259–261
    reaction to pregnancy, 290–291
    suicide, 285–294
    symptoms, 313–314
    with Frances Gelvin (photo), 10
    with Hill family (photo), 207
    with Tort Gelvin (photo), 14
    with Wilko Machetanz (photo), 7
    work and death, *i*
    yearbook description, 11
    yearbook photo, 12
Carothers family (photo), 3
Carothers Laboratory, memorial to inven-
    tor, *v*
Carr Fellowship, doctoral research, 33–34
Carty, J. J.
    Bell Labs, 66
    challenge to have coast-to-coast
        telephone in operation by 1914, 69
Catalysts, simple realization of shape and
    surface, 247
Catalytic hydrogenation, predictive
    mechanism, 38
Chain length for a useful fiber, 215
Chain terminator, acetic acid, 215

Change, difficulty in adapting, 82–83
Charch, Hale, tests on polyester fibers, 131
Chemical imbalance, Wallace Carothers, 319
Chemistry, introduction at Tarkio College, 15
Chlorobutadiene, structure, 110
Chloroprene, publications, 112
Cigarettes, father's reaction, 11
Civet, natural resources, 163
Cocaine
    medicinal uses, 53
    synthesis, 53–54
Coffman, Donald
    aminononanoic acid, 186–188
    new machinery for melt spinning the polymer, 214
    summer at DuPont, 94
Cofman, Victor, DuPont, 123–124
Cold drawing, polymer fibers, 115–116
Cold-drawn fibers, diagram, 116
Collins, Arnold
    synthetic drying oil, 107
    world's first synthetic rubber, 109–112
Colpitts, Edwin H., Bell Labs, 66
Conant, James B.
    chemical stability of two exotic ring structures, 282
    emphasis on research, 42–43
    response to research proposal, 57
Condensation polymerization
    presentation to Faraday Society, 244
    publication, 96–97
    structural unit vs. physical properties, 93
    worldwide stage, 236
Coolidge, Cole, 303–307
Coolidge, William D., ductile, high-resistance tungsten wire, 69
Cremation, Wallace Carothers, 293
Cupery, Martin
    Dutch Gold Dust Twins, 160
    structure of polymers, 161
Cyanide
    capsule attached to watch chain, 135
    cause of death, 291
    pills in his pocket, 41
Cyclic compounds
    alternatives to long-chain polymers, 162
    characteristic odors, 163–165

D

Davis, Albert G., GE industrial research laboratory, 65–66
Death, Wallace Carothers, 291
Delusterant, rayon, 183

Depression
    effect of alcoholism, 318
    effect on Carothers, *iv*
    overall discouragement, 145–156
    speculation, 119
    symptoms, 264
    teaching and research, 41–42
Depression, economic, *See* Hard Times
Despair, violent storms, 150
Diacids, synthesis of long-chain molecules, 92
Diazobenzene-imide, structure, 28–29
Difunctional monomers, long chains, 185–186
Disappearance
    hiking trip in the Alps, 275–276
    summer of 1934, 191–210
Divinylacetylene, catalyst, 106
Doctoral research, a form of slavery, 33
Dorough, Gus
    four-carbon succinic acid combined with ethylene glycol, 95
    research staff, 94
Double Bond, The, chemical reactions of the four-electron bond, 37–39
Dry-cleaning solvent, new polyamide fibers, 188
Duco automotive finish, application of science to a particular issue, 66–67
Duco lacquer, automotive finishes, 63
Duncan, Robert Kennedy
    imaginative chronicler and organizer, 67
    predictions, 13–15
    The New Knowledge, 3
DuPont
    decentralization of research, 64
    early proposed research, 55–56
    employment decision process, 77–85
    German chemists, 61–62
    growth from munitions to consumer chemicals, 60
    hiring practices, 123
    interest in Carothers, 52
    relationship with Ira Carothers, 296–307
    still-secret development of synthetic fiber, 236
    work at experimental station, 89
duPont
    family, churches built, 59
    Irénée, predictions for chemistry, 63
    Pierre Samuel, business development, 60

Duprene
    announcement, 132
    world's first synthetic rubber, 109–112
Dyestuffs, DuPont Company, 60–62
Dykstra, Harry, Dutch Gold Dust Twins,
    160

E

Electron pairs, planar illustration, 38
Electron polarization, chemical reactivity
    of the double bond, 39
Engine knock, tetraethyl lead, 65
Ethyl group, negatively charged, 53
Europische Hof
    Adams left and Carothers remained
        behind, 275
    Innsbruck, base for hikes, 274

F

Fabric, surface textures, 182
Family illness
    alcoholism and depression, 318
    Carothers family, 309–319
Faraday Society of London, international
    assembly of polymer scientists, 235–247
Fiber
    critical pathway discovery, 116
    increased molecular weight, 115
Fiber 66, fiber Wallace Carothers in-
    vented, 295
Financial worries, parents, 170
Fitzgerald, F. Scott
    crack-up of all values, 314–315
    treatment at Phipps Clinic, 200–201
Flory, Paul J.
    career, 230
    chain lengths of polymers, 215
Fortune magazine, discussion of nylon, *ii*
Friends, inability to understand Wallace
    Carothers, 233–234

G

Gelvin, Frances
    college friendships, 9
    Tarkio College, 6
    Tarkio College yearbook (photo), 8
    with Wallace Carothers (photo), 10
    See also Spencer, Frances
Gelvin, Tort
    married Wilko Machetanz, 85
    Tarkio College yearbook (photo), 8
    with Wallace Carothers (photo), 14
General Electric, first industrial research
    laboratory in America, 65–66

General Motors, control by DuPont, 60
German scientists, DuPont dye research,
    61–62
Germany, accomplishments in chemistry,
    21
Gifford, Alice, arrival at Tarkio College, 1
Glycols, synthesis of long-chain mole-
    cules, 92
Goiter, exemption from military service,
    18
Gordon Research Conferences, descrip-
    tion, 241
Graves, George, polyamide development,
    238, 249
Greenewalt, Crawford
    group manager under Bolton, 161
    married into duPont family, 221
    president of DuPont, 296
Grignard reagent, electric current, 53
Grimm, Leslie, Tarkio College, 15
Grimm, Marie, yearbook description, 11
Growth, personal and professional, 162

H

Hard Times
    end of Prohibition, 194
    family's financial plight, 179
    market crash effects, 143
    profitable opportunity in fibers, 182
    quiet marriage, 254
    staff with negligible turnover, 158
Harvard
    appointed instructor, 47
    comparison to University of Illinois, 51
    optimism about future, 55
    thoughts of returning, 170–171
Hasche, Leonard, Tarkio College, 15
Health, growing concern, 146
Healthy-mindedness, definition, 233
Hill, Julian
    barrier broken, 113–115
    best man, 253
    career and retirement, 228–229
    equilibrium of long-chain polymers and
        rings, 163
    fiber-forming properties of polymers,
        115
    guardian of Ira and Mary Carothers,
        305–307
    in 1990 (photo), 209
    memories of Kohler, 49
    nitroglycerin production problem, 99
    weekend on Martha's Vineyard, 206
    with family and Wallace Carothers
        (photo), 207

Hill, Polly
    description of Wallace Carothers,
        209–210
    with family and Wallace Carothers
        (photo), 207
Hohman, Leslie B.
    approach to patient depression,
        202–203
    lifestyle, 201–202
    professional background, 200
    psychiatric hospital in Baltimore, 197
Houses
    Arden, Delaware, 173–174, 196
    Whiskey Acres, 137–138, 140
Hutchins, Robert Maynard, offered
    Carothers position as chairman of Univ.
    Chicago Chemistry Department, 213–
    214
Hydrogenation, platinum catalyst, 31–32

I

Incandescent light bulb, research goal for
    GE laboratory, 68
Industrial application, research, 122
Industrial fellowships, system proposed
    by Duncan, 67–68
Industrial slave, thoughts about university
    work, 169
Innsbruck, hikes in the Tyrolean Alps,
    274
Insanity, history of treatment, 260

J

Jacobson, Ralph, research staff, 94
Jewett, Frank B.
    Bell Labs, 66
    signal amplification for cross-country
        voice transmission, 69
Johnson, Jack
    arranged hiking trip for Carothers, 267
    brassylic acid, 165
    career, 230–231
    College de France, 45
    invitation to visit Whiskey Acres, 290
    plagiarism dispute, 167–168
    University of Illinois, 40
Journal of the American Chemical Society
    associate editor, 136–137
    Carothers in need of treatment for
        nervous condition, 257
Jung, Carl, meaninglessness, 315–316

K

Ketene dimer, cyclic structure, 196
Ketones, natural odorants, 163

Kistiakowsky, George, injured in explo-
    sion, 126
Kohler, E. P.
    comparison to Carothers, 49
    Harvard Chemistry Department, 48–50
Kraemer, Elmer O., colloid research, 75
Kyle, Elizabeth Carothers (sister)
    alcoholism, 310–311
    family illness, 309–310
Kyle, Nancy (niece)
    drug dependence, 309–310
    family illness, 309–310
Kyle, Robert (brother-in-law), wedding,
    142
Kyle, *See also* Terrill, Suzie Wallace

L

Labovsky, Joe
    assistant to Carothers, 235
    mercury in a vibration-free
        environment, 161
    music and literature, 215–216
Lactam, failure to polymerize, 133
Langmuir, Irving, inert-gas-filled lamp, 69
Lenher, Sam
    DuPont career, 227–228
    explosion in laboratory, 126
    physical chemistry unit, 124–125
Lewis–Langmuir theory, octet theory,
    36–37
Lithium, control of bipolar depression,
    319

M

Machetanz, Wilko
    letter renewing old friendships, 85
    pessimistic outlook on life, 7–9
    roommate at Tarkio College, 7
    with Wallace Carothers (photo), 7
Mail room, rubber stamp, 166
Manic depression
    behavior pattern, 318–319
    family illness, 310
    support from friends, 41
Mapel, Evelyn, Tarkio graduate, 212
Mapel, William Latta
    Carothers wedding, 251–253
    friendship with Carothers, 211–212
    memories of Helen and Wallace
        Carothers, 253
Marital state, nosedive, 153
Mark, Herman
    atoms arranged in space, 241
    left Germany, 242–243

Marriage
  effect on career, 221
  Hohman's prescription, 206
Martha's Vineyard, summer holiday with
  friends, 206
Marvel, Carl S. (Speed)
  p-bromophenol, 25
  career, 231–232
  manufacturing scholarship, 26
  research consultant, 95
  synthetic rubber styrene–butadiene, for
    tires, 231
  University of Illinois, 40
Mazda, inert-gas-filled lamp, 69
McGinnis, Marjorie, yearbook descrip-
  tion, 11
McMullin, Mary (mother), marriage, 2
Memorial biography, childhood, 4
Memorial service, DuPont Company, 293
Metallic cocatalysts, effect on reduction,
  32
Meyer, Adolf
  Phipps Clinic, 200
  treatment of F. Scott and Zelda
    Fitzgerald, 200–201
Meyer, K. H., atoms arranged in space,
  241
Miles, John, purchase of new tuxedo, 239
Mistresses, common knowledge among
  friends, 168
Molecular biology, simple realization of
  shape and surface, 247
Molecular still
  barrier broken, 113–115
  diagram, 114
  Julian Hill, 108
Molecular structure, colloidal theory, 92
Molecular weight, breaking the 4200 barrier, 100, 113–115
Moore, Sylvia
  description as remembered, 221–222
  description in Carothers' letters, 147
  divorce, 169
  lobbying for Geneva Conference, 149
  married, Carothers' mistress, 168
  tight-lipped gaze of his parents, 178
Music
  Carothers' apartment, 212
  college years, 12
Musk, natural resources, 163

N

National Academy of Sciences
  election of Carothers, 256

notice of Carothers' election (photo),
  257
Neoprene, world's first synthetic rubber,
  109–112
Nernst lamp, metal oxide filament, 68
Neurocirculatory asthenia
  second hospitalization, 281
  Wallace Carothers, 261
Neurotics, elements of genius, 224
Nieuwland, Julius, acetylene chains, 106
Nitration process, trinitrotoluene (TNT),
  65
Nitroglycerin, production problem, 99
Norris, James Flack, hiking trip in Alps,
  273–275
Nose fatigue, perfume work, 164–165
Nurnberger Hutte, way station in Tyrolean Alps, 274
Nylon
  early production, 296
  early proposed research, 55–56
  first batch, 217
  ideas developed, 57–58
  introduction of new fiber, *i*
  invention, 181
  melting point, 218
  name of new fiber, 295
  results of invention, *iii–iv*
Nylon–6, publication and patent claims,
  134

O

Octet theory, carbon-containing double-
  bond systems, 37–39
Olefinic aldehyde, presence, 32
Organic silicon compounds, synthetic
  routes, 96

P

Pancreatitis, Wallace Carothers, 317
Pardee, Arthur, chemistry at Tarkio College, 15
Parents, decision to live with them, 178–
  179
Paris
  city of fantasy, 46–47
  visit alone, 246
Patent application
  Carothers' inventions, 129
  effect of publication, 133–134
  polymer and fiber specifications, 131
Penmanship, indication of sobriety level,
  287

Pennsylvania Hospital, Philadelphia Institute, 260
Pension, Ira and Mary Carothers, 296–299, 303
Perfume, odors of cyclic compounds, 163–165
Peterson, W. R., new way to make the 5–10 superpolymer, 214–215
Philadelphia Institute
Carothers admitted unwillingly, 259–269
commuting to work in Wilmington, 277
complaints from Carothers, 266–267
mild nervous disorders, 260
routine of admission, 262
Pinel Clinic, psychiatric hospital in Baltimore, 197
Pioneering applied research, distinguished from DuPont program, 72
Platinum catalyst, hydrogenation, 31–32
Poetry
college interest, 8–9
published, 172
Polyamide(s)
early history, 250
melting problem, 185
systematic study, 216
Polyamide 5–10
first knitted fabric samples, 218
goal, 214
Polyamide 6–6, DuPont's nylon, 217
Polychloroprene
accidental polymerization, 111
best the scientists ever get, 216
publication, 132
world's first synthetic rubber, 109–112
Polyester, cause of limited growth, 117–118
Polyester synthesis, publication, 118
Polyethylene, foreshadowed, 109
Polyethylene terephthalate, invention, 134
Polymer 5–10, W. R. Peterson, 189
Polymer theories, publication, 118
Polymerization
early proposed research, 55–56
goal of molecular weight greater than 4200, 90
slow pace, 95
Poly(tetrafluoroethylene) (Teflon), best the scientists ever get, 216
Powdered metal catalysts, hydrogenation, 31–32
Presbyterian Church, characteristics, 2
Prohibition, end, 193
Proteins, natural synthesis, 246

Psychiatric care, Philadelphia Institute, 259–261
Psychoanalysis, change and controversy, 152–153
Purity Hall
early experimentation, 95
Stine's new laboratory, 77

R

Railroad, tops of the mountains, 192
Rankin, Russell
Tarkio College, 6
yearbook description, 11
Rayon
DuPont, 62–63
modification, 182–183
publication policy, 129
Record collection, extensive, 213
Recruitment policies, DuPont chemical department, 83
Reese, Charles, research and development, 64
Relationships, judgments of friends, 223–224
Religious outlook on life, Carl Jung, 316
Research
business-oriented objectives, 157
maturing attitude, 54
progress self-evaluation, 159–160
proposals to DuPont, 52–57
Research management
chemical department, 98–99
lack of direction, 125
Rice, E. Wilbur, GE industrial research laboratory, 65–66
Rideal, Eric, rubber research association, 282–283
Rubber
evaluation, 144
synthetic, 105–112

S

Salzberg, Paul
file of Ira Carothers' letters to DuPont, 303
University of Illinois, 40
Scientific meetings, tonic for the mind, 185
Seabrook, William
chronic alcoholism and neurasthenic symptoms, 261
defining parable for Wallace Carothers, 314–315
fight with alcohol, 194
treatment and response, 265–266
Seaford, Del., world's first nylon plant, 296

Seven-membered ring, adipic anhydride, 99–100
Shape, ability to bond and fit, 247
Signal amplification, cross-country voice transmission, 69
Silk, history, 188
Smith, Merrill, yearbook description, 11
Spanagel, Ed, ethylene brassylate, 165
Spencer, Frances
  cache of 14 letters from Carothers, 103–105, 138–155, 172–179
  divorced and a mother, 85
  reconstructing a different past, 86
Spiritual phenomenon, William Wilson, 316–317
Staudinger, Herman
  arrangement of atoms, 242
  political restrictions, 243
Steinmetz, Charles
  GE industrial research laboratory, 66
  research operation based on German model, 68
Stine, Charles M. A.
  budget for "pure science or fundamental research work," 69–70
  DuPont research unit, 64
  introduction of new fiber, *i*
  lessons from other research facilities, 71
  process development, 65
Stock awards, employee bonuses, 299
Stock market, speculation, 97–98
Strangers, fear, 277
Succinaldehyde, synthesis, 53
Suicide
  printed justification in case of insanity, 294
  Wallace Carothers, 291
Summer Preps, preparation of organic chemicals, 24
Superpolyesters
  delayed publication, 131
  experimentation, 128
  publication, 133
Superpolymer, molecular weight of 10,000 or above, 187
Sweetman, Helen Everett (wife)
  attraction for Carothers, 224
  boarding train for New York wedding (photo), 252
  descriptions, 222
  DuPont patent applications, 219, 221
  marriage, 250–255
  See also Carothers, Helen
Synthesis, proof of polymer structure, 246

Synthetic rubber and fiber
  productive science, 227
  reaction to invention, 312–313

T

Tanberg, Arthur
  Carothers' correspondence, 54–55, 57–58, 256–258
  effect on DuPont's hiring, 76–77, 82–83, 94, 124–125
Tarkio College
  arrival of Wallace Carothers, 1
  chemistry laboratory, 16–17
  establishment, 6
  teaching chemistry, 18–19
  work for expenses and scholarship, 5
Taylor, Guy, catalysis group, 75
Teaching style, conflicting reports, 49–50
Terrill, Suzie Wallace Kyle (niece), family illness, 309–311
Tetraethyl lead, supplied by DuPont, 63
Thompson, J. A., attempts to build up Tarkio College, 1
Thomson, Elihu, GE industrial research laboratory, 65–66
Transesterification
  growing polymer chain, 113
  molecular still, 117–118
Tyrolean Alps, Innsbruck as base for hikes, 274

U

United Presbyterian Church, establishment of Tarkio College, 6
University of Chicago
  offer of position as chairman of Chemistry Department, 213–214
  summer of 1922, 30–31
University of Illinois
  Chemistry Department, 39
  chemistry laboratory, 23–24
  standards of the graduate school, 34
  work for doctorate, 22–43
University of South Dakota, instructor, 27–28
Urbana, work for doctorate, 22–43

V

Vagrant appearance, Chicago University campus, 30–31
van Natta, Frank
  polyesters from carbonic acid, 95
  research staff, 94
Viscose process, synthetic silks, 13

W

Wawaset Park
  Friday night gatherings, 255
  new apartment, 211
Weamer, Harry Claire, college chemistry
  teacher, 16
Wedding dinner, Wilmington's University Club, 250
Whiskey Acres
  lifestyle, 140

refuge for Carothers after leaving
  Philadelphia Institute, 277
shared house, 137–138
Whitney, Willis R., General Electric
  research laboratory, 65–66, 68
Will, January 1937, 301
Willsätter, Richard, visit to DuPont,
  177
World War II, use of nylon, *ii–iii*

Production: Margaret J. Brown
Acquisition: Barbara E. Pralle
Indexing: Colleen P. Stamm
Cover design: Amy J. O'Donnell

Composition: Vincent L. Parker
Printed and bound by Maple Press, York, PA

Other Titles from ACS Books

The ACS Style Guide: A Manual for Authors and Editors
Edited by Janet S. Dodd
264 pp; clothbound ISBN 0–8412–0917–0; paperback ISBN 0–8412–0943–X

Roger Adams: Scientist and Statesman
By D. Stanley Tarbell and Ann Tracy Tarbell
248 pp; clothbound ISBN 0–8412–0598–1

From Small Organic Molecules to Large: A Century of Progress
By Herman F. Mark
Profiles, Pathways, and Dreams Series
174 pp; clothbound ISBN 0–8412–1776–9

American Chemists and Chemical Engineers
Edited by Wyndham D. Miles
554 pp; clothbound ISBN 0–8412–0278–8

Chemistry and Crime: From Sherlock Holmes to Today's Courtroom
Edited by Samuel M. Gerber
135 pp; clothbound ISBN 0–8412–0784–4; paperback ISBN 0–8412–0785–2

Nobel Laureates in Chemistry 1901–1992
Edited by Laylin K. James
History of Modern Chemical Sciences Series
826 pp; clothbound ISBN 0–8412–2459–5; paperback ISBN 0–8412–2690–3

Stalin's Captive: Nikolaus Riehl and the Soviet Race for the Bomb
By Nikolaus Riehl and Frederick Seitz
History of Modern Chemical Sciences Series
240 pages; clothbound ISBN 0–8412–3310–1

Following the Trail of Light: A Scientific Odyssey
By Melvin Calvin
Profiles, Pathways, and Dreams Series
200 pp; clothbound ISBN 0–8412–1828–5

A Wandering Natural Products Chemist
By Koji Nakanishi
Profiles, Pathways, and Dreams Series
256 pp; clothbound ISBN 0–8412–1775–0

Seventy Years in Organic Chemistry
By Tetsuo Nozoe
Profiles, Pathways, and Dreams Series
292 pp; clothbound ISBN 0–8412–1769–6

For further information and a free catalog of ACS books, contact:
American Chemical Society
Distribution Office, Department 225
1155 16th Street, NW, Washington, DC 20036
Telephone 800–227–5558